Conference Proceedings of the Society for Experimental Mechanics Series

Series Editor

Kristin B. Zimmerman, Ph.D.
Society for Experimental Mechanics, Inc.,
Bethel, CT, USA

The Conference Proceedings of the Society for Experimental Mechanics Series presents early findings and case studies from a wide range of fundamental and applied work across the broad range of fields that comprise Experimental Mechanics. Series volumes follow the principle tracks or focus topics featured in each of the Society's two annual conferences: IMAC, A Conference and Exposition on Structural Dynamics, and the Society's Annual Conference & Exposition and will address critical areas of interest to researchers and design engineers working in all areas of Structural Dynamics, Solid Mechanics and Materials Research.

More information about this series at http://www.springer.com/series/8922

Shamim Pakzad
Editor

Dynamics of Civil Structures, Volume 2

Proceedings of the 37th IMAC, A Conference and Exposition on Structural Dynamics 2019

Editor
Shamim Pakzad
Department of Civil & Environment Engineering
Lehigh University
Bethlehem, PA, USA

ISSN 2191-5644 ISSN 2191-5652 (electronic)
Conference Proceedings of the Society for Experimental Mechanics Series
ISBN 978-3-030-12114-3   ISBN 978-3-030-12115-0 (eBook)
https://doi.org/10.1007/978-3-030-12115-0

© Society for Experimental Mechanics, Inc. 2020
This work is subject to copyright. All rights are reserved by the Publisher, whether the whole or part of the material is concerned, specifically the rights of translation, reprinting, reuse of illustrations, recitation, broadcasting, reproduction on microfilms or in any other physical way, and transmission or information storage and retrieval, electronic adaptation, computer software, or by similar or dissimilar methodology now known or hereafter developed.
The use of general descriptive names, registered names, trademarks, service marks, etc. in this publication does not imply, even in the absence of a specific statement, that such names are exempt from the relevant protective laws and regulations and therefore free for general use.
The publisher, the authors, and the editors are safe to assume that the advice and information in this book are believed to be true and accurate at the date of publication. Neither the publisher nor the authors or the editors give a warranty, express or implied, with respect to the material contained herein or for any errors or omissions that may have been made. The publisher remains neutral with regard to jurisdictional claims in published maps and institutional affiliations.

This Springer imprint is published by the registered company Springer Nature Switzerland AG.
The registered company address is: Gewerbestrasse 11, 6330 Cham, Switzerland

Preface

Dynamics of Civil Structures represents one of eight volumes of technical papers presented at the 37th IMAC, A Conference and Exposition on Structural Dynamics, organized by the Society for Experimental Mechanics and held in Orlando, Florida, January 28–31, 2019. The full proceedings also include volumes on Nonlinear Structures & Systems; Model Validation and Uncertainty Quantification; Dynamics of Coupled Structures; Special Topics in Structural Dynamics & Experimental Techniques; Rotating Machinery, Optical Methods & Scanning LDV Methods; Sensors and Instrumentation, Aircraft/Aerospace, Energy Harvesting & Dynamic Environments Testing; and Topics in Modal Analysis & Testing.

Each collection presents early findings from analytical, experimental, and computational investigations on an important area within structural dynamics. *Dynamics of Civil Structures* is one of these areas which cover topics of interest of several disciplines in engineering and science.

The Dynamics of Civil Structures Technical Division serves as a primary focal point within the SEM umbrella for technical activities devoted to civil structure analysis, testing, monitoring, and assessment. This volume covers a variety of topics including structural vibrations, damage identification, human-structure interaction, vibration control, model updating, modal analysis of in service structures, innovative measurement techniques and mobile sensing, and bridge dynamics among many other topics.

Papers cover testing and analysis of different kinds of civil engineering structures such as buildings, bridges, stadiums, dams, and others.

The organizers would like to thank the authors, presenters, session organizers, and session chairs for their participation in this track.

Bethlehem, PA, USA   Shamim Pakzad

Contents

1. **Improving an Experimental Test Bed with Time-Varying Parameters for Developing High-Rate Structural Health Monitoring Methods** 1
   D. T. Foley, B. S. Joyce, J. Hong, S. Laflamme, and J. Dodson

2. **Application of Electro-active Materials Toward Health Monitoring of Structures: Electrical Properties of Smart Aggregates** 7
   Patrick Manghera, Faiaz Rahman, Sankha Banerjee, Maryam Nazari, and Walker Tuff

3. **Output-Only Estimation of Amplitude Dependent Friction-Induced Damping** 17
   Karsten K. Vesterholm, Tobias Friis, Evangelos Katsanos, Rune Brincker, and Anders Brandt

4. **Modeling Human-Structure Interaction Using Control Models When Bobbing on a Flexible Structure** 27
   Ahmed T. Alzubaidi and Juan M. Caicedo

5. **Identification and Monitoring of the Material Properties of a Complex Shaped Part Using a FEMU-3DVF Method: Application to Wooden Rhombicuboctahedron** 35
   R. Viala, V. Placet, and S. Cogan

6. **Modal Tracking on a Building with a Reduced Number of Sensors System** 39
   Wladimir M. González and Rubén L. Boroschek

7. **Bayesian Damage Identification Using Strain Data from Lock Gates** 47
   Yichao Yang, Ramin Madarshahian, and Michael D. Todd

8. **Dynamic Tests and Technical Monitoring of a Novel Sandwich Footbridge** 55
   Jacek Chroscielewski, Mikolaj Miskiewicz, Lukasz Pyrzowski, Magdalena Rucka, Bartosz Sobczyk, Krzysztof Wilde, and Blazej Meronk

9. **Assessment and Control of Structural Vibration in Gyms and Sports Facilities** 61
   Aliz Fischer, Rob Harrison, Mark Nelson, François Lancelot, and James Hargreaves

10. **A Large Scale SHM System: A Case Study on Pre-stressed Bridge and Cloud Architecture** 75
    Gabriele Bertagnoli, Francescantonio Lucà, Marzia Malavisi, Diego Melpignano, and Alfredo Cigada

11. **Vibration Serviceability Performance of an As-Built Floor Under Crowd Pedestrian Walking** ... 85
    Jinping Wang and Jun Chen

12. **Identifying Traffic-Induced Vibrations of a Suspension Bridge: A Modelling Approach Based on Full-Scale Data** 93
    Etienne Cheynet, Jonas Snæbjörnsson, and Jasna Bogunović Jakobsen

13. **Floor Vibrations and Elevated Non-structural Masses** 103
    Christian Frier, Lars Pedersen, and Lars Vabbersgaard Andersen

14. **Vibration Performance of a Lightweight FRP Footbridge Under Human Dynamic Excitation** 111
    Stana Živanović, Justin M. Russell, and Vitomir Racic

15	**A Study of Suspension Bridge Vibrations Induced by Heavy Vehicles**	115
	Jonas Thor Snæbjörnsson, Thomas Ole Messelt Fadnes, Jasna Bogunovic Jakobsen, and Ove Tobias Gudmestad	
16	**Design and Performance of a Bespoke Lively All-FRP Footbridge**	125
	J. M. Russell, J. T. Mottram, S. Zivanovic, and X. Wei	
17	**Convolutional Neural Networks for Real-Time and Wireless Damage Detection**	129
	Onur Avci, Osama Abdeljaber, Serkan Kiranyaz, and Daniel Inman	
18	**The Influence of Truck Characteristics on the Vibration Response of a Bridge**	137
	Navid Zolghadri and Kirk A. Grimmelsman	
19	**Experimental Evaluation of Low-Cost Accelerometers for Dynamic Characterization of Bridges**	145
	Kirk A. Grimmelsman and Navid Zolghadri	
20	**Theoretical and Experimental Verifications of Bridge Frequency Using Indirect Method**	153
	Shota Urushadze, Jong-Dar Yau, Yeong-Bin Yang, and Jan Bayer	
21	**A Bayesian Inversion Approach for Site Characterization Using Surface Wave Measurements**	159
	Mehdi M. Akhlaghi, Babak Moaveni, and Laurie G. Baise	
22	**Estimating Fatigue in the Main Bearings of Wind Turbines Using Experimental Data**	163
	Giovanni M. Fava, Sauro Liberatore, and Babak Moaveni	
23	**Cointegration for Detecting Structural Blade Damage in an Operating Wind Turbine: An Experimental Study**	173
	B. A. Qadri, M. D. Ulriksen, L. Damkilde, and D. Tcherniak	
24	**System Identification of a Five-Story Building Using Seismic Strong-Motion Data**	181
	Rodrigo Astroza, Francisco Hernández, Pablo Díaz, and Gonzalo Gutierrez	
25	**Structural Property Guided Gait Parameter Estimation Using Footstep-Induced Floor Vibrations**	191
	Jonathon Fagert, Mostafa Mirshekari, Shijia Pan, Pei Zhang, and Hae Young Noh	
26	**Why Is My Coffee Cup Rattling: A Reassessment of the Office Vibration Criterion**	195
	Melissa W. Y. Wong and Michael J. Wesolowsky	
27	**Response of a SDOF System with an Inerter-Based Tuned Mass Damper Subjected to Non-stationary Random Excitation**	201
	Abdollah Javidialesaadi and Nicholas E. Wierschem	
28	**Experimental Study on Digital Image Correlation for Deep Learning-Based Damage Diagnostic**	205
	Nur Sila Gulgec, Martin Takáč, and Shamim N. Pakzad	
29	**Dynamic Response of the Suspended on a Single Cable Footbridge**	211
	Mikolaj Miskiewicz, Lukasz Pyrzowski, and Krzysztof Wilde	
30	**Event Detection and Localization Using Machine Learning on a Staircase**	219
	Blake Feichtl, Caleb Thompson, Tyler Liboro, Saad Siddiqui, V. V. N. Sriram Malladi, Tim Devine, and Pablo A. Tarazaga	
31	**Footbridge Vibrations and Their Sensitivity to Pedestrian Load Modelling**	225
	Lars Pedersen and Christian Frier	
32	**Recreating Periodic Events: Characterizing Footsteps in a Continuous Walking Signal**	231
	Ellis Kessler, Pablo Tarazaga, and Serkan Gugercin	
33	**On Wave Propagation in Smart Buildings**	237
	Mauro S. Maza, Mohammad I. Albakri, V. V. N. S. Malladi, and Pablo A. Tarazaga	
34	**Parameter Study of Statistics of Modal Parameter Estimates Using Automated Operational Modal Analysis**	243
	Silas S. Christensen and Anders Brandt	

35 Dynamic Bridge Foundation Identification 255
Nathan Davis and Masoud Sanayei

36 Damping Ratios of Reinforced Concrete Structures Under Actual Ground Motion Excitations 259
Dan Lu, Jiayao Meng, Songhan Zhang, Yuanfeng Shi, Kaoshan Dai, and Zhenhua Huang

37 Launching Semi-automated Modal Identification of the Port Mann Bridge 269
A. Mendler, C. E. Ventura, L. Nandimandalam, and Y. Kaya

Chapter 1
Improving an Experimental Test Bed with Time-Varying Parameters for Developing High-Rate Structural Health Monitoring Methods

D. T. Foley, B. S. Joyce, J. Hong, S. Laflamme, and J. Dodson

Abstract With the development of complex structures with high-rate dynamics, such as space structures, weapons systems, or hypersonic vehicles, comes a need for real-time structural health monitoring (SHM) methods. Researchers are developing algorithms for high-rate SHM methods, however, limited data exists on which to test these algorithms. An experimental test bed to simulate high-rate systems with rapid parameter changes was previously presented by the authors. This paper expands on the previous work. The initial configuration consisted of a cantilevered steel beam with a cart-roller system on a linear actuator to create an adjustable boundary condition along the beam, as well as detachable added masses. Experimental results are presented for the system in new configurations during various parameter changes. A clamped-clamped condition to increase the system's natural frequencies is studied, along with improvements in test repeatability and user control over parameter changes.

Keywords Time-varying systems · Testbed · Structural health monitoring · SHM · Damage detection · High-rate state estimation

1.1 Introduction

There are a growing number of advanced structures in dynamically harsh environments, such as hypervelocity air vehicles, space structures, and weapon systems. These structures can experience high speed impacts (>4 km/s) that result in damage propagating through the structures in microseconds [1, 2]. These applications have fueled interest in developing structural health monitoring (SHM) and damage prognosis methods for rapidly changing, time-varying systems [3–8]. These methods could calculate the location and severity of damage and determine what actions are required on timescales too small for human decision making.

Sufficient data pertaining to these rapidly changing systems is limited. Such data is needed for developing these SHM techniques and gain insight into structural damage. In order to address this absence, an experimental test bed is developed that allows for examination of multiple configurations for an example system. The DROPBEAR (Dynamic Reproduction of Penetrator Ballistic Environments for Advanced Research) is an experimental test bed capable of generating data for model-based estimators of rapidly changing, time-varying systems. The test bed is shown in Fig. 1.1. The system consists of a rectangular steel beam with several mechanical parameters that can be changed during the system response. The base cantilevered beam utilizes detachable electromagnets to add additional mass to any desired location along the beam's length. The electromagnets can be disengaged quickly to simulate a sudden detachment of a system component. The DROPBEAR

D. T. Foley
University of Florida, Gainsville, FL, USA

B. S. Joyce
University of Dayton Research Institute, Eglin AFB, FL, USA

J. Hong
Applied Research Associates, Inc, Valparaiso, FL, USA

S. Laflamme
Iowa State University, Ames, IA, USA

J. Dodson (✉)
Fuzes Branch, Air Force Research Laboratory, Munitions Directorate, Eglin AFB, FL, USA
e-mail: jacob.dodson.2@us.af.mil

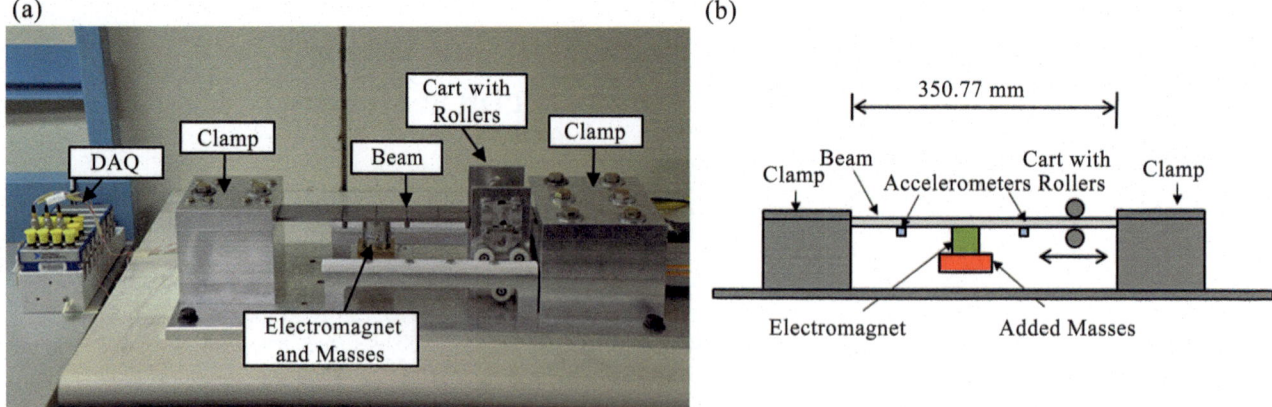

Fig. 1.1 Cantilever configurations for the DROPBEAR test bed. (**a**) and (**b**) show a photo and a schematic of the clamped-clamped beam with detaching electromagnet and rolling cart

also features a sliding cart on an actuator capable of acting as a pinned condition at a specific position on the beam. The sliding cart can be utilized in the same manner as a stationary pin or a moving pin after beam excitation. The electromagnet's attachment and the cart's position can be used as fixed parameters or as variable parameters at any time during testing. The versatility of both components coupled together provides an array of repeatable, fixed or variable testing configurations. Previous results have been presented on the beam in a clamped-free (cantilever) boundary condition [9]. The primary motivation for changing the boundary condition of the experiment is to increase the natural frequency of the system, and then illustrate faster changes in mechanical responses when the time-varying parameters occur. This extended abstract discusses the beam setup in a clamped-clamped boundary condition, which increased the natural frequencies of the system.

This paper focuses on the experimental setup, testing, and data analysis of the DROPBEAR in the clamped-clamped condition employing the lumped mass characterization. The frequency response functions of the beam are collected and analyzed in different experimental configurations to ultimately compare the response to analytical and finite element models. Comparison to these models verifies that the test bed is repeatable and reliable while providing a reference for any discrepancies experienced in the system's response. The objective is to capture data from the time-varying, rapidly changing configurations to analyze the response and develop models that can estimate system parameters and lead to damage detection.

1.2 Added Mass Experimental Setup and Test Procedure

The basic structure of the test bed features a large, rectangular aluminum plate fastened to a tabletop. This plate serves as the mounting base for the clamp housing in which the steel beam is secured. For these tests the cart with rollers was removed. The steel beam is 51 mm (2 in.) wide with a free length of 350.77 mm (13.81 in.) at a thickness of 6.3 mm (0.25 in.). Two single-axis PCB 353B17 accelerometers were attached to the beam at 87.63 mm (3.45 in.) and 219.2 mm (8.63 in.), from the primary clamp, as highlighted in Fig. 1.1b. The accelerometers were connected to a NI-9234 IEPE analog input module seated in a National Instruments (NI) cDAQ-9172 eight-slot chassis. The chassis was connected to a computer to acquire the generated signals via NI LabVIEW. A PCB 086C01 modal hammer was used to excite the beam at a desired location along the beam length. The hammer was equipped with various options for strike tips ranging from soft rubber to steel to vary the impulse provided to the beam. The hammer was connected to the chassis through the same NI-9234 module enabling impulse input data to be collected and analyzed. A LabVIEW program acquired and saved response data from all three sensors. MATLAB was employed for post-processing and generating frequency response functions (FRF).

The clamped-clamped configuration was tested in multiple configurations including pinned and added mass configurations. The clamped-clamped configuration with no added mass and with added mass near the tip of the beam is presented here. The added mass consisted of the electromagnet (weighing 0.259 kg). The beam was struck by the modal hammer in five locations measured from the beam base at the primary clamp at 87.6 mm (3.45 in.), 131 mm (5.16 in.), and 175.3 mm (6.9 in.), 219.2 mm (8.63 in.), and 262.9 mm (10.35 in.). For brevity, only the results from the impact at 262.9 mm (10.35 in.) and the tip accelerometer are used here. The beam was struck five times per test with ample time between each strike to observe the full decay in the response.

1.3 Added Mass Results

For mass configuration, the natural frequencies predicted by an analytical model and a finite element model are validated against natural frequencies calculated frequency response functions (FRFs) derived from experimental data. The natural frequencies were estimated against a finite element model capturing the clamped-clamped conditions of the beam.

A finite element of the beam with mass and variable cart position was derived from Euler's beam equation of motion using cubic Hermite polynomials [10]. A MATLAB script forms the mass and stiffness matrices, computes the natural frequencies and mode shapes, and calculates the frequency response functions for various impact and measurement locations. The finite element model accounts for the magnet as additional masses and rotational inertia terms at the degrees of freedom of the beam where the magnet attaches.

The experimental, time-domain data was processed in MATLAB to compute the FRF using the H_v estimation and the coherence [11, 12]. Figure 1.2 plots the FRFs and coherence near the first natural frequencies from experimental data alongside the FRF determined by the finite element model. The FRFs from the model and the data show good agreement. Adding mass decreases the natural frequencies as expected.

The first four natural frequencies are listed for no added mass in Table 1.1 and for the electromagnet mass in Table 1.2. For the bare beam with no added mass, the models show a good agreement (less than 2% error) with the experimentally obtained natural frequencies of the beam. Adding the electromagnet decreases all of the natural frequencies as expected. For the first three modes, the added mass model still had good agreement (less than 3% error), but there quickly started to be discrepancies at the higher modes (greater than 9% error). In the finite element model, the added mass is distributed along the elements of the beam occurred by the magnet (a diameter of 40 mm). The natural frequency estimates from the distributed mass of the finite element model produces a closer estimate to the experimental values compared to the results from the analytical model with lumped mass at the tip. It should be noted that model updating could be used to refine parameters and produces more accurate estimates of the natural frequencies. The small errors between the experimental and finite element natural frequencies indicate the model captures the behavior of the baseline system with fix parameters and could predict the beam's response once the mass detaches.

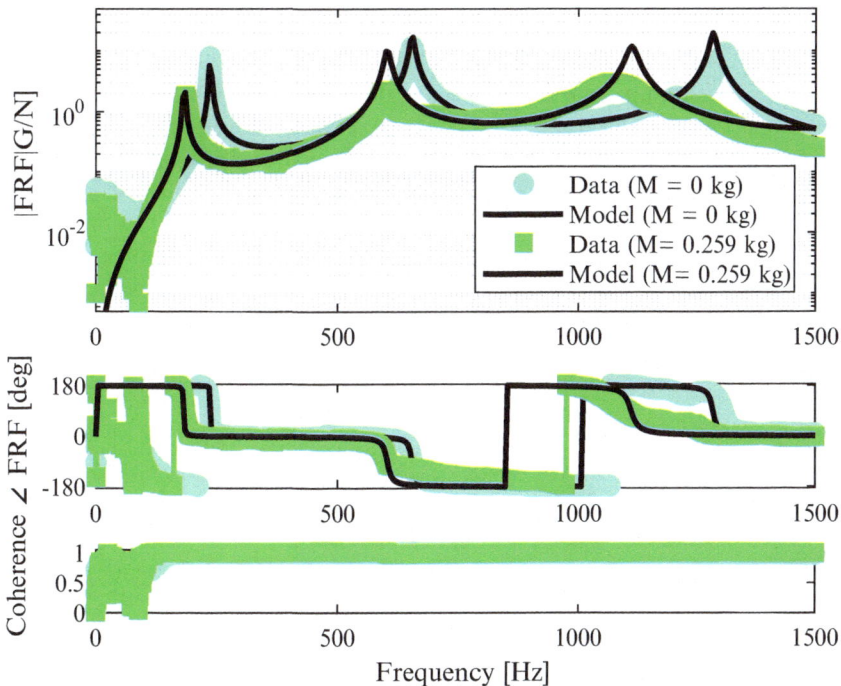

Fig. 1.2 Comparison of frequency response functions (FRF) between the beam without added mass ($M = 0$ kg), and with the mass of the electromagnet ($M = 0.259$ kg). Data is from the accelerometer at 87.63 mm (3.45 in.) and impacts at 262.89 mm (10.35 in.) from the base. The model FRFs are from the finite element model

Table 1.1 Comparison between theoretical and finite element frequencies for the bare beam with no added mass ($M = 0$ kg)

Mode	Experimental Natural frequency [Hz]	Finite element Natural frequency [Hz]	Percent error (%)
1	237.3	237.2	0.04
2	654.7	653.8	0.14
3	1306	1281.7	1.86
4	2255	2118.8	6.04

Percent errors are relative to the experimentally obtained natural frequencies

Table 1.2 Comparison between theoretical and experimental frequencies for the beam with the electromagnet near the beam tip with mass ($M = 0.259$ kg)

Mode	Experimental Natural frequency [Hz]	Finite element Natural frequency [Hz]	Percent error (%)
1	186.1	182.9	1.72
2	589.5	602.7	−2.24
3	1070	1111.1	−3.84
4	1922	1744.7	9.22

1.4 Static Cart Experiments

The cart with rollers was re-attached to the beam, as seen in Fig. 1.1a, b. The As the rollers moves along the beam, they create a moving pinned condition along the span of the beam. The beam was tested with several static configurations of the cart with rollers.

1.5 Conclusions and Future Work

There is a need for an experimental test bed for developing and demonstrating high-rate damage detection methods for the advancement of structural health monitoring. The DROPBEAR serves as a unique test bed capable of producing repeatable, time-varying system conditions that can assist in evolving state estimators for a highly dynamic environment. The finite element model showed good agreement to the experimentally obtained natural frequencies of the system with clamped-clamped boundary conditions and varying masses.

Future results presented will discuss the clamped-pinned-clamped conditions and the time varying conditions (mass drop and rolling pin) for the clamped-clamped boundary condition. This array of future experiments will capture response data for a multitude of conditions and parameter changes. This data will guide developing and testing algorithms for state estimation, system identification, and damage detection. The test bed's future in testing new, variable configurations offers encouraging potential for the progression of high-rate state and parameter identification methods.

Acknowledgements The material is based upon work supported by the Air Force Office of Scientific Research under award number FA9550-17RWCOR503. Any opinions, findings, and conclusions or recommendations expressed in this material are those of the authors and do not necessarily reflect the views of the United States Air Force.

References

1. Stein, C., Roybal, R., Tlomak, P., Wilson, W.: A review of hypervelocity debris testing at the Air Force Research Laboratory. Space Debris. **2**(4), 331–356 (2000)
2. Hallion, R.P., Bedke, C.M., Schanz, M.V.: Hypersonic Weapons and US National Security, a 21st Century Breakthrough. Mitchell Institute for Aerospace Studies, Air Force Association, Arlington, VA (2016)
3. Dodson, J., Inman, D.J., Foley, J.R.: Microsecond structural health montoring in impact loaded structures. In: Proceedings in SPIE, San Diego, CA (2009)
4. Lowe, R., Dodson, J., Foley, J.: Microsecond prognostics and health monitoring. In: IEEE Reliability Society Newsletter, p. 60 (2014)

5. Kettle, R., Dick, A., Dodson, J., Foley, J., Anton, S.R.: Real-time detection in highly dynamic systems. In: IMAC XXXIV A Conference and Exposition on Structural Dynamics, Orlando, FL (2015)
6. Dodson, J., Kettle, R., Anton, S. R.: Microsecond state detection and prognosis using high-rate electromechanical impedance. In: IEEE Prognostics and Health Management Conference, PHM'16 (2016)
7. Hong, J., Laflamme, S., Cao, L., Dodson, J.: Variable input observer for structural health monitoring of high-rate systems. In: Proc. AIP Conference Proceedings, p. 07003
8. Dodson, J., Joyce, B.S., Hong, J., Laflamme, S., Wolfson, J.: Microsecond state monitoring of nonlinear time-varying dynamic systems. In: ASME 2017 Conference on Smart Materials, Adaptive Structures and Intelligent Systems Snowbird, Utah, USA (2017)
9. Joyce, B., Greenoe, K., Dodson, J., Wolfson, J., Abramczyk, S., Karsten, H., Markl, J., Minger, R., Passmore, E.: An experimental test bed with time-varying parameters for developing high-rate structural health monitoring methods. In: IMAC XXXVII A Conference and Exposition on Structural Dynamics, Orlando, FL (2018)
10. Blevins, R.D.: Formulas for Natural Frequency and Mode Shape. Van Nostrand Reinhold Co, New York (1979)
11. Ewins, D.J.: Modal Testing: Theory, Practice, and Application. Research Studies Press Ltd., Hertfordshire, England (2000)
12. Allemang, R.J., Brown, D.L.: Experimental Modal Analysis and Dynamic Component Synthesis: Volume I Summary of Technical Work. Air Force Wright Aeronautical Laboratories, Wright-Patterson AFB, Ohio (1987)

Chapter 2
Application of Electro-active Materials Toward Health Monitoring of Structures: Electrical Properties of Smart Aggregates

Patrick Manghera, Faiaz Rahman, Sankha Banerjee, Maryam Nazari, and Walker Tuff

Abstract Electro-active polymers and piezoelectric energy harvesting materials offer enormous potential for developing smart damping devices for the health monitoring of civil structures during unknown excitations (e.g., earthquakes, blasts, etc.). These auto-adaptive and intelligent composites convert mechanical energy to electric energy by generating an electric field when subjected to mechanical excitation. Variation in the strength of the electric field response can detect any change in the structural properties due to damage in the host structure. An experimental research plan is developed to investigate the effectiveness of these smart materials in structural engineering applications with the purpose of damage detection and mitigation. To this end, two phases of this research study are outlined as follows: (1) materials fabrication: the piezoelectric composites were fabricated using a solution based wet lab fabrication methodology. Bulk sample geometries (such as bulk cylinders) of the two phase (such as $BaTiO_3$—Cement) composite electro-active materials were tested for their impedance and piezoelectric properties. The microstructure and elemental distribution of these materials were characterized using the Scanning Electron Microscope (SEM) and Energy-dispersive X-ray spectroscopy (EDS/EDX) to understand the process—structure—property relationships. The final composite product is labeled as electro-active smart aggregates in this project; (2) structural component testing: smart aggregates were embedded in a simply supported concrete beam to investigate their effectiveness in monitoring deflections of the beam under flexural testing. Electrical properties of smart aggregates will be presented herein.

Keywords Health monitoring · Concrete structures · Electro-active composites · $BaTiO_3$-Cement · Smart aggregates

2.1 Introduction

Electro-active polymers and piezoelectric energy harvesting materials offer enormous potential for developing smart damping devices for both the health monitoring and vibration control of *civil* structures during unknown excitations (e.g., earthquakes, blasts, etc.) [1–3]. These auto-adaptive and intelligent composites convert mechanical energy to electric energy by generating an electric field when subjected to mechanical excitation. Variation in the strength of the electric field response can detect any change in the structural properties due to damage in the host structure. These materials can also convert an applied electric field to mechanical vibrations and therefore act as active dampers to control the structural response. The integration

---

P. Manghera
Department of Mechanical Engineering, Lyles College of Engineering, California State University, Fresno, CA, USA

F. Rahman
Department of Civil and Geomatics Engineering, Lyles College of Engineering, California State University,
Fresno, CA, USA

S. Banerjee (✉)
Department of Mechanical Engineering, Lyles College of Engineering, California State University, Fresno, CA, USA
e-mail: sankhab@csufresno.edu

M. Nazari
Department of Civil and Geomatics Engineering, Lyles College of Engineering, California State University,
Fresno, CA, USA
e-mail: mnazari@mail.fresnostate.edu

W. Tuff
Department of Mechanical Engineering, Lyles College of Engineering, California State University, Fresno, CA, USA

of these two components results in the development of a smart structure which can accurately identify time-varying structural parameters during different excitations and optimally mitigate the structural vibrations [4].

Multiphasic piezoelectric and electro-active cement composites have been investigated over the last few years due to their potential for applications in structural health monitoring [5] and as sensors and actuators in civil structures [6, 7]. Quantitative and real-time monitoring of civil structures can be achieved by using a nondestructive impedance based measurement method [5, 8, 9]. The impedance method exhibits strong potential in the area of structural health monitoring and damage detection. This technique involves the use of a high frequency signal, such that any structural excitation or localized damage will lead to modifications in the local impedance characteristics and that of the smart aggregates. The concrete damage can be characterized using the local electromechanical coupling characteristics of the cement based piezoelectric materials [10]. In addition to the above methods, vibration based detection of structural damage has also been employed by other researchers [1, 8]. The vibration-based method is based on the comparison of characteristics such as damping parameters, modal shape and natural frequency [5–7]. These electro-active composites can be used as surface bonded attachments [11] or as embedded sensors of different geometries such as cubic, cylindrical and spherical [12, 13]. The surface bonded 2D structures, which include thin and thick films, have certain limitations due to their location and orientation [5, 6]. Similarly, embedded sensors can provide only localized information since they are fixed and cannot be moved in the structure [6, 13].

The current work includes the impedance and piezoelectric characterization of multiphasic $BaTiO_3$—Cement based smart aggregates. The microstructural properties are analyzed using a scanning electron microscope to characterize the fabrication process—microstructure—property relationships. These aggregates are embedded in concrete structural elements as smart sensors to monitor their deflections under flexural testing.

2.2 Background and Methodology

The following work involves the fabrication of cylindrical sensors as shown in Fig. 2.1. An overview of the fabrication process of two-phase $BaTiO_3$ bulk composite based smart aggregate is described below. Sigma Aldrich $BaTiO_3$ with mean particle size of approximately 3 μm is mixed with Portland Cement with different volume fractions ranging from 5 to 50% in steps of 5%. The fabrication process starts with the measurement of appropriate amounts of $BaTiO_3$ and Portland Cement, which were combined, stirred by hand and then ball milled for 8 h. Next, a specified amount of water (40% by weight of the cement) was added. The mixture was stirred to make a composite slurry that was then poured into an aluminum mold and pressed at 55.2 MPa for 15 min in a compression molder. After compression, the sample was removed from the mold and cured in an environmental system at a relative humidity of 98% for 24 h. Once the curing was complete, the sample was polarized applying an electric field of 0.6 kV/mm with a high voltage power supply at a temperature of 120 °C. Two different poling methods were used: the contact based poling method and the contactless poling method (in which corona discharge based micro-plasma were employed). After cooling inside the fume hood, the sample was wrapped in aluminum foil and allowed to sit for the next 12 h prior to measurements. The temperature and humidity of the fume hood were monitored during this time. The micro-structure of the samples was characterized using a scanning electron microscope and the electrical and electro-mechanical properties were measured using an Impedance Analyzer and a Piezometer.

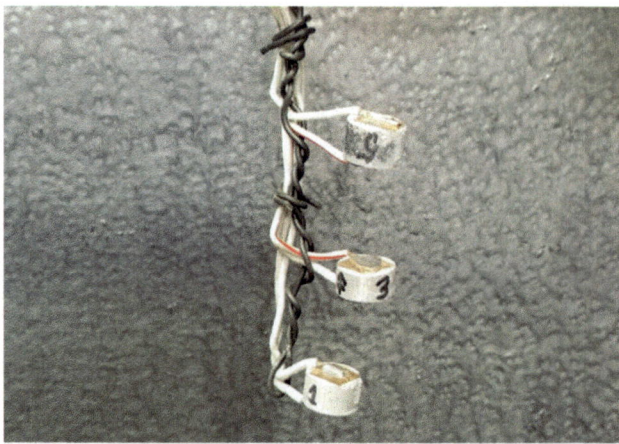

Fig. 2.1 Three cylindrical smart aggregates prior to being embedded inside the concrete sample

2.3 Analysis

The SEM micrographs of the smart aggregates are shown in Fig. 2.2 below. Taken at magnifications of 26.7 kX, 13.3 kX, 8.00 and 2.7 kX, these micrographs depict the distribution of $BaTiO_3$ inclusions in the cement matrix. A uniform distribution of the $BaTiO_3$ clusters was found, along with the agglomeration of larger cement particles in the microstructure.

The micrographs also show the interaction of $BaTiO_3$ particles with each other and also with cement agglomerations at a $BaTiO_3$ volume fraction of 40%. The micrographs from other volume fractions show the same trends as above. The $BaTiO_3$ clusters enhance the electron transfer mechanisms of direct electron transfer and electron hopping toward developing electro-mechanical coupling characteristics with the host structure.

Figure 2.3 depicts the variation of the piezoelectric strain coefficient (d_{33}) in pC/N with changes in the volume fraction of $BaTiO_3$ and also with variation in applied static force for corona poled bulk composites. The d_{33} decreases from a maximum value of about 0.08 pC/N at a $BaTiO_3$ volume fraction of 5% to a value of 0.01 pC/N for a $BaTiO_3$ volume fraction of 50%. This change is possibly due to the increase in the contact resistance in the microstructure with the increase in the $BaTiO_3$ volume fraction. This can be further analyzed by SEM micrographs at different volume fractions. The variation in

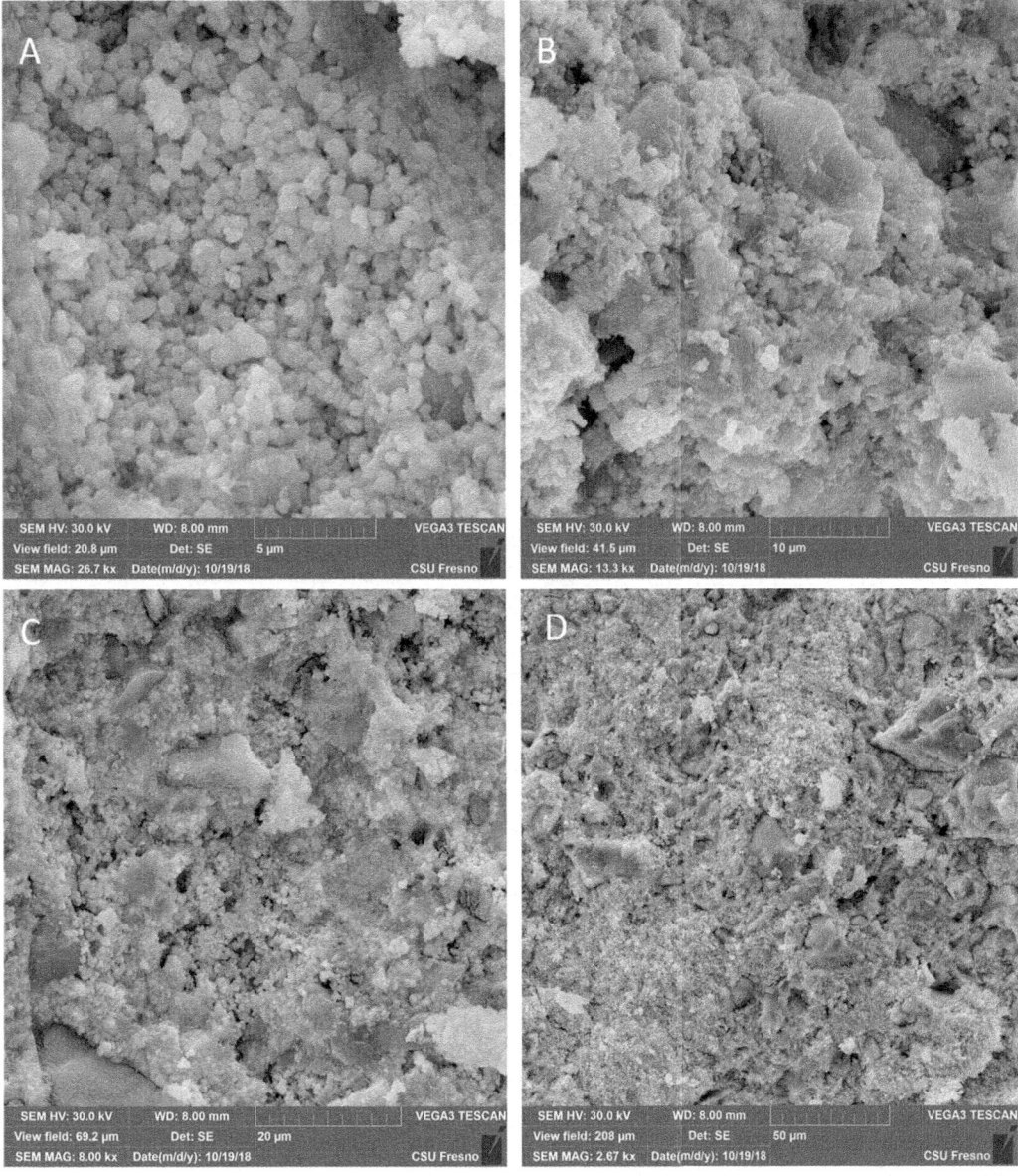

Fig. 2.2 The distribution of $BaTiO_3$ and Portland Cement particles at magnifications of (**a**) 26.7 kX, (**b**) 13.3 kX, (**c**) 8.00 kX, and (**d**) 2.7 kX

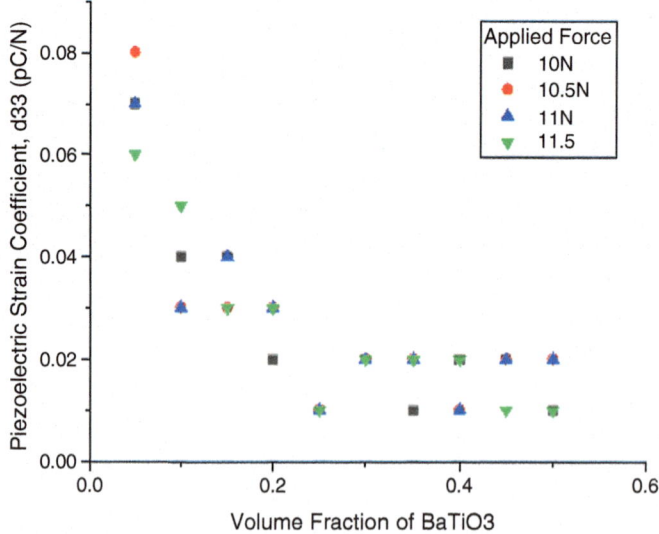

Fig. 2.3 The variation in the electromechanical performance, as demonstrated by the piezoelectric strain coefficient with variation in BaTiO$_3$ volume fractions. The d$_{33}$ decreases an increase in BaTiO$_3$ volume fractions

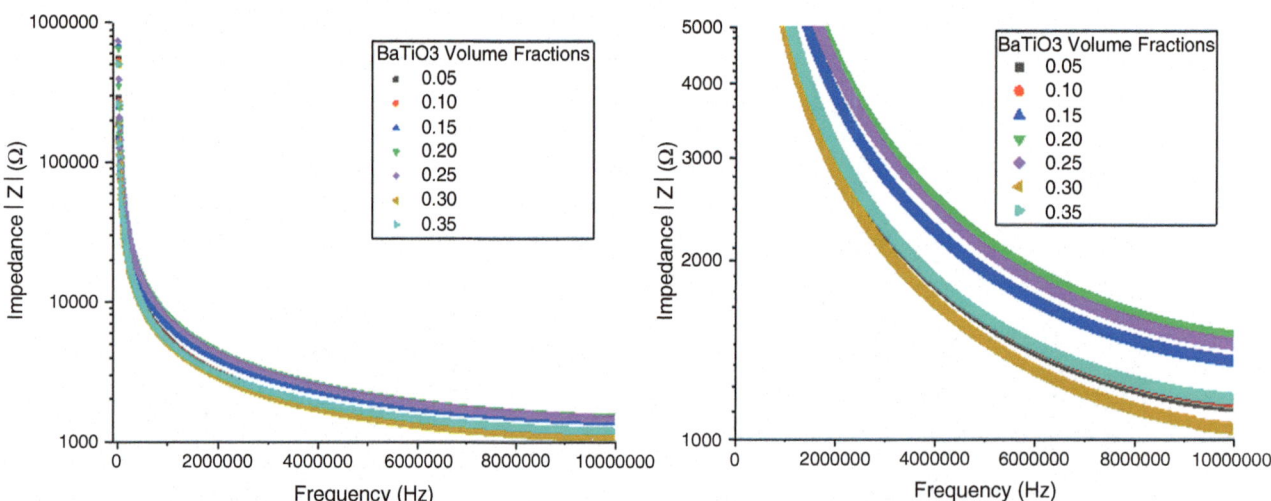

Fig. 2.4 The variation in impedance characteristics with change in BaTiO$_3$ volume fractions, which shows the increase in impedance values in volume fractions greater than 5% before reaching a drop in impedance values in volume fractions greater than 20%

applied force does not show any specific trend when compared to d$_{33}$. This indicates that the electro-mechanical coupling characteristics do not change with the change in loading values.

The variation in impedance, |Z| impedance values, IZI, with a frequency variation from 200 Hz to 10 MHz shows that the impedance decreases with an increase in BaTiO$_3$ volume fraction except for a volume fraction of 5%. The impedance spectrum is shown in Fig. 2.4. Also, the impedance values from 10 kZ to 500 kHz are approximately in the range from 100 to 10 kΩ. This range will be used for the quantitative and real-time structural health monitoring and damage detection for impedance based measurements. The impedance values decrease to approximately 1 kΩ at a frequency of 10 MHz. The phase angle measurements will also be taken to determine the resonant frequencies of the smart aggregates in the above range.

2.4 Structural Engineering Application

The sensors, as shown in Fig. 2.1, were placed in the middle of a concrete beam with the dimensions of 6 in. × 6 in. × 18 in. and were subjected to four-point flexural testing, using a Universal Testing Machine (UTM). The experimental setup is shown in Fig. 2.5. As presented in this figure, a strain gauge was used to measure deformations of the beam at its center. Additionally, the vertical displacement of the middle of the beam was recorded by the UTM. The deformation data from this testing was measured up to the flexural failure of the concrete beam; this data will be compared with those obtained from the smart aggregates to show the effectiveness of these electro-active sensors. The electro-active properties of the smart aggregates (i.e., impedance, resistance, and capacitance) were measured during loading with a sampling rate of 60s (spectrum data acquisition rate of 30s) and a frequency range of 20 Hz–10 MHz.

The strain gauge data, which is presented in Fig. 2.6, shows failure of the beam specimen after approximately 12 min of loading, with the maximum deflection occurring before fracture at about 11–12 min from the start of loading. Measured simultaneously with the strain gauges, the smart sensors' electrical data spectra are recorded in Figs. 2.7, 2.8 and 2.9, color-coded by the time at which the values were measured from the start of loading. The electrical impedance spectrum, as shown in Fig. 2.7, depicts peaks at different frequency ranges of 0.5–3.0 and 4.5–5.5 MHz. These peak impedance values of 340 Ω and 112.5 Ω, at frequencies of approximately 2.4 MHz and 4.6 MHz, respectively, occur at the same timeframe at which the strain gauge data recorded maximum deflection. For smaller deflections of the beam at about 10–11 min from the start

Smart aggregates placed in the middle of a concrete beam

Four-point flexural testing setup

Strain gauge attached to measure the vertical deformations

Fig. 2.5 Smart aggregates are used in a concrete beam to monitor its deflection

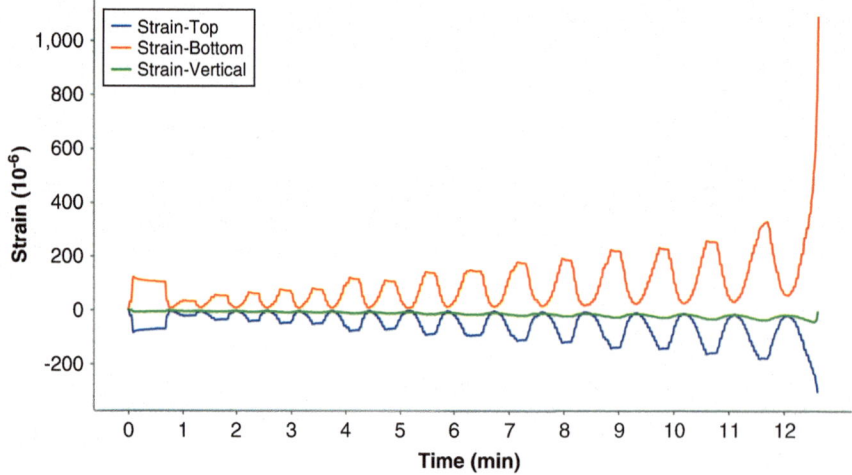

Fig. 2.6 Strain data measured at the center of the beam specimen during four-point flexural testing

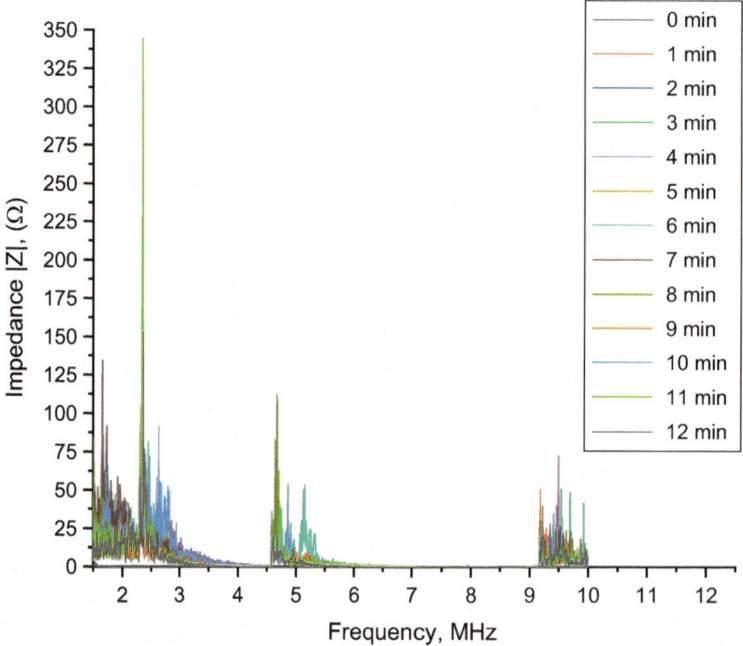

Fig. 2.7 Electrical impedance spectrum of the smart aggregates

of loading, the peaks heights in the same frequency ranges are approximately 105 and 62.5 Ω. A similar trend is observed with the other data spectra at lower deflections. The resistance spectrum, which is presented in Fig. 2.8, is consistent with the impedance data.

The correlation between electro-active properties and structural health monitoring was demonstrated by Sirohi and Chopra [14]. The smart aggregates can be treated as a parallel plate capacitor using the following equation:

$$C_P = \frac{e_{33}A}{t} \qquad (2.1)$$

where C_P is the parallel plate capacitance, e_{33} is the permittivity in the 3–3 direction, A is the cross-sectional area of the cylindrical aggregate, and t is the thickness of the aggregate. Due of the anisotropic nature of these materials, the electroactive properties depend on the direction. To identify directions, the axes 1, 2, and 3 is shown below (corresponding to X, Y, Z of

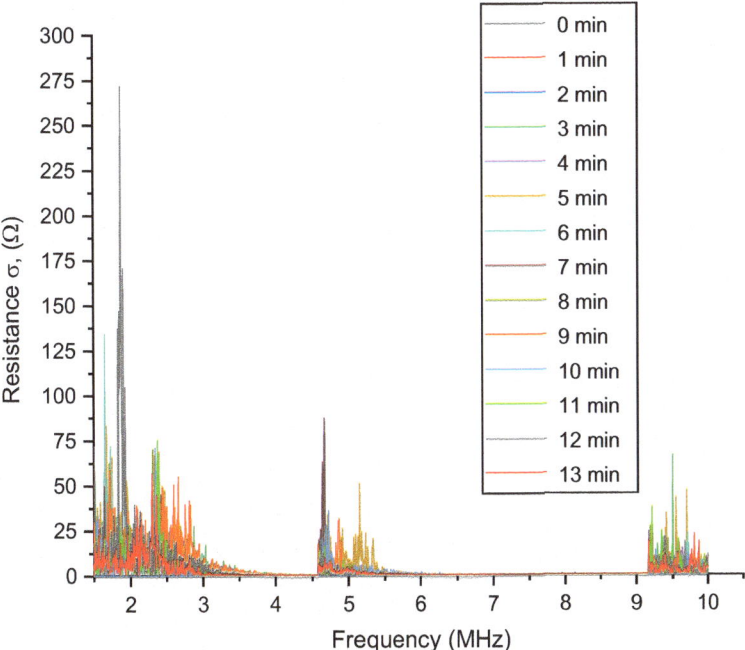

Fig. 2.8 Resistance spectrum of the smart aggregates, which is consistent with the impedance data

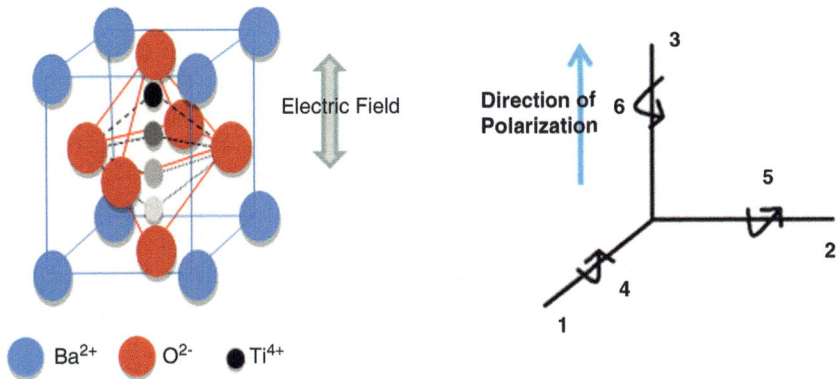

Fig. 2.9 In polycrystalline piezo-ceramics the dipole is aligned by the application of an external electric field. The orthogonal system describes the properties of a poled piezoelectric ceramic. Axis 3 is the poling direction

the classical right-hand orthogonal axis set) in Fig. 2.9. The axes 4, 5 and 6 identify rotations (shear), $\theta_X, \theta_Y, \theta_Z$ (also known as U, V, W). The axes and polarization (direction of dipole alignment), and the tetragonal $BaTiO_3$ crystal structure are also shown in the figure.

The capacitance spectrum is shown in Fig. 2.10 over a frequency range of 20 Hz to 10 MHz. A maximum peak height of 350 μF is observed for the spectrum at 11–12 min in the frequency range of 6–9 MHz. The capacitance values in the frequency ranges from 0.5–3 to 4.5–5.5 MHz are the lowest for the maximum beam deflection, which is due to lower electrical impedance and resistance values. The change in the electro-active properties can be attributed to the modification of the aggregates' microstructure with an increase in the beam deflection. Furthermore, the microstructural contact resistance is increased with increase in the beam deflection, which is depicted in Fig. 2.8. The magnitude of electrical impedance is the root mean square of the magnitude of the electrical resistance and the reactance. The increase in impedance with increase in beam deflection is observed at two different frequency ranges in Fig. 2.7. With the increase in contact resistance, the electron density is reduced due to lower electron transport, which leads to higher energy loss. As a result of this higher energy loss, the permittivity is reduced, which can be explained by Eq. (2.2) below with the real and imaginary components of permittivity,

$$e = e' - je'' \tag{2.2}$$

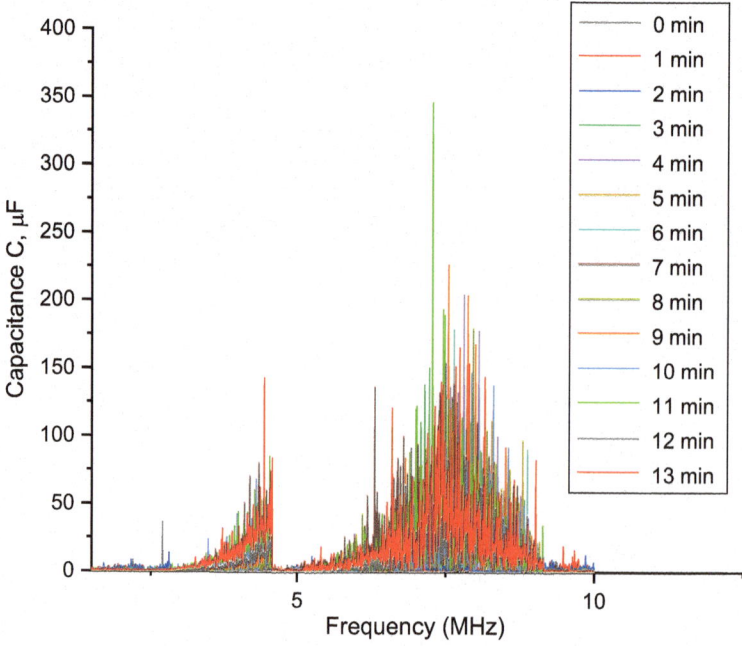

Fig. 2.10 Capacitance spectrum of the smart aggregates

The imaginary part of the permittivity is linked to the bound charge and dipole relaxation of the electro-active material. This increases the energy loss and in turn increases the magnitude of permittivity. The capacitance is directly proportional to the permittivity, as shown in Eq. (2.1). Reduction in permittivity due to higher energy loss will lead to lower capacitance values, as observed in the frequency ranges from 0.5–3 to 4.5–5.5 MHz, while the reverse phenomenon is observed in the rest of the spectrum, where lower energy loss leads to higher capacitance values.

The strain can also be related to the parallel plate capacitance, as shown in the equation below [14],

$$\varepsilon_1 = \frac{V_c C_p}{S_q} \tag{2.3}$$

2.5 Conclusion

The above work includes the detailed fabrication process for the $BaTIO_3$—Cement composite based smart aggregates. The change in electro-mechanical properties and in impedance characteristics of the smart aggregates at different volume fractions were analyzed. Experimental data suggests that variation in volume fractions of the active material in the composite affects the electro-mechanical coupling performance, while the applied force does not. The range of measurement for the impedance based monitoring method has also been determined, while phase angle measurements remain requisite for analysis to further determine the resonant frequencies of the smart aggregates in the above range. Electrical impedance based excitation of the smart aggregates was used for monitoring structural deformations in the host structure. Both the impedance and capacitance shows consistent changes with the beam deflection. Further quantitative analysis needs to be performed to establish relationships between strain and electroactive properties.

Acknowledgements The above work was partially funded by the Department of Defense, Southern California Edison, CSU Transportation Consortium, and Fresno State Transportation Institute.

References

1. Park, G., Inman, D.J.: Impedance-based structural health monitoring. In: Damage Prognosis: For Aerospace, Civil and Mechanical Systems, pp. 275–292 (2005)
2. Sodano, H.A., et al.: Comparison of piezoelectric energy harvesting devices for recharging batteries. J. Intell. Mater. Syst. Struct. **16**(10), 799–807 (2005)
3. Min, J., et al.: Impedance-based structural health monitoring incorporating neural network technique for identification of damage type and severity. Eng. Struct. **39**, 210–220 (2012)
4. Kim, M.-H.: Simultaneous structural health monitoring and vibration control of adaptive structures using smart materials. Shock. Vib. **9**(6), 329–339 (2002)
5. Qingzhao, K., et al.: A novel embeddable spherical smart aggregate for structural health monitoring: part I. Fabrication and electrical characterization. Smart Mater. Struct. **26**(9), 095050 (2017)
6. Yuan, M.Z., et al.: Processing method and property study for cement-based piezoelectric composites and sensors. Mater. Res. Innov. **19**(1), S1-134–S1-138 (2015)
7. Zhao, P., et al.: Investigation of cement-sand-based piezoelectric composites. J. Intell. Mater. Syst. Struct. **27**(12), 1666–1672 (2016)
8. Weijie, L., et al.: Feasibility study of using smart aggregates as embedded acoustic emission sensors for health monitoring of concrete structures. Smart Mater. Struct. **25**(11), 115031 (2016)
9. Xu, K., et al.: Real-time monitoring of bond slip between GFRP bar and concrete structure using piezoceramic transducer-enabled active sensing. Sensors. **18**(8), 2653 (2016)
10. Yang, Y., et al.: Sensitivity of PZT impedance sensors for damage detection of concrete structures. Sensors (Basel, Switzerland). **8**(1), 327–346 (2008)
11. Talakokula, V., Bhalla, S.: Reinforcement corrosion assessment capability of surface bonded and embedded piezo sensors for reinforced concrete structures. J. Intell. Mater. Syst. Struct. **26**(17), 2304–2313 (2015)
12. Cédric, D., Arnaud, D.: A study on the performance of piezoelectric composite materials for designing embedded transducers for concrete assessment. Smart Mater. Struct. **27**(3), 035008 (2018)
13. Yongli, M., et al.: A cement-based 1−3 piezoelectric composite sensor working in d 15 mode for the characterization of shear stress in civil engineering structures. Smart Mater. Struct. **27**(11), 115013 (2018)
14. Sirohi, J., Chopra, I.: Fundamental understanding of piezoelectric strain sensors. J. Intell. Mater. Syst. Struct. **11**(4), 246–257 (2000)

Chapter 3
Output-Only Estimation of Amplitude Dependent Friction-Induced Damping

Karsten K. Vesterholm, Tobias Friis, Evangelos Katsanos, Rune Brincker, and Anders Brandt

Abstract Identification of modal parameters, when a structure is under operational conditions is termed Operational Modal Analysis (OMA). Current OMA techniques are based on the assumption of linear time-invariant systems, and thus have limited applicability when applied to structures known to violate these assumptions. The present study investigates how the Random Decrement (RD) technique can improve robustness of OMA methods when friction-induced nonlinear damping is present in a system. This is done by estimating the amplitude dependent damping. A friction mechanism is introduced in a model of a structure, and by applying the RD technique at different amplitudes of simulated responses, RD signatures are produced, that represent the system vibrating with these amplitude levels. This allows the modal parameters to be estimated based on RD signatures computed with each amplitude level, using time domain parameter estimation methods, and the amplitude dependency of the damping is identified.

Keywords Random decrement · Random vibrations · Friction-induced Nonlinear damping · Operational modal analysis · Identification of nonlinearity

3.1 Introduction

The Random Decrement (RD) technique was invented in the late 1960s and the early 1970s by Cole [1–3], as a way to extract the 'signature' of a vibration signal, while the signal was being measured. It was invented to detect when damage in a vibrating structure occurred. The idea was, that damage was difficult to detect from observing the raw vibration signal, but easy to see in the signature. The RD technique is now an established output-only method for evaluating modal parameters of structures under random force inputs. It is possible to apply the RD technique, with a certain triggering condition, at multiple amplitude levels in the same response signal, and producing RD signatures that represent the system vibrating at these amplitude levels. The RD signatures can be treated as correlation functions, and amplitude specific modal parameters of the system can be estimated using time domain parameter estimation methods. When the modal parameters, representing different amplitude levels, are estimated, it reveals amplitude dependent nonlinearities in the system. The system investigated here has the stick-slip friction nonlinearity. The purpose of this study is to propose a new method of estimating the amplitude dependency of the damping caused by this type of nonlinearity, using only the measured response of the system. Amplitude dependent nonlinearities are present in many systems, and can cause difficulties, and erroneous results, since the current OMA methods are designed for linear systems [4].

The idea of applying the RD technique at multiple amplitude levels in one response signal was described by Jeary [5], where it successfully uncovered the amplitude dependent damping in tall buildings. The damping was estimated by fitting a decay curve to the envelope of the auto RD signatures.

The local extremum (LE) triggering condition was first presented by Tamura and Suganuma [6], as a method of evaluating amplitude dependent frequency and damping in buildings. The triggering condition was derived for a single-degree-of-freedom (SDOF) system, and experimental investigations of the amplitude dependent damping and frequency of 3 different towers was performed. In their methodology, the first step was to bandpass the response signals around the fundamental frequency with a narrow bandpass filter, to eliminate unwanted components. This analysis step can lead to erroneous results if the investigated system has closely spaced modes around the fundamental frequency. It is very plausible that this error

---

K. K. Vesterholm (✉) · A. Brandt
Department of Technology and Innovation, University of Southern Denmark, Odense, Denmark
e-mail: kav@iti.sdu.dk

T. Friis · E. Katsanos · R. Brincker
Technical University of Denmark, Danish Hydrocarbon Research and Technology Centre, Lyngby, Denmark

© Society for Experimental Mechanics, Inc. 2020
S. Pakzad (ed.), *Dynamics of Civil Structures, Volume 2*, Conference Proceedings
of the Society for Experimental Mechanics Series, https://doi.org/10.1007/978-3-030-12115-0_3

is present in their results, given that two of the three towers are symmetric, and the third is symmetric to some extent, and are therefore highly likely to have closely spaced modes. Furthermore, the frequency and damping was estimated using the logarithmic decrement technique, where the peak values from the first two periods in the RD signature were used.

A resent study in 2018 by Bajrić and Høgsberg [7] investigated the output-only estimation of the system parameters in an SDOF system with hysteretic damping, using Covariance Driven Stochastic Subspace Identification (COV-SSI) and the linearization of a Bouc–Wen hysteresis model. The proposed method identified the system parameters robustly, but the method requires that the mass is known. Hysteresis damping and stick-slip friction have similar properties, where hysteresis could be considered a more versatile description of the physics in an actual damper [8].

The methods proposed in the present study are meant to improve robustness in Operational Modal Analysis (OMA), when applied to nonlinear systems. The numerical system being investigated consists of an underlying linear system, with a localized stick-slip friction present, and is therefore a nonlinear system. This type of nonlinearity is present in many places, for instance in the bridge bearings, connecting two off-shore platforms. One of the great challenges of this nonlinearity is that it depends not only on the present state and input of the system, but also on the previous state of the system.

To the authors' knowledge, no one has estimated the modal parameters of a numerical multiple-degree-of-freedom (MDOF) system with a localized stick-slip friction with the purpose of identifying the amplitude dependent damping as a function of the response, using the RD technique. The purpose of this study is therefore to investigate a novel approach of performing a RD analysis together with a time domain modal parameter estimation, that can estimate the amplitude dependent damping as function of the response. Matlab R2018a is used for simulation, analysis and visualization of results.

3.2 Theory

This section describes the techniques and methods used during this study.

The present study applies the RD technique at multiple amplitude levels of simulated responses, and calculates the RD signatures using the LE triggering condition presented by Tamura and Suganuma [6], but the signal processing consisting of a bandpass filtering will be omitted to avoid the issues described above. The damping is estimated using a time domain parameter estimation method, where the auto- and cross RD signatures forms the equivalent of a correlation function matrix. With this method, many values in the RD signatures are used in the estimation, compared to the logarithmic decrement method used in [5, 6], where only the peak values from the first couple of periods were used. This means the damping estimates in the present study will have a higher statistical certainty, since it is based on more information. Also, since the RD signatures describe nonlinear vibrations, it cannot be assumed that the auto signature behaves entirely like an exponential decay, which is the assumption when using logarithmic decrement [8]. The limit of the approach proposed by Bajrić and Høgsberg is the assumption of a known mass. This assumption is not necessary for the approach proposed in the present study, which can result in an easier implementation. There is however, a trade-off between the two approaches, in the fact that Bajrić and Høgsberg estimated the mechanical parameters, and it is therefore possible to reconstruct the system from the estimated parameters. The present study estimates modal parameters without scaling, where it is not possible to reconstruct the system, without additional information.

For a linear system, these two assumptions apply. One: the modal parameters do not change when estimating them with RD signatures calculated from different amplitude levels in the response signal [9]. Two: the modal parameters do not change when the amplitude of the zero mean Gaussian force signal exciting the system, is changed. The second assumption is based on the principle of superposition. These assumptions do not apply for the nonlinear system investigated in the present study. The RD analysis is used to describe the amplitude dependent damping of the system. For this to be successful, the case where the system is excited by one force level and analyzed at a specific response level, the damping estimate must be exactly the same as for a case where the system is excited by a different force level, while RD is applied at the same specific response level. This means that the modal parameters are found to be the same for a specific response level, regardless of the force level.

3.2.1 Random Decrement

The RD signature is calculated as the average of N segments extracted from a discrete stochastic process. The segments used in the averaging are found by identifying certain points, called triggering points, in the signal, that satisfy some condition, known as the triggering condition (TC). There are two types of RD signatures, the auto signature, where the TC is applied to, and the segments are extracted from the same signal. The cross signature is where the TC is applied to one signal, and the segments are extracted from another signal.

The cross RD signature is calculated by

$$\hat{D}_{X,Y}(\tau) = \frac{1}{N} \sum_{i=1}^{N} x(t_i + \tau) | T_{y(t_i)} \tag{3.1}$$

where x is the signal, that N segments are extracted from, and $T_{y(t_i)}$ is the triggering condition applied to signal y, identifying N triggering points t_i. For the auto RD signature, x and y denote the same signal. The number of lags in the RD signature is controlled by the parameter M.

The triggering condition used in this study is the LE, where a triggering point is identified to be the local peak value, within a band of the response. This triggering condition was first proposed by Tamura and Suganuma to investigate amplitude dependent damping and frequency, where it was described as RD technique, ranked by peak amplitude. The TC was described more comprehensively by Asmussen [9], where the name LE was adopted, and formally defined as

$$T_{y(t)}^{LE} = \{a_1 \leq y(t) < a_2, \dot{y}(t) = 0\} \tag{3.2}$$

where a_1 and a_2 determine the band in the response, in which to find a peak, which occurs when $\dot{y}(t) = 0$.

The analysis is performed on both the positive and negative parts of the response signal. Doing so, approximately doubles the number of triggering points, which helps limit the random error in the RD signature.

For the further analysis, RD signatures can be treated as correlation functions, and form the equivalent of a correlation function matrix, as done in [4, 9], and be used in the time domain modal parameter estimation.

3.2.2 Multiple Reference Ibrahim Time Domain

The time domain modal parameter estimation method used in this study is the Multiple Reference Ibrahim Time Domain (MITD) method.

The Ibrahim Time Domain (ITD) method, first published by Ibrahim and Mikulcik [10], was developed to estimate modal parameters from free decays of linear systems, and is associated with the RD technique. ITD was extended to multiple references as described by Allemang and Brown [11]. MITD is very similar to COV-SSI, the identification method used in [7]. The implementation of MITD is done in matlab with the ABRAVIBE toolbox [12].

3.2.3 Equivalent Linear Damping

A simplified approach to calculating the relative damping of a mode affected by a Coulomb type friction damping is to consider one period of the harmonic response of an SDOF system at resonance, i.e., a modal period, at a specific amplitude. Equalizing the dissipated energy of a combined linear and friction damped system with the dissipated energy of a linear system, the equivalent linear, viscous damping at a specific amplitude for a mode oscillating at its natural frequency becomes [13]:

$$\zeta_{eq(n)} = \frac{1}{4\pi} \frac{E_{d(n)}}{E_{t(n)}} = \frac{\pi \omega_n c_n Q_n + 4Fd(n)}{2\pi k_n Q_n} \tag{3.3}$$

where $E_{d(n)}$ and $E_{t(n)}$ are the modal dissipated and total energy, respectively, and ω_n, k_n, c_n, Q_n, $Fd(n)$ are the nth modal components of the natural angular frequency, stiffness, viscous damping, displacement amplitude and friction force, respectively.

The equivalent linear damping is calculated to obtain a reference, with which to compare the results from the RD analysis.

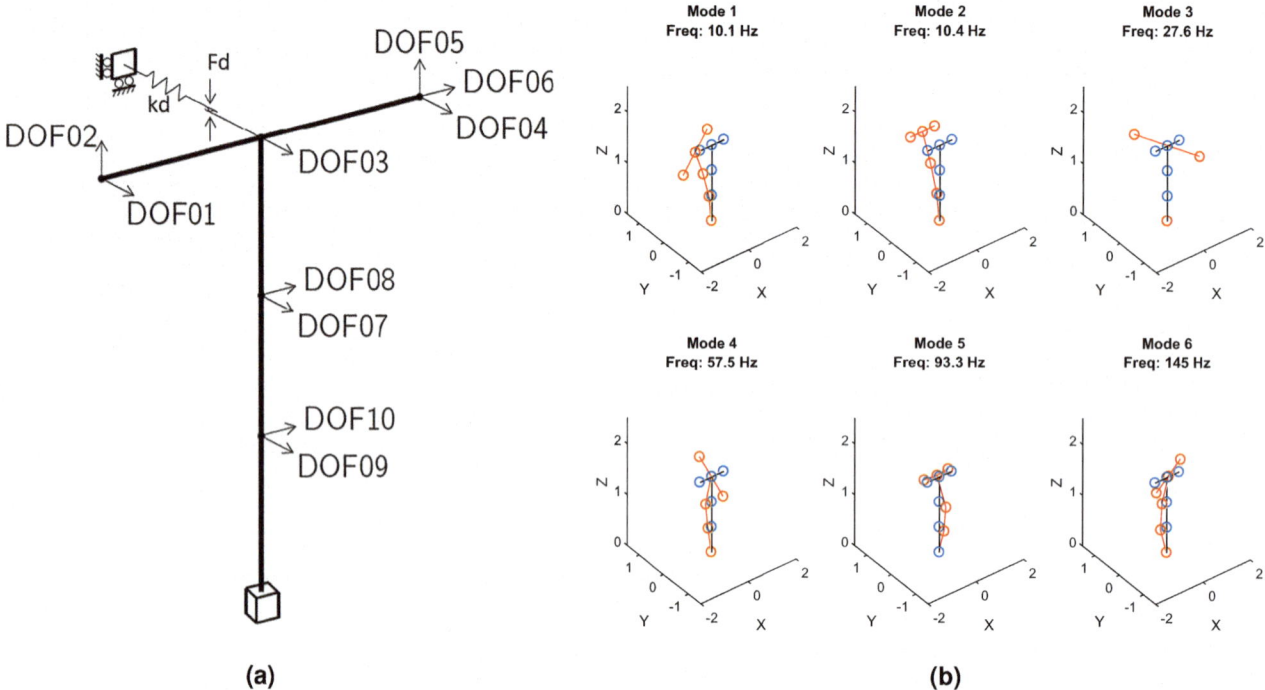

Fig. 3.1 Illustration of T-shaped structure. Notice that mode 1 and 2 are closely spaced, and that only mode 2 and 5 have in-phase motion in the y-direction. (**a**) 10 DOF model of T-structure, with stick-slip friction located at DOF03. (**b**) First 6 mode shapes and frequencies

3.3 Numerical Case Study

3.3.1 System Investigated

The system investigated in the present study is a numerical model, presented in Fig. 3.1a, where the first 6 modes of the system are shown in Fig. 3.1b. It is a T-shaped structure, that is clamped in its base, and has otherwise free boundary conditions, and is a 10 DOF system reduced from 156 DOFs [14]. The linear damping in the structure is implemented as modal damping, such that all modes have 1% relative damping. A stick-slip friction is positioned at DOF03, making the structure nonlinear.

The stick-slip friction describes a situation where two surfaces are in contact, for instance when a bridge is resting on a bridge bearing. There is friction between the two surfaces, meaning that when the bridge has a force acting on it, the resulting force has to overcome the friction, in order to move the bridge relative to the bearing. When the resulting force is not large enough to move the bridge, the system is in the 'stick' phase, and when the force overcomes the friction and moves the bridge, it is now in the 'slip' phase. In the slip phase, the friction force dissipates energy, meaning it acts as a damper. In the stick phase, the friction acts as a spring, which can have any stiffness, depending on the situation. This spring stiffness can describe different situations, and in this study, a very high stiffness is used together with a relatively low friction force compared to the external excitation, to maintain the mode shapes, and eigenfrequencies as in the linear case. This is done, since the purpose of this study is only to investigate an amplitude dependent damping alone.

For the T-structure, the stick-slip friction is implemented as Jenkins element going from DOF03 to ground. It is described by the parameters Fd, the friction force, and kd, the spring stiffness. In this study the stick-slip parameters $Fd = 2.5\,\mathrm{N}$ and $kd = 10^9\,\frac{\mathrm{N}}{\mathrm{m}}$ are used. Moreover, with the chosen parameters of the friction element and the later defined magnitudes of the external excitation, the friction-induced damping is assumed to behave similarly to Coulomb friction. This means the friction force Fd is the one appearing in Eq. (3.3). kd is so high that we can use the simplified approach in Eq. (3.3), and it does not appear in the Equation.

Since the stick-slip friction is acting in the direction of DOF03, it is expected that only modes that have motion in this direction are affected by the nonlinearity, meaning mode 2 and 5 in Fig. 3.1b. It is expected that the only parameter of these modes that has an amplitude dependency is the relative damping.

3.3.2 Analysis

Simulation Parameters

The numerical simulation is performed in matlab using a state space formulation as presented in [15], using a sampling frequency $f_s = 2048$ Hz, and with a time history length of 30 min. A zero mean Gaussian signal is used as a force signal, and all DOFs are excited by an independent force signal, all with the same amplitude.

Random Decrement Analysis Parameters

The number of lags in the RD signature, defined by the parameter M, must be large enough such that the number of lags in the RD signature passed to MITD can be adjusted during the investigations, and 512 is chosen. This leaves plenty of room for the analyst to tweak the number of lags in the RD signature passed to MITD.

Next, the triggering levels are defined. Firstly, the width of this triggering band, then at what response level it should be applied to is discussed. In Eq. (3.2), a_1 and a_2 determine the band in the response, where the local peaks are found. The width of the triggering band is chosen to be constant for all the triggering levels. This ensures, that when the RD analysis is applied to different response levels, the amplitude is equally well defined for all triggering levels. This comes at a cost, in the form that the number of triggering points in each band is not equal, since the number of peaks in the signal, is not the same for all amplitudes. It has the consequence that the RD signatures calculated from different triggering levels, are not based on the same number of averages. It is assumed that this will not influence the results in this study, because of the long time histories. The width of the triggering band is defined as 3% of the signal range of the response from each DOF.

Recommendations for the optimal range of the response signal, in which to perform the RD analysis, are given by Asmussen [9]. These recommendations apply for a linear system, where the desire is to find the best level to trigger the signal to obtain the best RD signature. The recommendation is that triggering should not be performed below the standard deviation of the triggering signal σ_y, and not above $2\sigma_y$.

When choosing triggering levels with the purpose of investigating the amplitude dependency, it is desired to have the largest range possible, and triggering levels between $1.2\sigma_y$ and $2.5\sigma_y$ are used in the present analysis. For higher triggering levels, the number of triggering points in the triggering band $[a_1, a_2]$ decrease to a point where the random error has a significant influence on the RD signatures. This random error propagates through the analysis and will increase the random error of the modal parameter estimates. For triggering levels below $1.2\sigma_y$, a Monte Carlo simulation showed that the variance of damping estimates increase significantly, and few stable poles were obtained from MITD. This was discovered in the preliminary tests, where the parameters of the RD analysis were tweaked until enough stable poles were repeatedly obtained in the parameter estimation.

Multiple Reference Ibrahim Time Domain Parameters

Once the RD signatures are obtained, the modal parameters are estimated using the MITD method. There are no general recommendations for the number of lags that should be used in MITD, and reasoning behind the number chosen in the present study is discussed here.

The RD signature with LE, can be thought of as a method of investigating what happens, on average, a number of time lags after the system has experienced a local peak in the response, that has some specific amplitude. For a nonlinearity that is dependent on the systems' present and past states, it is not trivial or simple to explain, what happens a number of lags after a local peak in the response. The RD signature can therefore not be described as a linear model of the free decay, as the assumption for RD signatures is in MITD. Measures must be taken to mitigate this deviation from the linear assumption, in order to obtain good modal parameter estimates. An idea to solve this issue was proposed by Brincker et al. [16], where a nonlinear rocking system was investigated. The idea is to only use a number of lags covering the very first part of the RD signature, as an effort to minimize a bias in the RD signatures. This idea is adopted in the present study, and 110 lags from the RD signature are used in MITD. 110 lags cover a little more than half a period with the lowest frequency in the RD signature.

As a method of removing noise from RD signatures, in case any was present, the first 5 lags are omitted from all RD signatures, before passing them to the MITD method [17].

In summary, this means that the first 5 lags are skipped, and then the next 110 lags of the RD signatures are used in the MITD.

Monte Carlo Simulation

To establish a baseline for the damping estimates $\hat{\zeta}$, a Monte Carlo simulation, using 50 runs, was performed with various excitation force levels. From each case with a different force level, the RD analysis was performed at 3 triggering levels, $1.2\sigma_y$, $1.85\sigma_y$, and $2.5\sigma_y$. The specific triggering levels were different for each DOF, since the response level is different, and triggering levels used in the final Monte Carlo simulation was calculated from the triggering levels $1.2\sigma_y$, $1.85\sigma_y$, and $2.5\sigma_y$ for each force level, as an average of the triggering levels from 50 runs. This ensures that the same triggering level was used in the final Monte Carlo simulation. From the 50 runs, the standard deviation, and the mean of the damping estimates were calculated, and a region spanning from the mean ± 3 standard deviation of the damping estimate $3\sigma_{\hat{\zeta}}$ was formed. This was done for all the triggering levels in the different cases with different excitation force levels. The outcome of the Monte Carlo simulation will constitute the final results of this study, and be compared with the equivalent linear damping.

Equivalent Linear Damping

The equivalent linear damping in Eq. (3.3) is defined for a modal period, meaning the quantities needed must first be calculated, which is described in the following. The modal stiffness and modal damping, are found by pre- and post multiplying the stiffness and damping matrices from Sect. 3.3.1 with the real part of the estimated mode shape matrix Ψ, that has mode shape vectors as columns. To calculate the modal displacement values Q_n, a vector of linearly spaced displacement values going from 0.0005 to 0.01 is created and multiplied with mode shape 2 in DOF 3 $\Psi_{3,2}$, meaning the third row in second column of Ψ. DOF 3 because it is chosen as the reference DOF, and mode 2 is used because it is the mode of interest regarding the results of the analysis.

3.4 Results and Discussion

The results presented will have a main focus on mode 2, since it is most affected by the nonlinearity.

When both a friction damper and a viscous damper are affecting a mode, which is the case for mode 2, the relative damping ζ will decrease when the response level is increasing. The explanation is that the dissipation force in a friction damper is determined by a constant F_d, and the dissipation force for the linear term is the damping coefficient c times velocity $\dot{y}$. For a low velocity, F_d is the governing term regarding the dissipation of energy. This results in a high ζ. When the response level increases, the linear dissipation force increases, while the friction damping force is constant. This results in lower ζ, compared to when the response was lower.

For the various force levels investigated in the Monte Carlo simulation, the minimum number of triggering points in a response signal from the DOF's related to mode 2, was 930 in DOF03 when using a triggering level of $2.5\sigma_y$. 930 triggering points is considered sufficient, but not excellent, to calculate an accurate RD signature. The number of triggering points changed from 1800 to 930 for the force levels 15.2 N and 55.6 N, a phenomenon that does not occur in the purely linear case, and serves as an indicator that the system is nonlinear.

In Fig. 3.2, the mean of the ζ estimates $\pm 3\sigma_{\hat{\zeta}}$ for modes 1 and 2 from the Monte Carlo simulation are compared with the equivalent damping, given in Eq. (3.3). The response from DOF03 is chosen to be the reference, and the triggering level on the 1st axis is from DOF03. The Monte Carlo simulation, represented by Fig. 3.2 shows that for mode 1, $\hat{\zeta}$ is between 0.8 and 1.2 %. This is the expected result, and can be seen by two blue dashed lines enveloping the solid green line, representing the equivalent linear damping. This shows that the closely spaced modes 1 and 2 are successfully separated in the analysis. $\hat{\zeta}$ for mode 1 are as accurate as if the system had been linear.

$\hat{\zeta}$ of mode 2, is found to be amplitude dependent, and inversely proportional to the triggering level. The two dashed black curves envelop the equivalent linear damping, the solid red curve. This means that the RD analysis with the MITD parameter estimation falls within $3\sigma_{\hat{\zeta}}$ of the equivalent linear damping. This is considered a satisfactory result. However, there is still room for improvements. The width of the $\mu_{\hat{\zeta}} \pm 3\sigma_{\hat{\zeta}}$ is very wide for some triggering levels, for instance at 10^{-3} m. Here $\hat{\zeta}$ is estimated to be anywhere between 11% and 15.5%, which is a very large margin.

The desire to have a RD analysis that estimates the same ζ for a particular triggering level, regardless of the force level applied to the system has not been satisfied in the present study. As the force level decreases, the damping estimates increase relative to triggering level. The phenomenon is most pronounced at the low triggering levels, where the damping is high. As the response level increases, the damping estimates for the different force levels start to meet each other, and eventually overlap when the damping estimate is around 4%.

In Fig. 3.2, $\mu_{\hat{\zeta}} \pm 3\sigma_{\hat{\zeta}}$ for mode 2 has an increasing uncertainty as $\hat{\zeta}$ increases. Some of the increased uncertainty is caused by the nonlinearity, but not all of it. When a linear system has a large ζ, the uncertainty of the estimate is also large. This is illustrated in Fig. 3.3, where Monte Carlo simulations with 50 runs of linear systems with ζ of mode 2 ranging from 1% to 15%, were performed. The analysis is otherwise exactly the same, as for the nonlinear system. The graph in Fig. 3.3, shows the mean ± 3 standard deviations of the 25 estimates, against the different $\hat{\zeta}$ of mode 2, and a trend is clear. The uncertainty of the damping estimate is increasing when the damping is increasing for linear systems. This increase in uncertainty should be taken into account, when interpreting the results in Fig. 3.2.

Mode 5 is also affected by the stick-slip friction, but the influence is much more subtle. $\hat{\zeta}$ is found to be around 1.2%, for the lowest triggering levels investigated, and drops quickly to 1% which is what the linear damping is.

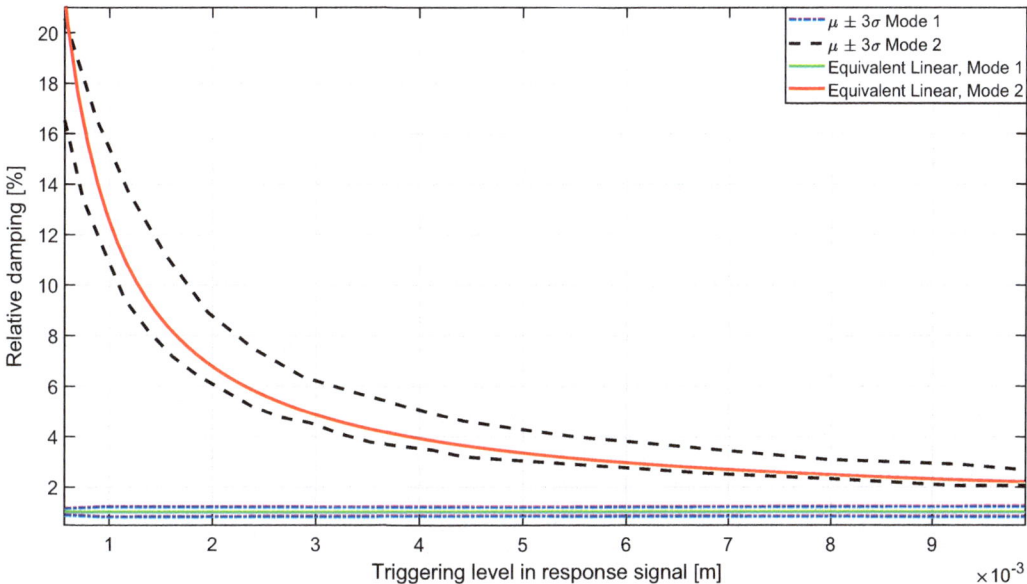

Fig. 3.2 Regions representing the damping estimates for modes 1 and 2 are illustrating the results from the Monte Carlo simulation. The region for mode 1 is between the two dot-dashed blue lines, representing $\mu_{\hat{\zeta}_1} \pm 3\sigma_{\hat{\zeta}_1}$. The region for mode 2 is between the two dashed black lines representing $\mu_{\hat{\zeta}_2} \pm 3\sigma_{\hat{\zeta}_2}$. Inside the estimation regions for modes 1 and 2, are the equivalent linear damping, the solid green and solid red lines respectively

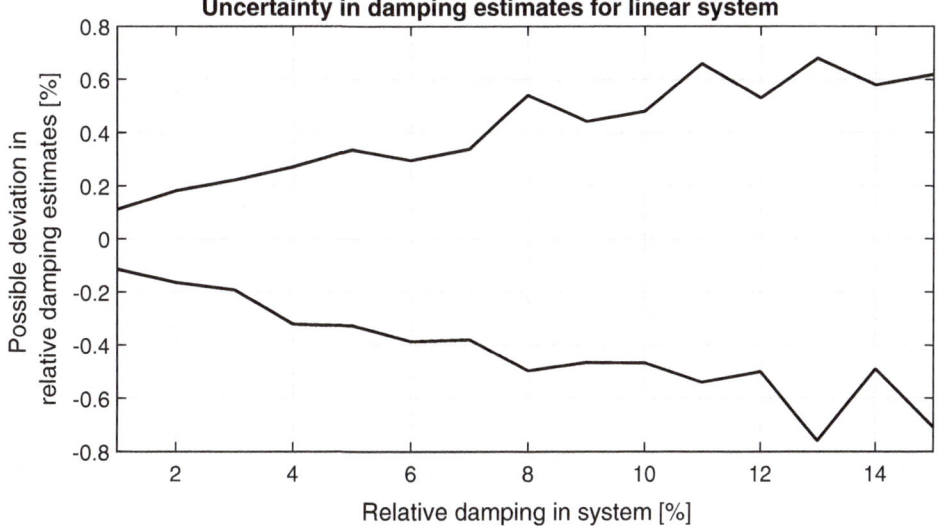

Fig. 3.3 From a Monte Carlo simulation with 25 runs, $3\sigma_{\hat{\zeta}}$ is calculated and plotted against the linear damping in system with ζ ranging from 1% to 15%, representing a possible deviation in the estimate

Table 3.1 Results from a Monte Carlo simulation of a standard OMA analysis using 50 runs

RMS force level [N]	$\mu_{\hat{\zeta}}$ %	$3\sigma_{\hat{\zeta}}$ %
15.2	13.7	1.0264
17.7	10.9	1.0734
20.2	8.9	0.8195
22.8	7.6	0.7512
25.3	6.5	0.3815
27.8	5.8	0.4716
30.4	5.2	0.3859
32.9	4.7	0.3171
35.4	4.3	0.4214
37.9	3.9	0.3134
40.5	3.7	0.2935
43.0	3.5	0.3641
45.5	3.3	0.3223
48.1	3.1	0.2949
50.6	2.9	0.2167
53.1	2.8	0.2197
55.7	2.7	0.3018

Excitation force in the left column, with corresponding mean and 3 standard deviations of damping estimate for mode 2 in middle and right column respectively

A standard OMA analysis has been performed to compare with the results from the RD analysis. Everything is the same for the numerical model and Monte Carlo simulation. The differences between the analyses are that a correlation function matrix is calculated instead of a RD matrix, and the first 300 lags of the correlation functions are used in MITD. This is considered a standard OMA setting, and the results are presented in Table 3.1. It is not possible to show the results from the standard OMA analysis in the same plot as the RD analysis, because these damping estimates cannot be associated with a specific response level. This also means that the results are difficult to compare with the equivalent linear damping, meaning it is difficult to determine if the results are correct or not. The tendency of the standard OMA estimates is the same as for the RD analysis, that for a low force level, the relative damping is high, and it lowers when the force is increased.

3.5 Conclusion

It can be concluded that the friction mechanism in DOF3 is causing the relative damping of modes 2 and 5 to exhibit amplitude dependent behavior.

The suggested method is successful in identifying that only modes 2 and 5 have amplitude dependent relative damping, even when modes 1 and 2 are closely spaced. The functional form of the damping estimates of mode 2 as a function of triggering level, is very similar to the equivalent linear damping. From this it can be concluded that the proposed approach to investigating the amplitude dependent damping of a system with a stick-slip friction mechanism is close to uncovering the parameters of the amplitude dependency, but further investigations are needed. Nothing about the system has to be known in order to perform this analysis, which means it could be used in structural dynamics output-only setting, to investigate the amplitude dependency of the relative damping. It can also be concluded that a standard OMA analysis cannot determine how the relative damping of mode 2 depends on the response amplitude with the same level of detail as the RD analysis is capable of.

Acknowledgement The authors acknowledge the funding received from Centre for Oil and Gas – DTU/Danish Hydrocarbon Research and Technology Centre (DHRTC).

References

1. Cole, H.A.: On-the-line analysis of random vibrations. In: AIAA/ASME 9th Conference on Structural Dynamics Materials. Palm Springs, USA (1968)
2. Cole, H.A.: Failure detection of a space shuttle wing flutter by random decrement. In: NASA, TMX-62,041 (1971)
3. Cole, H.A.: On-line failure detection and damping measurement of space structures by random decrement signatures. NASA CR-2205 (1973)
4. Brincker, R., Ventura, C.: Introduction to Operational Modal Analysis. Wiley, London (2015)
5. Jeary, A.P.: Establishing non-linear damping characteristics of structures from non-stationary response time-histories. Struct. Eng. **70**, 61–66 (1992)
6. Tamura, Y., Suganuma, S.Y.: Evaluation of amplitude-dependent damping and natural frequency of buildings during strong winds. J. Wind Eng. Ind. Aerodyn. **59**(2), 115–130 (1996)
7. Bajrić, A., Høgsberg, J.: Estimation of hysteretic damping of structures by stochastic subspace identification. Mech. Syst. Signal Process. **105**, 36–50 (2018)
8. Singiresu, S.R.: Mechanical Vibrations. Prentice Hall, Upper Saddle River (2011)
9. Asmussen, J.C.: Modal analysis based on the random decrement technique: application to civil engineering structures. PhD Thesis, University of Aalborg, 1997
10. Ibrahim, S.R., Mikulcik, E.C.: A method for the direct identification of vibration parameter from the free responses. Shock Vib. Bull. **47**(4), 183–198 (1977)
11. Randall J. Allemang and David L. Brown. Experimental modal analysis and dynamic component synthesis. Modal Parameter estimation, vol. 3. Technical report, University of Cincinnati, Department of Mechanical and Industrial Engineering, 1987
12. Brandt, A.: ABRAVIBE–A MATLAB toolbox for noise and vibration analysis and teaching (2018)
13. Chopra, A.K.: Dynamics of Structures. Pearson Education, London (2012)
14. O'Callahan, J.: System equivalent reduction and expansion process. In: Proceedings of the 7th International Modal analysis Conference on Society of Experimental Mechanics, pp. 29–37 (1989)
15. Lu, L.Y., Chung, L.L., Wu, L.Y., Lin, G.L.: Dynamic analysis of structures with friction devices using discrete-time state-space formulation. Comput. Struct. **84**(15–16), 1049–1071 (2006)
16. Brincker, R., Demosthenous, M., Manos, G.C.: Estimation of the coefficient of restitution of rocking systems by the random decrement technique. In: Proceedings of the 12th International Conference on Modal Analysis (1994)
17. Orlowitz, E., Brandt, A.: Influence of noise in correlation function estimates for operational modal analysis. In: Topics in Modal Analysis & Testing, vol. 9, pp. 55–64. Springer, Berlin (2019)

Chapter 4
Modeling Human-Structure Interaction Using Control Models When Bobbing on a Flexible Structure

Ahmed T. Alzubaidi and Juan M. Caicedo

Abstract High performance materials has enabled engineers to design civil structures with smaller dead loads than in the past. However, lower dead loads results in a higher live to dead load ratio and the possibility of excessive vibrations due to human loading. This paper extends a controller theory based model to model the human-structure-interaction (HSI) problem. Prior work focused on modeling a standing individual with bent knees using a proportional, integrative and derivative (PID) controller model. This work extends this idea to a person bobbing or performing short movement up and down by bending his or her knees at the frequency provided by a metronome. Prior work considered the input to the human-structure system was a force applied to the structure. This work consider the bit produced by a metronome as the input to the overall human-structure system. The force applied to the structure is modeled as the output of the human, while the structure's acceleration is fed back into the control human system. Experiments performed at the University of South Carolina using a flexible platform that behaves as a single degree of freedom system are used to test the model. A force plate is installed in the platform to measure the forces exerted by the person on the platform as he or she moves. Model parameters and their corresponding uncertainty are quantified in a probabilistic fashion using Bayesian inference with the force plate forces as well as the acceleration measurements of the structure as observations. The model performance is evaluated by comparing probabilistic predictions with force and acceleration measurements found experimentally.

Keywords Human-structure interaction · Control theory · Human activity · Structural dynamics · Bayes inference

4.1 Introduction

New high performance strength materials have allowed structural engineers to design and construct structures using less material. Slender and lightweight structural members in grandstand, dance floor, malls, and fitness center have the potential to have vibration problems due to a higher live to dead load ratio [1, 2]. There are many examples of excessive vibration induced by human walking, and/or dancing [2]. The Millennium bridge is arguably one of the most well known examples. This structure caught the attention of researchers due to excessive vibration created by people walking [3]. Other examples are the 1831 footbridge failure in Broughton, UK due to unison marching of soldiers [4], and the excessive vibrations of the 30 floor TechnoMart building caused by aerobic exercise in Seoul, South Korea [5]. These excessive vibrations occurred not only because of the small structural damping typical of civil structures but because of the low mass and large spans of the structure.

The interdependence of the structural and human sub-systems cause what some researchers call the human-structure interaction (HSI) phenomenon, where the overall dynamic system might have a new properties [6]. Two remarkable issues appear in human-structure interaction problems: the combined human structure system, such as natural frequency and damping ratio can make the structure prone to excessive vibration due to human dynamic load. The second issue is the synchronization among the people because of the interaction with structure and crowd dynamics that occurred in the London Millennium Bridge[3, 6–8].

These excessive vibrations have captured the attention of scientists and engineers who have developed models to study and explain the combined human-structure system. Classical models such as the mass, damper, spring (MDS) has been widely used. Lasprilla et al. [9], Sachse et al. [10], Wei and Griffin [11], Brownjohn [12]. However, MDS models cannot add energy to the system and therefore, cannot consider other sources of excitation to the overall human-structure system such as sound (or music) excitation. Recently, Ortiz and Caicedo proposed a new model based on the control theory where they

---

A. T. Alzubaidi (✉) · J. M. Caicedo
Department of Civil and Enviromental Engineering, University of South Carolina, Columbia, SC, USA
e-mail: alzubaa@email.sc.edu; caicedo@cec.sc.edu

used the Proportional, Integrative and Derivative (PID) controller to represent the effect of an individual or a group standing on the flexible structure [1, 9]. The contribution of their work was to frame the human-structure interaction HSI problem as a control problem where the plant is the structure and the controller is modeling the person or people.

In this paper, we expanded the model proposed by Ortiz-Lasprilla and Caicedo [1] where we propose the other use of a controller to include sound (or the bits created by a metronome) as an input to the overall HS system. Experimental data have been collecting for people bobbing (flexing their knees) on a flexible structure behaving as a single degree of freedom system. This experimental data is used to update the parameters of the controller in a deterministic fashion and create a probabilistic model that models the HSI.

This paper is organized as follows. Section 4.2 describes provides an overview of the models overview. A brief discussion of the experimental setup and instrumentation is given in Sect. 4.3. Section 4.4 a comparison between the experimental tests and the predictions of the model is provided. Finally, a brief set of conclusions are discussed.

4.2 Background

Several applications such as robotics, radar antenna, and automobile steering control work based on control theory. Control systems can be classified as open and closed loop control system [13]. Closed loop control of linear system fundamentally uses the concept of feedback. A measure from the plant is used as input to the controller and control device which provides an additional input to the plant. In this study, we modeled HSI using the same concepts where the structure is modeled as the plant of the system and the human is modeled as the controller.

4.2.1 Human-Structure Model

It is important to highlight that the objective of closed loop controllers are to obtain a desired response of the plant by using feedback and reduce the system error [13, 14]. While in most applications the objective is to reduce the plant output (e.g. reduce vibrations), in this particular application the objective is to mimic the behavior of the combined human-structure system. The block diagram used by Ortiz-Lasprilla and Caicedo [1] consists of a plant $G(s)$ and controller $H(s)$. In this paper we expand this idea by including an additional block to model the response of a person to sound. Figure 4.1 shows the proposed model where $H_1(s)$ represents the dynamics of the person due to floor motion, and $H_2(s)$ represents dynamics of the person due to sound excitation. The excitation force on the structure is represented by the term $B(s)$ whereas the term $C(s)$ represents the output acceleration. $M(s)$ represents the input in form of sound. The combined structure and human system can then be described by the equation

$$TF(s) = \frac{G(s)H_2(s)}{1 + G(s)H_1(s)} \qquad (4.1)$$

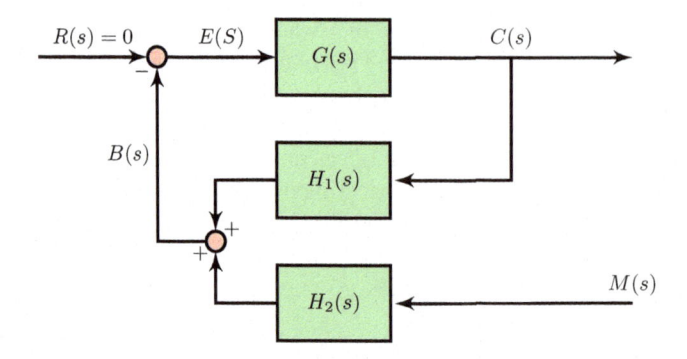

Fig. 4.1 Block diagram of a closed-loop control system

4.2.2 Human Model

Two PID controllers were used in this study, although the idea of close loop control does not limit the models to be PID controllers. The first controller takes acceleration of structure as input and the second takes the sound of a metronome as input. The output of both controller are the force applied to the structure and are added as input to the structure $G(s)$. Each PID controller transfer function has three parameters which are k_p, K_i, and K_d as described in the following equation

$$H_1(s) = K_p + \frac{K_i}{s} + K_d s \tag{4.2}$$

4.2.3 Structural Model

A cantilever structure was built at the Department of Civil and Environmental Engineering at the University of South Carolina and has been used to experimentally test the proposed model. This structure consists of a steel frame and concrete blocks as shown in Fig. 4.2. This structure has a mobile support and masses which are used to change the dynamic properties of the system and the live to dead load ratio. In this study, the length of the cantilever is 124 inch. The structure can be modeled as single degree of freedom and described by the equation

$$G(s) = \frac{\frac{1}{m}}{s^2 + 2\zeta\omega_n s + \omega_n^2} \tag{4.3}$$

where m, ζ, ω_n are the mass, damping ratio, and natural frequency of system. The poles which are the root of denominator of transfer function are expressed in terms of the natural frequency (ω_n) and damping ratio ζ of the structure as shown in the following equation [15]

$$p_{1,2} = -\zeta\omega_n \pm \sqrt{(\zeta\omega_n)^2 - \omega_n^2} \tag{4.4}$$

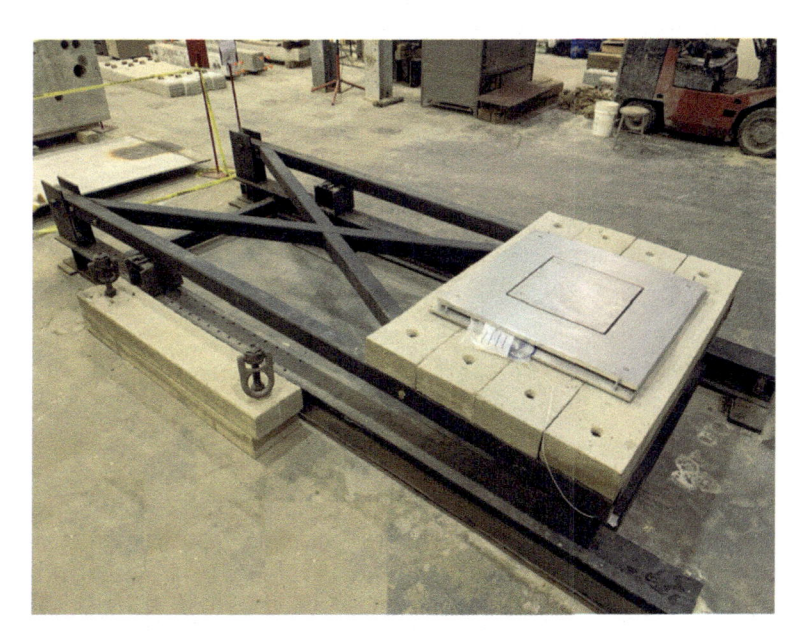

Fig. 4.2 Experimental structure

4.3 Experimental Testing and Model Updating

A PCB 096D50 impact hammer with a sensitivity of 0.2198 mV/N and a PCB 333B50 accelerometer with a sensitivity of 1019 mV/g was utilized for experiments. The accelerometer was used to measure the vertical acceleration at the tip of the cantilever and the impact hammer was used to excite the empty structure. In addition, a PCB 130F20 microphone with a sensitivity of 40.2 mV/Pa and a force plate developed in house were utilized to measure the sound created by a metronome and the forces exerted by the person on the structure. The data was collected using an NI 9234 data acquisition system. Data was collected in 20 second records at a sampling frequency of 6400 Hz. While this frequency is high for the structure, it was required to correctly describe the pressure captured by the microphone. The frequency response function of the system was estimated using the equation [16]

$$\text{TF}(f) = \sqrt{h_1(f) * h_2(f)} \tag{4.5}$$

where

$$h_1(f) = \frac{S_{xx}(f)}{S_{xy}(f)}, \qquad h_2(f) = \frac{S_{yx}(f)}{S_{yy}(f)} \tag{4.6}$$

S_{xx} and S_{yy} represent the auto spectral densities for the force and acceleration respectively and S_{xy} is the cross spectral density between the output acceleration and the input force. S_{yx} is the cross spectral density between the input force and the acceleration.

Three different type of experiments were performed. The first test used the empty structure and it was performed with the objective of investigating the structural parameters only. The second and third tests were performed with a person standing with bent knees and a person bouncing (bending their knees) with a specific beat provided by a metronome.

4.3.1 Bayes Inference

Bayesian inference is used to update the parameters of the human and structural models. Bayes theorem is expressed by the equation

$$P(\Theta|D) = \frac{P(D|\Theta)P(\Theta)}{P(D)} \tag{4.7}$$

where $P(\Theta|D)$ is the posterior probability density function of the parameters Θ given the observed data D. $P(D|\Theta)$ is the likelihood, and $P(\Theta)$ is the prior probability density function of the parameters. The prior expresses our knowledge or beliefs about the parameters before updating. Chain Monte Carlo methods (MCMC) are used to sample the posterior [17–19].

4.3.2 Empty Structure

In the empty structure experiments the impact hammer was used to excite the structure. Figure 4.3a shows the input force of the hammer and Fig. 4.3b show the acceleration response.

The prior probability density function PDF distribution of structure mass is assumed to be a normal distribution P(mass) $\sim N(600, 25)$. The prior PDF of the natural frequency of the empty structure is assumed to be a normal distribution P(ω_n) $\sim N(18.85, 1.0)$. The prior PDF for the damping ratio was assumed as P(ζ) $\sim N(0.006, 0.004)$. The likelihood function was assumed as a Normal distribution with a standard deviation modeled by an Inverse Gamma with shape $\alpha = 4$ and scale $\beta = 0.2$. A Markov Chain Monte Carlo MCMC algorithm was utilized to sample P($\Theta = \{$mass$, \omega_n, \zeta\}|D$). The posterior distributions for mass, ω_n and ζ were used as prior distributions for the model of the human-structure system of the person bouncing.

4.3.3 Person Standing with Bent Knees

The structure was tested again with a single occupant standing over the structure with bent knees and not moving. The structure was excited by the impact hammer similar to the excitation of the empty structure. The experimental transfer function of occupied structure was calculated using Eq. (4.5) and it is shown in Fig. 4.4. The parameters of the closed

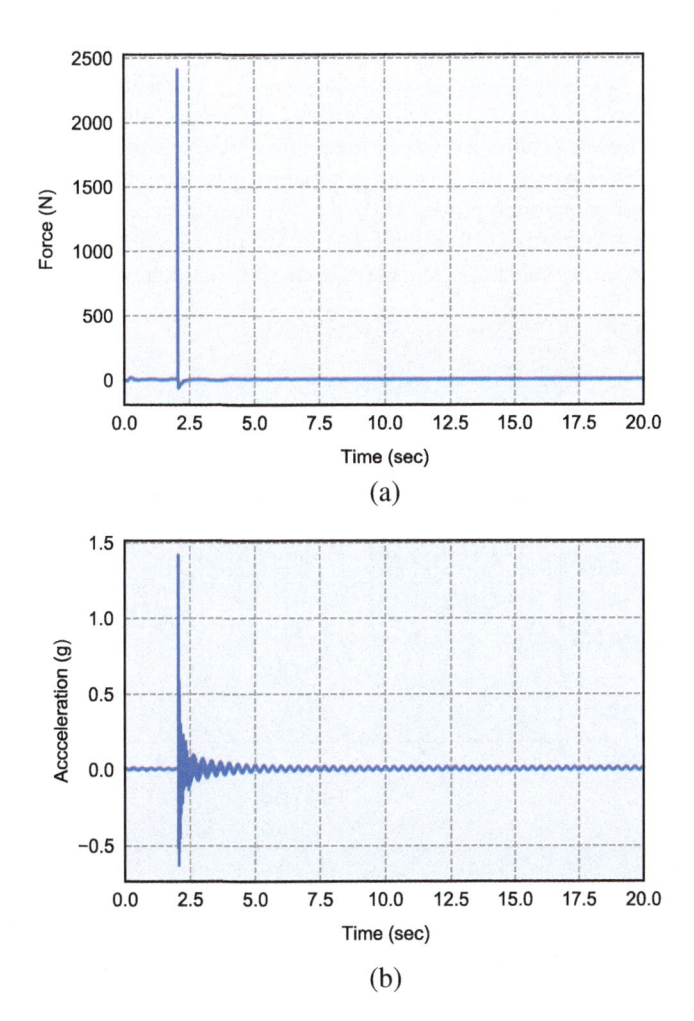

Fig. 4.3 Acceleration and impact force-time history of empty structure. (**a**) Impact force-time history. (**b**) Acceleration-time history

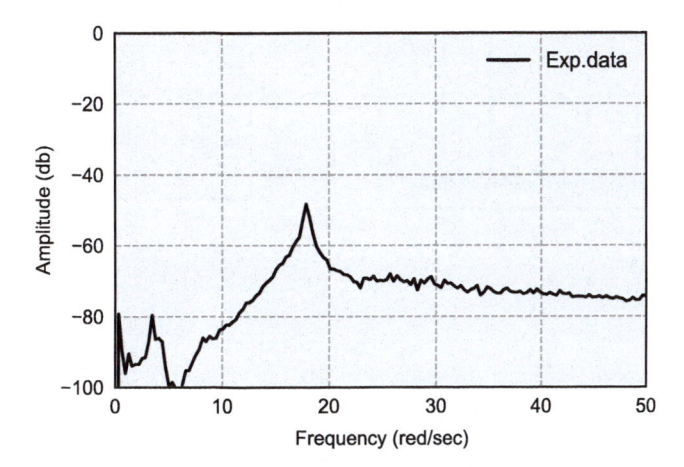

Fig. 4.4 Experimental transfer function of occupied structure

loop control system were updated, finding the posterior distribution of the parameters $\Theta = \{mass, \omega_n, \zeta, K_p, k_d, k_i\}$. The posterior $P(\Theta|D)$ from the previous experiment was used to inform the structural parameters.

4.3.4 Individual Bouncing on Flexible Structure

A test using the force plate was used to understand the interaction between the human and the flexible structure. Here the force generated by the person bobbing was measured to update the second PID controller ($H_2(s)$ in Fig. 4.1). During bouncing the subject move her/his body up and down with his/her knees bent and keeps fully contact with the floor. In other words, the person is not jumping. Bouncing is a simpler activity to model than jumping and it is the focus of this paper [20, 21]. A metronome was set to 120 bpm and the person standing in the structure was asked to excite the structure with his/her feet at the frequency of the sound. The sound pressure produced by the metronome, acceleration, and load applied by the person were acquired using data acquisition system and shown in Fig. 4.5. The prior probability density function PDF distributions of a person bouncing model are shown in Table 4.1. The parameters of the structure and both PID controllers were updated using experimental data.

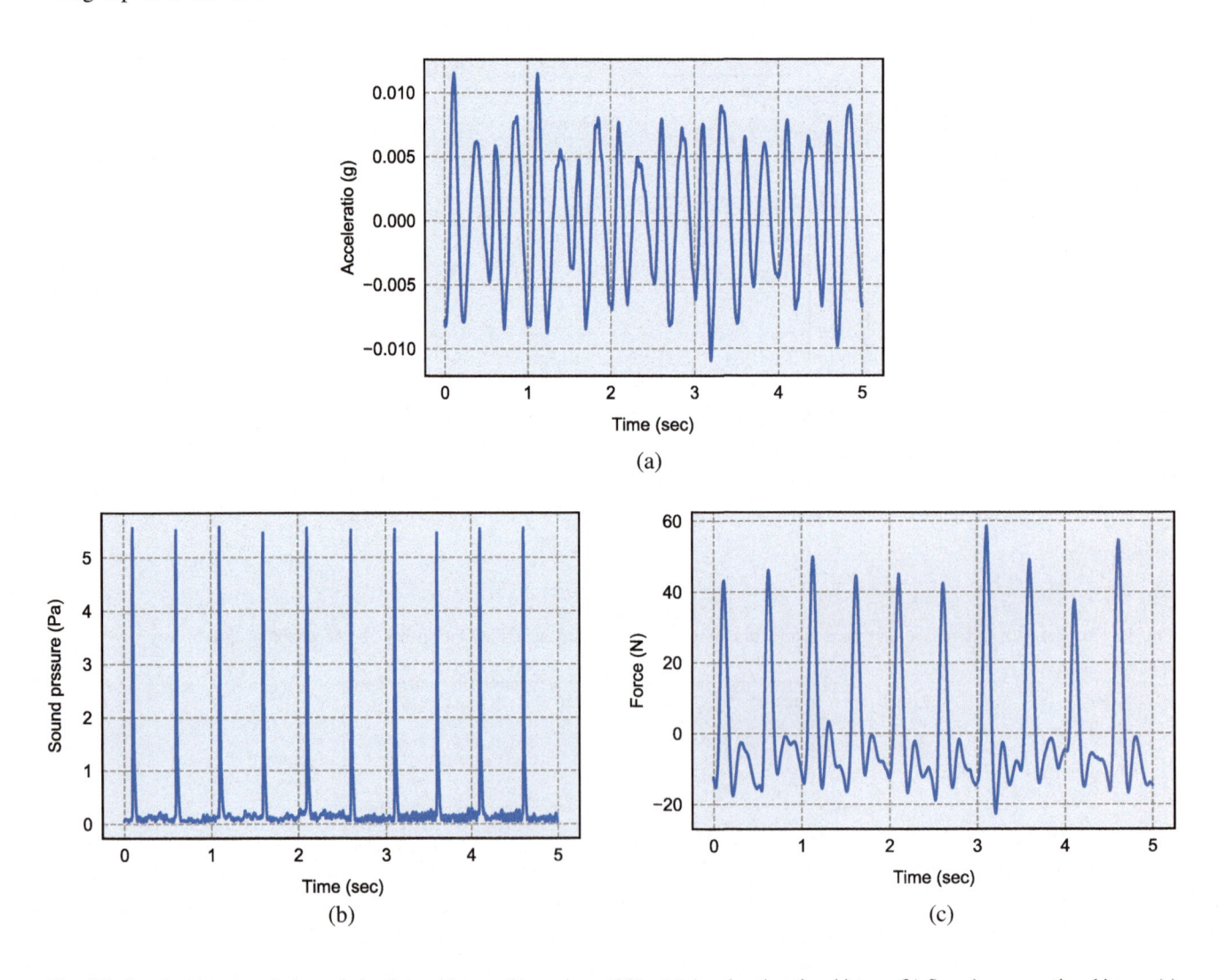

Fig. 4.5 Acceleration, sound, dynamic load-time history of bouncing at 2 Hz. (**a**) Acceleration-time history. (**b**) Sound pressure-time history. (**c**) Person dynamic load -time history

Table 4.1 Prior PDF for model of structure excited by the person

Parameter	PDF	Mean	Standard deviation
ω_n	Normal	$\mu = 18.85$	$\sigma = 1.0$
ζ	Normal	$\mu = 0.005$	$\sigma = 0.00025$
Mass	Normal	$\mu = 600$	$\sigma = 25$
K_p	Normal	$\mu = 1299$	$\sigma = 130.0$
K_d	Normal	$\mu = 42$	$\sigma = 8.0$
K_i	Normal	$\mu = 495.0$	$\sigma = 25.0$
K_{p1}	Normal	$\mu = 75,240$	$\sigma = 7524$
K_{d1}	Normal	$\mu = 3179.0$	$\sigma = 320.0$
K_{i1}	Normal	$\mu = 802,800$	$\sigma = 80,280$

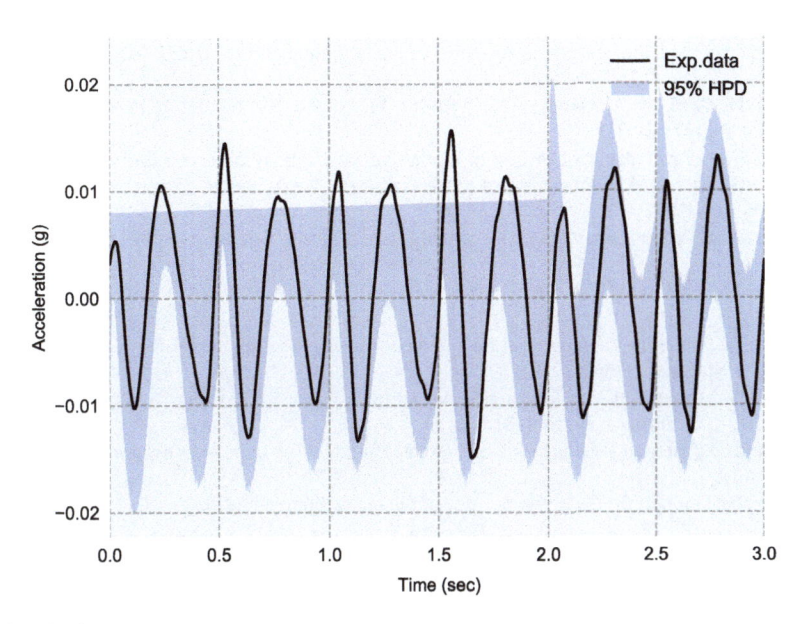

Fig. 4.6 Posterior predicative check

Table 4.2 Moments of random variables describing the parameters of the human, music and structure

Parameter	ω_n [red/s]	ζ	Mass[kg]	K_p	K_d	K_i	K_p1	K_d1	K_i1
Mean	18.21	0.006	634	−1341	115.979	1128	84624	8217	2,838,127
St. Deviation	0.0.198	0.003	11.08	54.4	2.19	175.5	4119	182.8	63,490
95% HPD	(17.98, 18.89)	(0.001, 0.009)	(609.5, 654.6)	(−1465, −1269)	(112.7, 120.09)	(789.5, 1503	(70,666, 89,047)	(6091, 7262)	(2,739,917, 2,952,269)

4.4 Results and Discussion

Because of the type of experiments performed, different parts of the model were updated with each data set. The parameters of the structure (plant) were updated with the empty structure. Then, the parameters for $H_1(s)$ were updated with the experiment of the person standing with bent knees. Finally, the complete system was updated with the sound excitation. The posterior PDF of the parameters were sampled using Metropolist Hasting algorithm. A posterior prediction check was performed by using the model to predict the acceleration of a structure with a different experiment as shown in Fig. 4.6. Table 4.2 shows the posterior of mean, standard deviation, and 95% of HPD interval of each parameter of the final model.

4.5 Conclusions

This paper presents a new model to represent the human-structure interaction phenomenon to simulate the complete system when an individual flexes their knees on a flexible structure. The proposed model uses two Proportional, Derivative, and integrative (PID) controllers to model both the feedback from the structure and the excitation due to sound. Model parameters were updated using Bayesian inference. Posterior predictions of the model match well with experimental results for a single person. However, additional research needs to be performed to verify the validity of the model with other individuals.

Acknowledgement The author would like to acknowledge the higher education and scientific research ministry of Iraq to support this research.

References

1. Ortiz-Lasprilla, A.R., Caicedo, J.M.: Comparing closed loop control models and mass-spring-damper models for human structure interaction problems. In: Dynamics of Civil Structures, vol. 2, pp. 67–74. Springer, Berlin (2015)
2. Racic, V., Pavic, A.: Mathematical model to generate near-periodic human jumping force signals. Mech. Syst. Signal Process. **24**(1), 138–152 (2010)
3. Dallard, P., Fitzpatrick, T., Flint, A., Low, A., Smith, R.R., Willford, M., Roche, M.: London millennium bridge: pedestrian-induced lateral vibration. J. Bridge Eng. **6**(6), 412–417 (2001)
4. Tilly, G.P., Cullington, D.W., Eyre, R.: Dynamic behaviour of footbridges. In: IABSE Surveys S-26/84, pp. 13–24 (1984)
5. Lee, S.H., Lee, K.K., Woo, S.S., Cho, S.H.: Global vertical mode vibrations due to human group rhythmic movement in a 39 story building structure. Eng. Struct. **57**, 296–305 (2013)
6. Jones, C.A., Reynolds, P., Pavic, A.: Vibration serviceability of stadia structures subjected to dynamic crowd loads: a literature review. J. Sound Vib. **330**(8), 1531–1566 (2011)
7. Madarshahian, R., Caicedo, J.M., Zambrana, D.A.: Benchmark problem for human activity identification using floor vibrations. Expert Syst. Appl. **62**, 263–272 (2016)
8. Alzubaidi, A.T., Caicedo, J.M.: Modeling human-structure interaction using control models: external excitation. In: Dynamics of Civil Structures, vol. 2, pp. 183–190. Springer, Berlin (2019)
9. Lasprilla, A.R.O., Caicedo, J.M., Ospina, G.A.: Modeling human–structure interaction using a close loop control system. In: Dynamics of Civil Structures, vol. 4, pp. 101–108. Springer, Berlin (2014)
10. Sachse, R., Pavic, A., Reynolds, P.: Human-structure dynamic interaction in civil engineering dynamics: a literature review. Shock Vib. Dig. **35**(1), 3–18 (2003)
11. Wei, L., Griffin, M.: Mathematical models for the apparent mass of the seated human body exposed to vertical vibration. J. Sound Vib. **212**(5), 855–874 (1998)
12. Brownjohn, J.M.: Energy dissipation in one-way slabs with human participation. In: Proceedings of the Conference on Asia-Pacific Vibration, pp. 13–15. Nanyang Technological University, Singapore (1999)
13. Dorf, R.C., Bishop, R.H., Modern Control Systems. Pearson, London (2011)
14. Sim, J., Blakeborough, A., Williams, M.: Modelling of joint crowd-structure system using equivalent reduced-DOF system. Shock Vib. **14**(4), 261–270 (2007)
15. Ogata, K.: Modern control engineering. Book Rev. **35**(1181), 1184 (1999)
16. Ewins, D.: Modal testing: theory, practice and application (mechanical engineering research studies: engineering dynamics series), 2003
17. Hastings, W.K.: Monte carlo sampling methods using markov chains and their applications. Biometrika **57**(1), 97–109 (1970)
18. Beck, J.L., Katafygiotis, L.S.: Updating models and their uncertainties. I: Bayesian statistical framework. J. Eng. Mech. **124**(4), 455–461 (1998)
19. Madarshahian, R., Caicedo, J.M.: Reducing mcmc computational cost with a two layered bayesian approach. In: Model Validation and Uncertainty Quantification, vol. 3, pp. 291–297. Springer, Berlin (2015)
20. Ji, T.: Floor vibration. Struct. Eng. **72**(3/1), 37 (1994)
21. Ellis, B., Ji, T.: Floor vibration induced by dance-type loads: verification. Struct. Eng. **72**, 37–37 (1994)

Chapter 5
Identification and Monitoring of the Material Properties of a Complex Shaped Part Using a FEMU-3DVF Method: Application to Wooden Rhombicuboctahedron

R. Viala, V. Placet, and S. Cogan

Abstract In this study, a non-destructive method, FEMU-3DVF, based on velocity fields measurements and finite element model updating is used to identify the material properties of a complex shaped wooden part. The part studied is a rhombicuboctahedron made of spruce wood. Its dynamical response is measured with a laser-3D vibrometer and the experimental modal basis is used to identify by inverse method the elastic properties of the spruce wood. Considering the experimental modes, the shear modulus in the RT plane of spruce (G_{RT}) can be reliably measured. The part undergoes three different relative humidity cases which lead to three moisture content levels for which the material properties are identified. It is shown that G_{RT} decreases and (η_{RT}) increases when the moisture content increases. This method can be used to monitor complex shaped structures in a non-destructive way.

Keywords Non-destructive testing · Material characterization · Wood materials · Structure monitoring · Dynamical domain

5.1 Introduction

In the material characterization domain, both direct and indirect methods can be used to determine the mechanical parameters of a part. Direct methods are generally destructive or at least with contact and are used to measure the properties of a wide variety of materials. Indirect methods are usually model-based and identify the material properties by minimizing an error on the model predictions. Non-destructive methods used for material characterization can be full-field measurements [1] and, especially for wood, resonant [2] and non-resonant [3] ultrasonic methods. Inverse methods based on dynamic experimental data and Finite Element Model Updating (FEMU) can be particularly useful. The fitting of a numerical simulation with experiments using minimization methods [4], enables the determination of the properties of materials exhibiting a complex behavior. This type of method is non-invasive, fast and easy to set-up and it requires a reduced preparation of the samples, compared to other traditional methods that require calibrated specimens. In addition, dynamical responses like eigenmodes are more representative of the global behavior of the material when compared to more usual local measurements.

These benefits have led to the development of numerically based identification methods from the early works like [5]. The efficiency of the FEMU method in the case of complex specimen shapes has been assessed by studying layered materials [6]. Many studies have focused on other types of structures or materials such as spruce wood plates for guitar making undergoing moisture content changes [7]. Nevertheless, only a few studies concern non-planar shapes, such as tubes [8]. More recently, in 2016, a review [9] made a list of the numerous works concerned with the identification of material properties using vibrational approaches. This paper highlighted the fact that, despite its potential, this method has mostly been applied to flat plate specimens and to not complex shape parts. These parts are generally made of materials whose mechanical properties depend on the manufacturing process.

In 2018, a study of complex shaped bio-based soundboards using both finite element model updating and 3D velocity field measurements in the dynamical domain has been proposed [10]. It illustrates the efficiency of such a method for the study of complex shaped non-planar parts. The method used in this above-mentioned paper, FEMU-3DVF, is applied for the study of a bulk part with complex shape, a wooden rhombicuboctahedron made of spruce wood. The shape proposed for modal analysis is inspired by the shapes used in the resonant ultrasonic method to determine all the parameters of the rigidity

R. Viala (✉) · V. Placet · S. Cogan
Department of Applied Mechanics, Univ. Bourgogne-Franche-Comté, FEMTO-ST Institute, CNRS/UFC/ENSMM/UTBM, Besançon, France
e-mail: romain.viala@univ-fcomte.fr

matrix [3]. In the case of bulk structures, sensitivity to anatomic discrepancies in wood is alleviated and the results obtained are more representative of a global behavior of the part. Moreover, the studied part undergoes different relative humidity (RH) levels and the identification is applied for each RH level and its corresponding moisture content (MC). This study aims at highlighting the fact that this method has a strong potential for the monitoring and survey of complex shape and material structures and parts, in a non-destructive, contact-free way.

5.2 Material and Method

The studied part is made of spruce (*Picea abies*) which is a softwood whose mechanical behavior is considered to be orthotropic and elastic in the dynamical domain and strain level considered. The shape of the part is a rhombicuboctahedron, an Archimedean solid, which has been machined from a log of the wood chosen for musical instruments making, without defects like nodes and with a straight grain. The rhombicuboctahedron faces are square and equilateral triangle shaped whose sides are equal to 37.3 mm, and its bounding box is equal to 90 mm. The sample has been suspended on nylon wires for the measurements and excited with a speaker in the frequency bandwidth between 2000 and 8000 Hz. The measurements have been performed using a 3D laser vibrometer by POLYTEC, as displayed in the Fig. 5.1.

The modal basis evaluated with a POLYMAX method is compared with the one computed by a numerical model, using the software NASTRAN. The part is modelled using the finite element method, with tetrahedral elements, as shown in the Fig. 5.2. The computation of the first ten modes above 1 Hz lasts 30 s. The minimization of a cost function is used to reduce the discrepancy between the numerical modal basis and the experimental one. The cost function is defined by the sum of the absolute relative error between experimental and numerical matched frequencies. The frequencies are matched with a modal assurance criterion, as proposed in [10]. The MAC value considered for the pairing is equal to 0.7. At the end of the procedure, the cost function is minimized, and the material properties are identified.

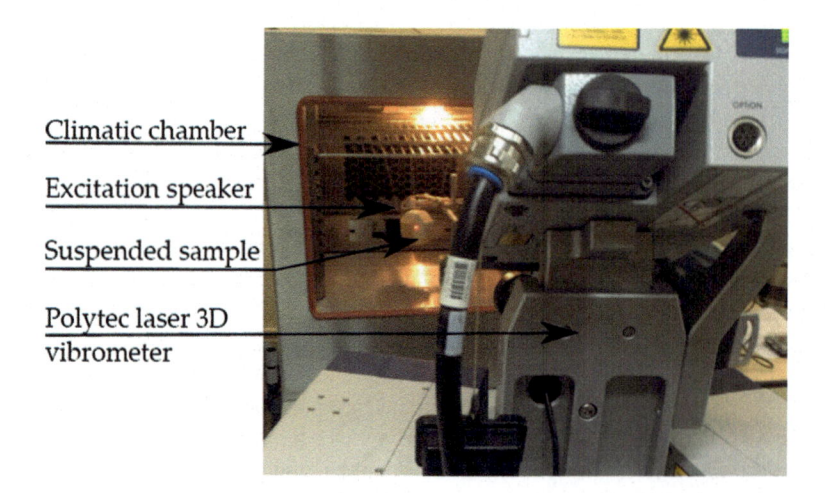

Fig. 5.1 Experimental protocol for the measurement of the velocity fields used for the modal analysis

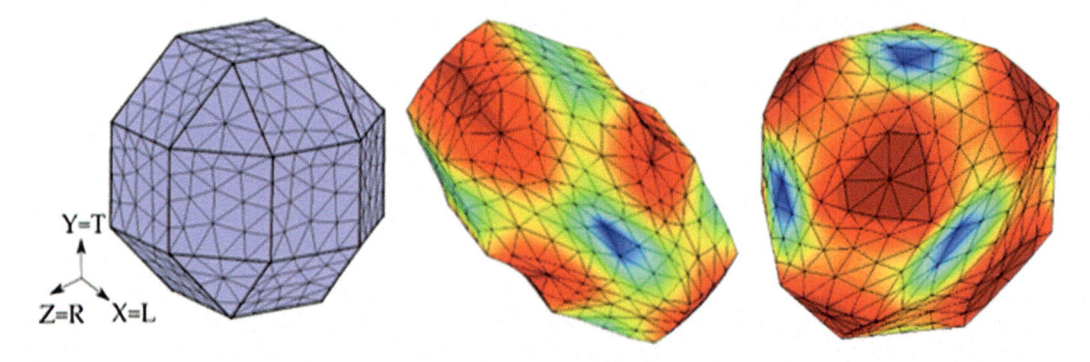

Fig. 5.2 Mesh and first two modes of the model of the rhombicuboctahedron

In addition, the sample is prepared during 7 days in a climatic chamber at three different RH steps: 30, 50 and 75% at a constant temperature equal to 23 °C. This is used to measure the dynamical response for three different levels of moisture content MC in usual room conditions. At the end of the dynamical tests, the sample is oven dried for 3 days at 103 °C and is weighed to obtain the dry mass m_0 used to calculate the moisture content of the specimen.

5.3 Results

The physical properties of the sample are given in the Table 5.1. The reference value is considered for RH = 50%. For other levels of RH, the dimensions of the sample change and the bounding box side is equal to either 89 mm or 91 mm for RH equal to 30% and 75%, respectively. The m_0 mass measured is equal to 176.3 g and is used for the calculation of the MC levels. The moisture content increases from 8.9 to 15.4% when the relative humidity varies from 30 to 75%. The specific gravity variation is very small, as the sample undergoes both dimensional and mass changes which counterbalance.

The evolution of the first two eigenfrequencies is given in the Table 5.2 and the shape of the modes is given in the Fig. 5.2. It is shown that the experimental frequencies drop by 127 and 274 Hz for respectively mode 1 and 2 when the relative humidity increases from 30 to 75%, (the moisture content increases from 8.9 to 15.4%). For each case, a numerical model is created with corresponding geometry evolution and material properties change, as evaluated in [11] and [12] for spruce. The Table 5.3 lists the material properties used for each case of RH step.

Each numerical model is used to identify the material properties for each RH step, and the cost function is minimized. The average value of the eigenfrequencies error is given in the Table 5.4, as well as the average MAC value. It is shown that the error is close to 1.5% at the end of the minimization procedure which highlights a good correlation between the experimental and computed modal bases.

Based on these results, the shear modulus in the RT plane is identified for each RH (or MC) case, as given in the Table 5.5. The Table 5.5 also gives the loss factor which has been determined as twice the modal damping of the first mode of the experimental modal basis. The values obtained for G_{RT} are in accordance with the one obtained with both ultrasonic test [13] and shear static [14] tests. The evolution of the material properties as a function of the moisture content has led to empirical laws. These laws express the relative variation of both the shear modulus in RT plane (G_{RT}) and the loss factor (η_{RT}) for a variation of 1% MC. Based on the given results, it gives $-1.1\%/\%$MC for **G_{RT}** which is higher than [13] ($-0.5\%/\%$MC)

Table 5.1 Physical properties and moisture content of the sample for three different RH conditioning

RH step (%)	Mass [g]	Volume [cm^3]	Specific gravity [−]	MC (%)	Variation MC (absolute)	Variation MC (relative)
30	192	436.6	0.440	8.9	−3.4	−28
50	198	451	0.439	12.3	0	0
75	203.4	466	0.436	15.4	3.1	25

50% RH is considered as the reference value

Table 5.2 Evolution of the first two modes displayed in the Fig. 5.2 as a function of the RH step level

RH step (%)	Mode 1 [Hz]	Mode 2 [Hz]
30	2562	3539
50	2533	3318
75	2435	3265

Table 5.3 Initial material properties of spruce used for the simulations, for each MC case, relations taken from [11] and [12]

MC (%)	E_L [MPa]	E_T [MPa]	E_R [MPa]	ν_{LT} [−]	ν_{TR} [−]	ν_{TL} [−]	G_{LT} [MPa]	G_{RT} [MPa]	G_{LR} [MPa]	Specific gravity [−]
8.9	12,410	670	975	0.4	0.3	0.03	795	33	825	0.440
12.3	12,120	615	900	0.4	0.3	0.03	720	32	740	0.439
15.4	11,810	550	820	0.4	0.3	0.03	645	31	645	0.436

Table 5.4 Average eigenfrequency error between the computed and experimental modal bases before and after calibration

RH step (%)	Average error before calibration (%)	Average error after calibration (%)	Average MAC after calibration (%)
30	15	1.3	75
50	12	1.5	85
75	8.2	1.8	72

Table 5.5 Evolution of G_{RT} and η_{RT} as a function of the moisture content of the sample

MC (%)	G_{RT} [MPa]	η_{RT} (%)
8.9	34.4	1.1
12.3	33.5	1.2
15.4	31.9	1.6

and lower than [12] (−3%/%MC). However, only one sample was considered, and more experiments must be conducted to provide more detailed and accurate laws. The loss factor η_{RT} evolution is also evaluated as +11%/%MC, which is an evolution more important than for the rigidity but correspond to usual values.

5.4 Conclusion

In this study, an inverse method based on finite element model updating and 3D measurements in the dynamical domain enables the determination of at least two material parameters that are usually hard to determine in a non-destructive way. The results are representative of a global behavior of the studied part. The shear modulus value obtained is correlated with the usual ones. The method is also used to evaluate the relative changes of elastic and damping parameters when the studied part undergoes relative humidity changes. The results obtained are used to create empirical laws that link the elastic and damping properties of wood with respect to moisture content changes. The application field of the FEMU-3DVF method proposed here is a potential tool for the monitoring of the properties of complex shaped parts and structures in a non-destructive and contact free way, and for which the global behavior is considered.

Acknowledgements Authors acknowledge M. Vincent TISSOT, machinist, who realized the studied sample.

References

1. Hild, F., Roux, S.: Digital image correlation: from displacement measurement to identification of elastic properties—a review. Strain. **42**(2), 69–80 (2006)
2. Longo, R., Delaunay, T., Laux, D., El Mouridi, M., Arnould, O., Le Clézio, E.: Wood elastic characterization from a single sample by resonant ultrasound spectroscopy. Ultrasonics. **52**(8), 971–974 (2012)
3. François, M.L.M.: Vers une mesure non destructive de la qualité des bois de lutherie. Revue Des Composites et Des Matériaux Avancés. **10**, 261–279 (2009)
4. Mottershead, J.E., Link, M., Friswell, M.I.: The sensitivity method in finite element model updating: a tutorial. Mech. Syst. Signal Process. **25**(7), 2275–2296 (2011)
5. Sol, H.: Identification of Anisotropic Plate Rigidities Using Free Vibration Data, Free University of Brussels (VUB) (1996)
6. Lauwagie, T., Heylen, W., Sol, H., Van Der Biest, O.: Validation of a vibration based identification procedure for layered materials. In: Proceedings of ISMA 2004, pp. 1325–1336 (2004)
7. Pérez, M.A., Poletti, P., Gil Espert, L.: Vibration testing for the evaluation of the effects of moisture content on the in-plane elastic constants of wood used in musical instruments. In: Vasques, C.M.A., Rodrigues, J.D. (eds.) Vibration and Structural Acoustics Analysis, pp. 21–57 (2001)
8. Cunha, J., Piranda, J.: Identification of stiffness properties of composite tubes from dynamic tests. Exp. Mech. **40**(2), 211–218 (2000)
9. Viala, R., Placet, V., Cogan, S.: Identification of the anisotropic elastic and damping properties of complex shape composite parts using an inverse method based on finite element model updating and 3D velocity fields measurements (FEMU-3DVF): application to bio-based composite violin soundboard. Compos. A: Appl. Sci. Manuf. **106**, 91–103 (2018)
10. Allemang, R.J., Brown, D.L.: A correlation coefficient for modal vector analysis. In: First International Modal Analysis Conference, pp. 110–116 (1982)
11. Viala, R.: Towards a Model-Based Decision Support Tool for Stringed Musical Instrument Making, Université Bourgogne Franche-comté (2018)
12. Guitard, D.: Mécanique Du Matériau Bois et Composites. Cépaduès, Toulouse (1987)
13. Keunecke, D., Sonderegger, W., Pereteanu, K., Lüthi, T., Niemz, P.: Determination of Young's and shear moduli of common yew and Norway spruce by means of ultrasonic waves. Wood Sci. Technol. **41**(4), 309–327 (2007)
14. Dahl, K.B., Malo, K.A.: Linear shear properties of spruce softwood. Wood Sci. Technol. **43**(5–6), 499–525 (2009)

Chapter 6
Modal Tracking on a Building with a Reduced Number of Sensors System

Wladimir M. González and Rubén L. Boroschek

Abstract Typical modal tracking algorithms make use of frequency, mode shape and environmental conditions, among other information to identify, classify and follow the modal response. To correctly characterize mode shapes, well distributed and located sensors are required. In civil strictures, the use of large number of sensors is limited by the size of the structure, the difficulty to install cables or wireless connections and limitations due to cost, technical, usage or aesthetics requirements. In this article, we explore the effects of a limited set of sensors to perform modal tracking in a building that has been monitored continuously for approximately 5 years and 8 months. We implement a methodology that relies mostly on the modal frequency and a poorly defined modal shape. The reduced number of sensors are used to discriminate between closely spaced modes and modes that are highly sensitive to environmental conditions. In order to increase the robustness of the tracking algorithm, the global environmental conditions, particularly temperature, are monitored and used. It is concluded that in order to obtain a reliable tracking, a minimum number of sensors is required if closely spaced modes are present. Also, the need of highly defined mode shapes can be reduced if the environmental conditions are considered. As expected, we have found that tracking is possible with very limited number of sensors, if the objective is to capture the global response contained in the lower modes of the system. In order to capture low excited higher modes, the number of sensors should be increased and location selected cautiously.

Keywords SHM · Modal tracking · Civil engineering · Sensor density · Environmental effects

6.1 Introduction

Typically, modal tracking algorithms use frequencies and modal shapes to identify, classify, evaluate and track the modal response through time. The inclusion of the modal shape allows the process to distinguish between close frequencies, increasing the process's robustness. However, this inclusion implies more demands for the monitoring system, since the proper characterization of the modal shape requires an array of sensors large enough and well distributed in the structure.

In civil structures, particularly in buildings, the use of a large number of sensors is limited by the size of the structure, the difficulty to install cables or wireless connections and limitations due to cost, usage or aesthetics requirements. Therefore, in numerous occasions a poor characterization of some physical modes will happen, especially if they are poorly excited, as is the case with higher modes.

This paper explores the effect of a limited characterization of the modal shape on the modal tracking, due to a monitoring process done with a few sensors. For this purpose, two methodologies are applied in a medium-rise building monitored by six accelerometers and a weather station, for a continuous period of 5 years and 8 months. The first, proposed by the authors in [1], which uses the external temperature measurements to build predictive models for the frequency, and the second, presented in [2], based exclusively on the temporal regularity of the modal shape. The former is evaluated four times, with six, three, two and one channel acceleration sensor(s) for the task.

W. M. González (✉) · R. L. Boroschek
Department of Civil Engineering, University of Chile, Santiago, Chile
e-mail: wladimir.gonzalez.v@ing.uchile.cl; rborosch@ing.uchile.cl

6.2 Methodology

6.2.1 System Identification

The system identification considers 15 min windows without overlap, from April 8, 2011 to December 31, 2016. The identification algorithm, proposed in [3], is an automatic interpretation of stabilization diagrams based on OPTICS [4]. The stability diagrams generated with the SSI-COV algorithm [5]. For this evaluation, no special care is taken to remove spurious modes from the time window identification. So, the data to be used for modal tracking presents considerable noise. This noisy data is labeled as "raw identification data" and it is presented as background information in Figs. 6.1, 6.2, 6.3, 6.4, 6.5, 6.6, 6.7, 6.8, and 6.9.

Since the modal tracking is done for four spatial configurations of sensors (each with a different number of active sensors), the identification is also performed four times; one for each configuration.

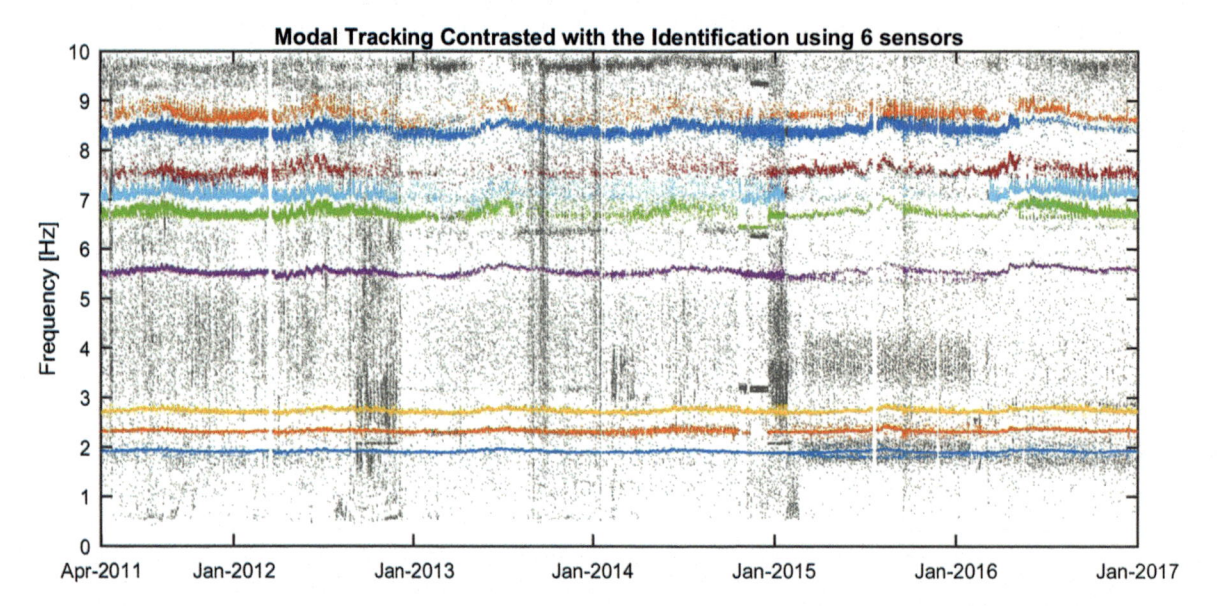

Fig. 6.1 MAT carried out with six available accelerometers. The raw identification data set using six sensors in the background

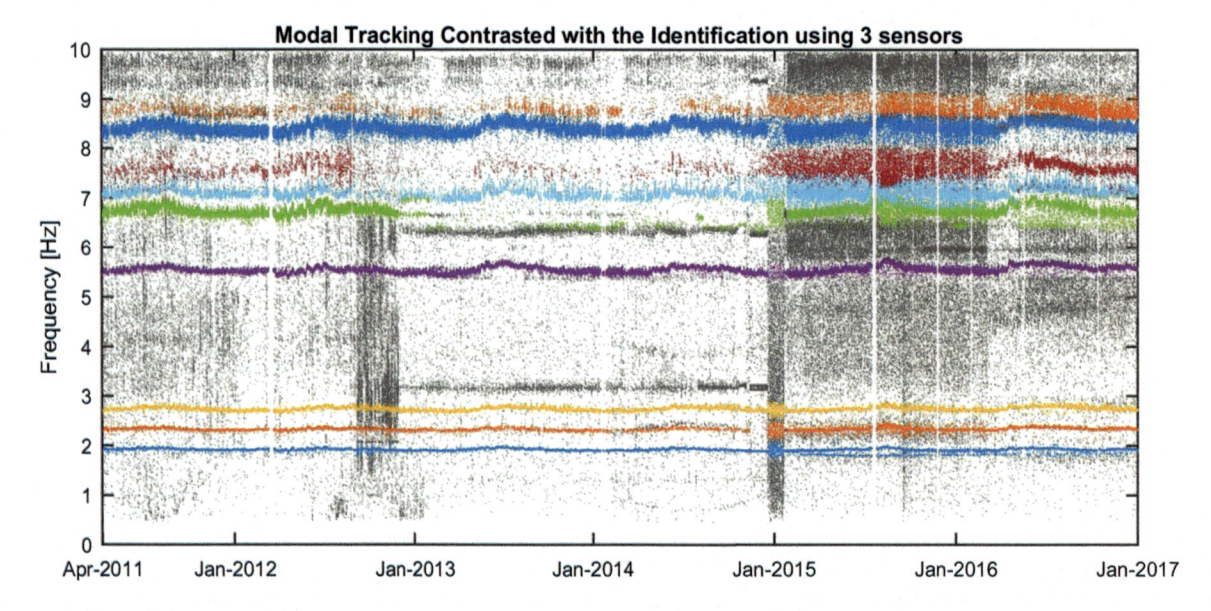

Fig. 6.2 MAT carried out with three available accelerometers. The raw identification data set using three sensors in the background

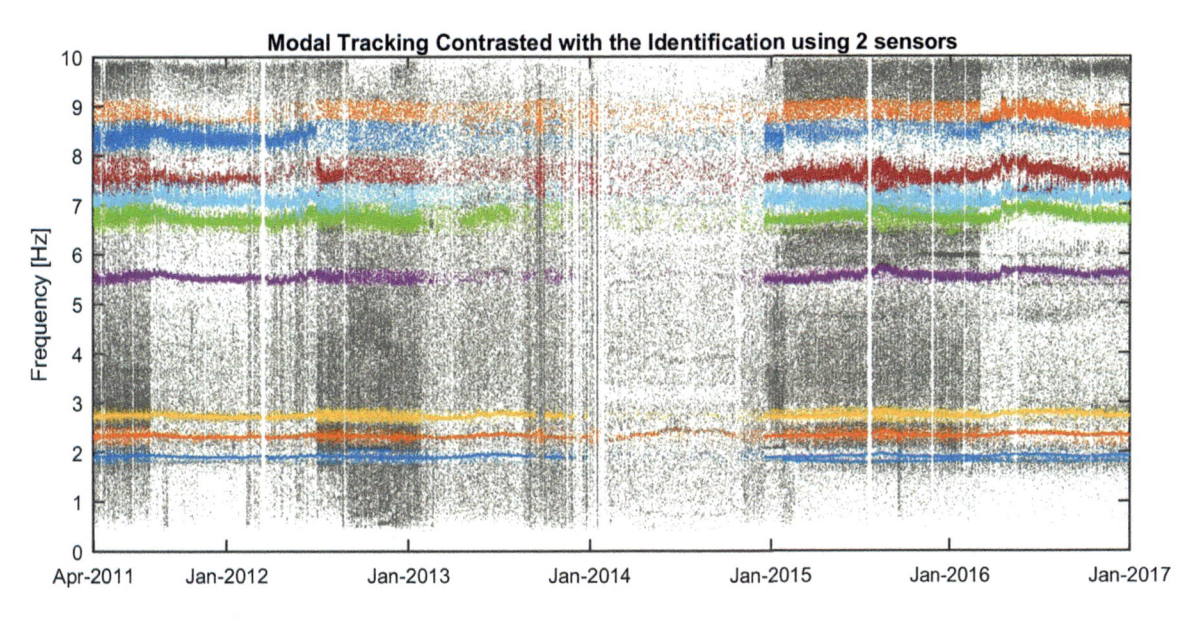

Fig. 6.3 MAT carried out with two available accelerometers. The raw identification data set using two sensors in the background

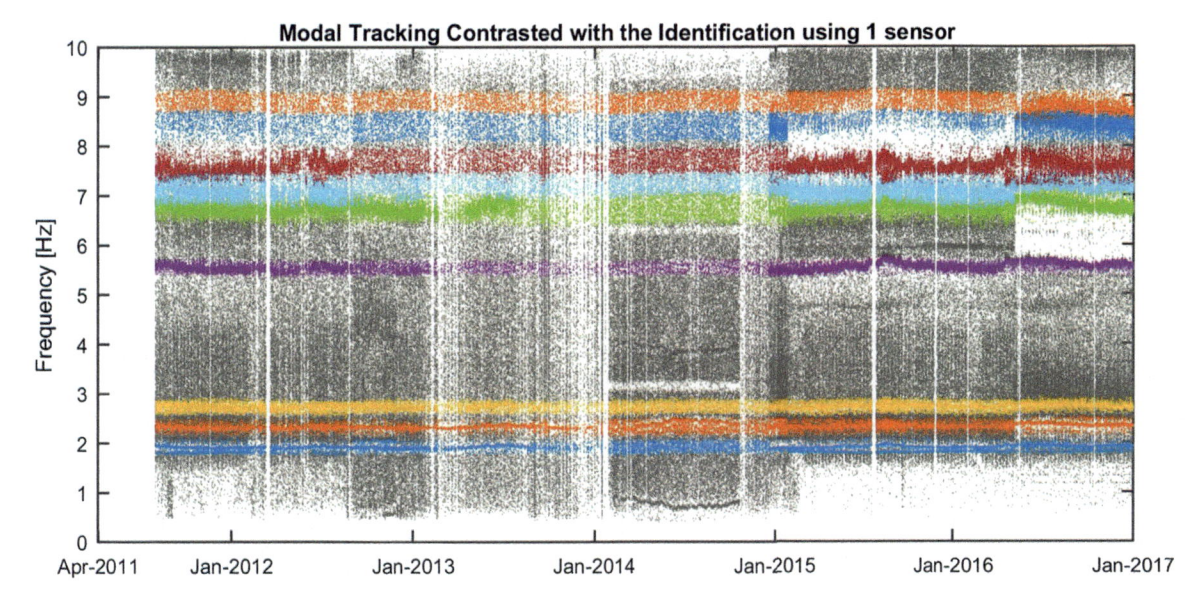

Fig. 6.4 Modal Tracking carried out with one available accelerometer. The raw identification data set using one sensor in the background

6.2.2 Model Assisted Tracking (MAT)

The Model Assisted Tracking algorithm [1] associates the identification with the reference properties of each physical mode, at each time step, to find the modes' instantaneous state (based on the work by [6], among others by the same author).

Considering temporal variation of the frequency as a result of variations in environmental conditions, the methodology automatically updates the reference frequency from previously trained models based on meteorological information. These models correspond to linear models of the frequency depending on the past temperature:

$$f_i = \sum_j a_j \cdot T_{i-j} + b \tag{6.1}$$

where f_i indicates frequency at the i-th time step and T_{i-j} represents the external temperature at the (i-j)-th time step, indicating any of the temperature measurements of the past 12 h, not necessarily consecutive.

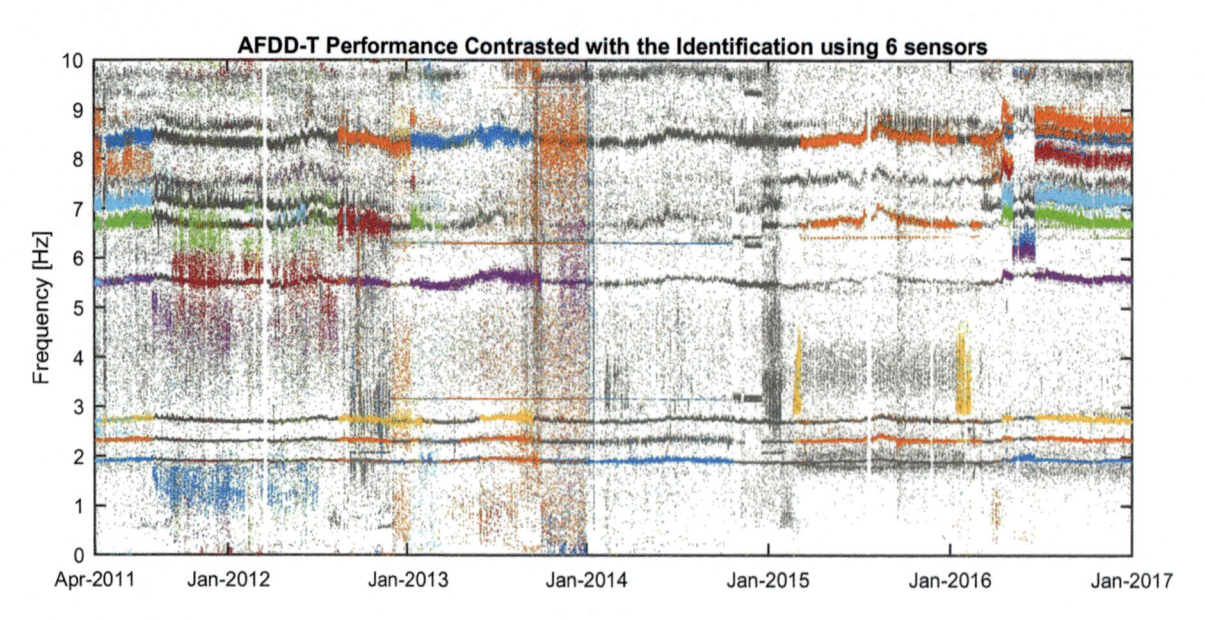

Fig. 6.5 Implementation of the AFDD-T algorithm with six available accelerometers. The raw identification data set using six sensors in the background

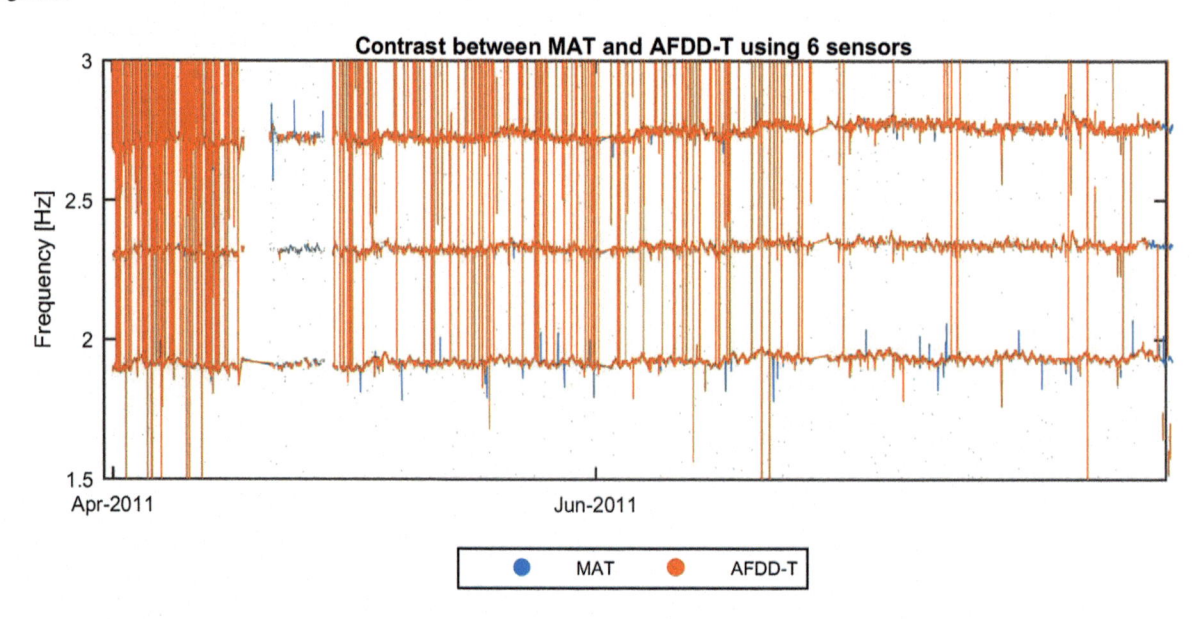

Fig. 6.6 Comparison between the AFDD-T and MAT, with six sensors available in both cases. In black, the system identification with six sensors is represented. In Fig. 6.5, the resulting frequencies of AFDD-T are represented in a dot pattern, unlike the figure above, where they are represented in a line pattern. This is made to differentiate them, since the vertical lineal seen above would obstruct them

To characterize the similarity of the modes identified with each reference mode, the following metric is used:

$$d_{ref,i} = (1-\alpha) \cdot \frac{|f_{ref} - f_i|}{\max_j (f_i)} + \alpha \cdot \left(1 - MAC\left(\phi_{ref}, \phi_i\right)\right) \tag{6.2}$$

where ϕ_{ref} and f_{ref} denotes the modal shape and frequency of the reference, while ϕ_i and f_i denotes the modal shape and frequency of the i-th mode for the analyzed time window. $MAC(\phi_{ref}, \phi_i)$ corresponds to the MAC index between the two modal shapes [7]. $\max_j (f_i)$ is the maximum identified modal frequency in the analysis window, whose only purpose is to normalize $|f_{ref} - f_i|$ to make it comparable to the distance term associated with the modal shape. In addition, the function of α is to incorporate the spatial definition of the modal shape characterization, and it is given the following form:

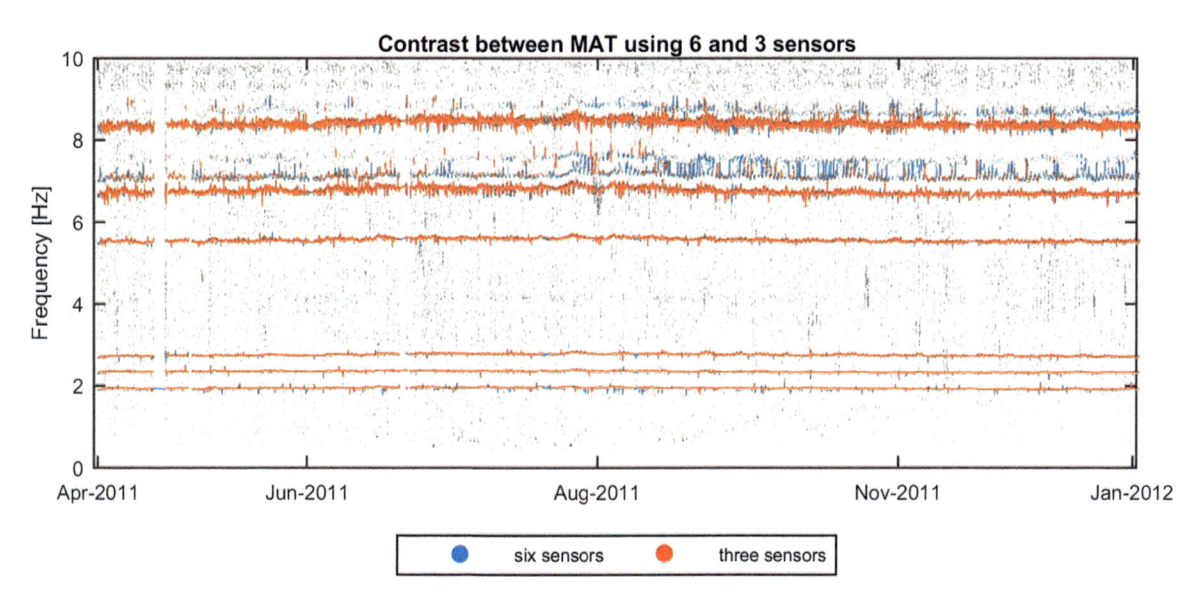

Fig. 6.7 Contrast between the MAT algorithm with six and three available sensors, for the year 2011. In black, the identification of the system made with six sensors is represented. The selection of year 2011 to compare both approaches responds solely to the need to use a shorter window of time to appreciate in detail the performance of the algorithms

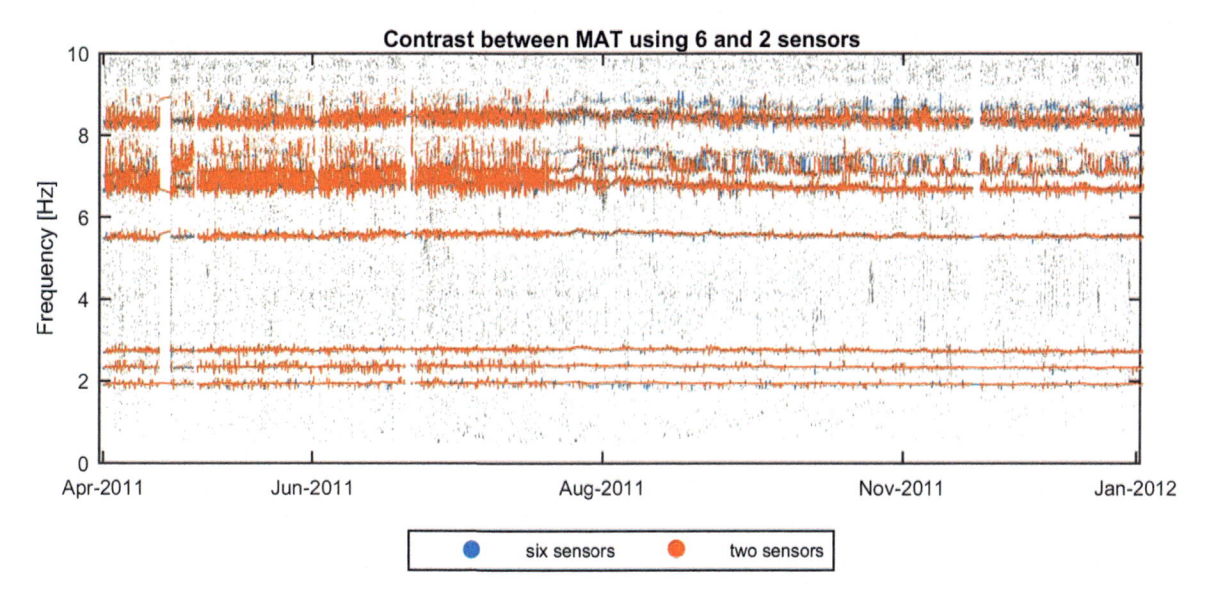

Fig. 6.8 Contrast between the MAT algorithm with six and two available sensors, for the year 2011. In black, the identification of the system made with six sensors is represented

$$\alpha = \begin{cases} 0.5 \text{ if number of shared active channels} \geq 2 \\ 0 \qquad\qquad\qquad \text{otherwise} \end{cases} \quad (6.3)$$

Finally, the identified mode that is closest to the reference is considered as a valid sample of the physical mode only if, given that no extreme event like and earthquake has occurred, it is within a neighborhood defined by fixed thresholds to the similarity in frequency and modal shape, whose values for the case study are explicit in Table 6.1.

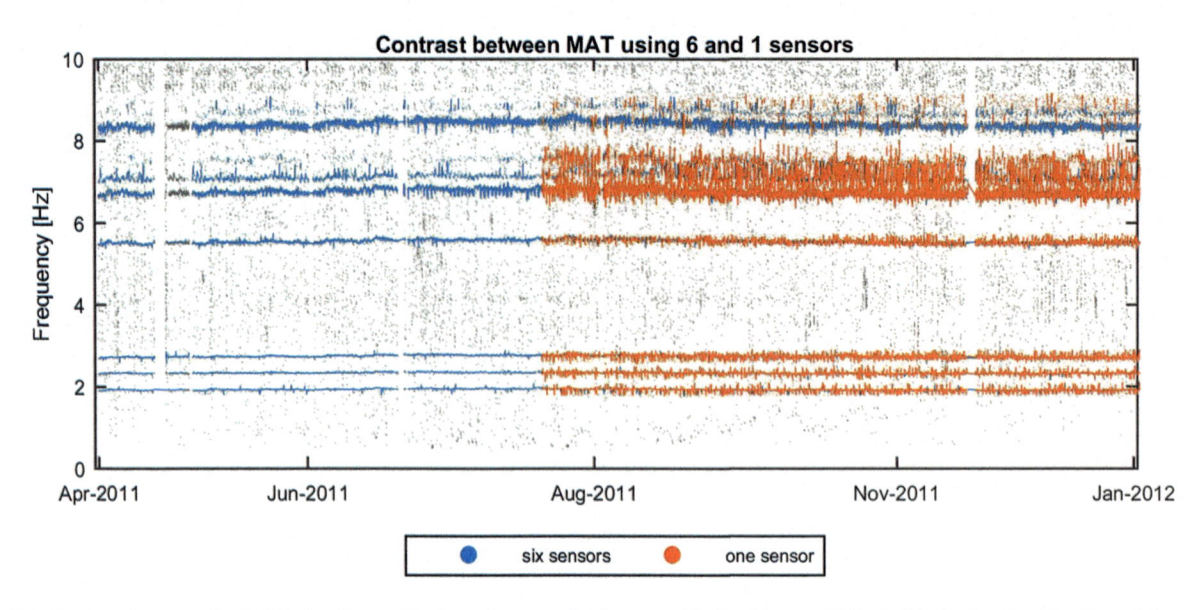

Fig. 6.9 Contrast between the MAT algorithm with six and one available sensor(s), for the year 2011. In black, the identification of the system made with six sensors is represented. The single sensor used during this evaluation was discarded the first half of the year due to noise in the signal, explaining the absence of tracked frequencies for this period of time

Table 6.1 Thresholds used for each distance term, for each physical mode

Mode	1st	2nd	3rd	4th	5th	6th	7th	8th	9th
Frequency	1.75%	2%	1.75%	2%	4%	4.5%	5%	3.5%	4%
Modal shape	35%	35%	35%	35%	35%	35%	35%	35%	35%

6.2.3 AFDD-T

AFDD-T [2], based on the FDD identification algorithm [8], uses the reference modal shape to apply a filter on the signal's periodogram to automatically isolate the modal response. Then, the modal characterization proceeds just like FDD. In this way, AFDD-T identifies and follows the modal response over time based exclusively on the modal shape, ignoring the frequency.

For this methodology, unlike the algorithm mentioned above, its performance is evaluated using all six sensors available in the structure.

6.3 Case Study

The analyzed structure corresponds to the Torre Central building at the Faculty of Physical and Mathematical Sciences of the University of Chile, located in Santiago, Chile (33°27′27″S, 70°39′44″O). It is an eight-story medium-rise building with two underground stories, which has been used mainly for administrative purposes throughout the period of analysis, between April 8, 2011 and December 31, 2016.

The monitoring system consists of six uniaxial accelerometers, three of them located on the eighth story and the other three on the third. For the evaluation with three available channels, the three on the eight story are used, whereas for the evaluation with two available channels, a pair of perpendicular sensors is chosen among these. Finally, for the evaluation with one available channel, the one on the longest direction is chosen. As is characteristic of long-term monitoring, some channels are sporadically affected with noise and therefore must be discarded, resulting in a variable number of active sensors.

The external temperature measuring system consists of a meteorological station located approximately 40 m from the structure, administered by the Department of Geophysics of the University of Chile.

6.4 Application to the Case Study

As has already been mentioned, two methodologies are implemented, with five different approaches, for the same case study: the modal tracking aided by models with six, three, two and one available channels, and AFDD-T. For the first four (see Sect. 6.2.1), the thresholds defined for each of the metric's components (see Eq. 6.2) are the following:

For AFDD-T, as in the article where it is proposed, the threshold defined for the MAC value is 0.95.

6.4.1 MAT with Six Available Channels

The MAT algorithm applied on the identification with six available acceleration sensors is shown in Fig. 6.1. In this figure, it is shown that this tracking methodology is capable of recognizing the modal response over time. Although it is observed some missing frequencies in time, they are mostly attributed to the absence of observable modes in the identification window. The modes with frequencies above 9 Hz are not tracked due to its low energy and presence in the raw data.

6.4.2 MAT with Three Available Channels

The MAT algorithm applied on the identification with three available acceleration sensors is shown in Fig. 6.2. Compared to Fig. 6.1, there is a greater number of spurious modes present in the identification. This is due to a more limited spatial distribution of the monitoring system in the structure; making harder to identify the stable poles in the records.

Additionally, it is noticed that the tracking algorithm is able to capture the modal response evolution but has a less accurate performance for the higher modes, especially for the year 2015, in which spurious modes are labeled as samples of physical modes. As for the four lower modes located below 6 Hz, there is no noticeable decay in performance, when compared to the six sensors array approach.

6.4.3 MAT with Two Available Channels

The MAT algorithm applied on the identification with two available acceleration sensors is shown in Fig. 6.3. The trend seen in Fig. 6.2 for the three sensors approach is repeated here, as more spurious modes are present in the identification.

Regarding the algorithm's performance, the lower modes located below 3 Hz appear to be correctly tracked for most of the analyzed period, though they are absent from the identification during almost all the year 2014. For the higher modes, the decay with respect to the three sensors approach is evident.

6.4.4 MAT with One Available Channel

The MAT algorithm applied on the identification with one available acceleration sensor is shown in Fig. 6.4. In this case, even for the lower modes it is seen a decay in performance with respect to all previous approaches. The amount of spurious modes in the identification, combined with the absence of the modal shape as a descriptor of the identified modes, makes it impossible to track any physical mode correctly. The higher modes above 8 Hz are not even identified by the identification algorithm, making them impossible to track.

6.4.5 AFDD-T

Figure 6.5 shows the modal tracking developed with the AFDD-T algorithm. Due the noise level in the acceleration registers and the use of the modal shape as the single descriptor, it is noted this methodology has a poorer performance than previous methodology for 6, 3 and 2 sensors.

At the beginning of the study period, from April 8 to August 1, 2011, the AFDD-T algorithm performs better than in any other time window. In Fig. 6.5, this result is compared with that shown in Fig. 6.1, also done with all six available sensors, at the aforementioned time period.

In Fig. 6.6 it is seen that, even at this time window, AFDD-T fails to correctly capture lower modes, repeatedly labeling spurious modes as physical. Vertical lines shown in Fig. 6.6 correspond to frequencies that are very far from the typical values for these modes. The algorithm still catalogs these frequencies as samples of the modal response. As explained in the Methodology section, this is a consequence of both the fact that the algorithm discriminates exclusively based on modal shapes, and the relatively low number of sensors and the presence of noisy signals for the case study.

Figure 6.7 compares the results obtained from the MAT algorithm with six and three available sensors, for the year 2011.

As can be seen, both approaches have a very similar performance for the first four modes. For higher modes, on the other hand, the performance of the six sensors approach is better, being the only one able to capture all the modes detected.

Figure 6.8 shows the compared results obtained from the MAT algorithm with six and two available sensors, for the same period of time as above.

The trend seen in Fig. 6.7 regarding the lower modes below 6 Hz is repeated in this figure, as the results of both approaches have a similar performance. For the higher modes, it can be seen how closely spaced frequencies are confused with each other and with spurious modes.

Figure 6.9 shows the compared results obtained from the MAT algorithm with six and one available sensor(s), for the same period of time as all the above.

In this figure, it is seen a large amount of dispersion in the tracked frequencies using one sensor, due to both a high presence of spurious modes in the identification and the absence of a modal shape characterizing each mode. For the evaluation with one sensor, clearly the only frequencies tracked in time are those well spaced from each other; the four lower modes below 6 Hz.

6.5 Conclusions

Two modal tracking methodologies have been implemented, with a variable number of sensors, with the intention to characterize the effect of a small monitoring system on the process of tracking the modal response. The performance of the Model Assisted Tracking algorithm with six sensors shows that it is able to track the response of all the modes of interest, when they are characterized by the identification algorithm. When applying MAT with three sensors, it is still possible to track the first four modes below 6 Hz, with an accuracy that's similar to the evaluation using six sensors. For higher modes, though, the tracking is considerably poorer (but possible). This loss identifiability intensifies as less sensors are used, reaching the extreme case of an evaluation using a single sensor, where only well spaced identified frequencies are possible to track.

The second tracking methodology evaluated, AFDD-T, has limited results, being unable to consistently track even the lower modes. Such an algorithm requires a more complete and less noisy monitoring system than the available in this case. Supporting the monitoring on the frequency and environmental measurements allows working with a small monitoring system.

If the objective of monitoring is to capture the overall behavior of the structural system, which is characterized mainly by the lower modes, very few well located sensors are sufficient. If, on the other hand, the objective is to capture the higher modes, which are more difficult to distinguish from noise in the signal, then more and well located sensors are necessary. Even so, it is remarkable that only six sensors installed in the building are enough to satisfactorily track all the identified modes, showing that in this case a small monitoring system is sufficient for the task.

References

1. Gonzalez, W., Boroschek, R., Bilbao, J.: Modal tracking of an instrumented building assisted by temperature measurements. Unpublished manuscript, University of Chile, Santiago, Chile (2018)
2. Rainieri, C., Fabbrocino, G., Cosenza, E.: Near real-time tracking of dynamic properties for standalone structural health monitoring systems. Mech. Syst. Signal Process. **25**(8), 3010–3026 (2011)
3. Boroschek, R., Bilbao, J.: Interpretation of stabilization diagrams using density-based clustering algorithm. Engineering Structures (approved for publication) (2018)
4. Ankerst, M., Breunig, M.M., Kriegel, H.P., & Sander, J.: OPTICS: ordering points to identify the clustering structure. In: ACM Sigmod Record, vol. 28, no. 2, pp. 49–60. ACM (1999, June)
5. Peeters, B., De Roeck, G.: Reference-based stochastic subspace identification for output-only modal analysis. Mech. Syst. Signal Process. **13**(6), 855–878 (1999)
6. Alessandro, C., Filipe, M., Carmelo, G., Álvaro, C.: Automatic operational modal analysis: challenges and practical application to a historical bridge. In: 6th ECCOMAS Conference on Smart Structures and Materials, pp. 1–20 (2013)
7. Allemang, R. J., Brown, D. L.: A correlation coefficient for modal vector analysis. In: Proceedings of the 1st International Modal Analysis Conference, vol. 1, pp. 110–116. SEM Orlando (1982, November)
8. Brincker, R., Zhang, L., Andersen, P.: Modal identification of output-only systems using frequency domain decomposition. Smart Mater. Struct. **10**(3), 441 (2001)

Chapter 7
Bayesian Damage Identification Using Strain Data from Lock Gates

Yichao Yang, Ramin Madarshahian, and Michael D. Todd

Abstract Damage identification plays a significant role in the maintenance of navigation locks, which are part of the United States' $500B replacement value in inland waterway infrastructure; maritime transport disruption for closed lock gates causes substantial economic and utility losses. Lock gates are normally instrumented with strain gauges, and one of the critical failure modes is the development of a gap between the supporting wall and gate, initiating from quoin and/or pintle part wear. This gap leads to undesirable load distributions that can induce gate failure by overload. The probability of damage exceedance from different values affects repair strategies. This work uses Bayesian inference to identify that damage. The input features are raw strain data, while the loading is assumed unknown. The inherent uncertainty in measurements and model assumptions result in a posterior distribution of parameters of damage such as gap size and location. The results show that the true parameter of damage, which is used to generate simulated data, could be predicted using the posterior.

7.1 Introduction

Navigation locks play an important role in waterway transportation system in the US and all over the world [1, 2]. The unexpected closure of these assets, e.g. due to malfunction of structural components, is very costly because it prevents many shippers from fulfilling their scheduled transport missions [3]. To prevent the unscheduled closure of navigation locks in the United States, US Army Corps of Engineers (USACE) designed a discrete, indexed-based condition rating system for components of a navigation lock. Through this rating system, decision-makers are informed about repair and maintenance scenarios [4, 5]. Structural health monitoring of these assets can reduce the uncertainty about assigning a condition to different components of navigation locks.

One of the most common damage scenarios in miter gates is the occurrence of gaps in the quoin block, which is an interface between the gate and the supporting wall [6]. When this gap size increases, the area which connects the gate to the supporting wall decreases, leading to stress redistribution that possibly exceeds design or performance limit states. For example, the cyclic nature of loading on the structure may cause fatigue failures in such high-stress zones [7]. Early prediction of gap development in miter gates informs the planning of repair strategies at the time of scheduled closure and to eliminate unscheduled (and much more costly) closures due to an unexpected failure of components of the structure. The most common loading scenario for miter gates is hydrostatic loading due to the different water levels on both sides of the lock chamber. Hydrostatic loading at upstream and downstream will create an internal moment on vertical girders of a lock [8]. Therefore, distribution of stress and subsequently strains on the structure is a function of boundary conditions, hydrostatic loading, and temperature. Several navigation locks are instrumented using strain gauge sensors by USACE.

In this paper, we are going to perform a Bayesian inverse analysis utilizing strain gauge data to obtain gap and loading parameters. Here, we used only simulated data using a previously-validated finite element model in ABAQUS. We also ignored the effect of the temperature on distribution of strains. Using the Bayesian framework, we can investigate how the accuracy of strain gauge data results in uncertainty in parameter estimation. Also, we can consider the experience of the modeler and field engineers using priors. For example, based on the available information, we can estimate the water level on both sides of the miter gates, but we are not certain of the exact value. So we may define a normal distribution reflecting our initial belief about the level of water, and then let the Bayesian model update our belief by observing the data.

This work presents a preliminary investigation of Bayesian inverse finite element modeling of Navigation locks. A Markov chain Monte Carlo (MCMC) algorithm is used to obtain the posterior distributions for gap size and hydrostatic loading. We used PyMC3 package of python to implement MCMC [9]. One of the challenges in this project is compiling both ABAQUS

Y. Yang (✉) · R. Madarshahian · M. D. Todd
University of California at San Diego, La Jolla, CA, USA
e-mail: yiy018@ucsd.edu; mdtodd@ucsd.edu

and a python package named Theano [10]. To run ABAQUS from python script through Theano is a long and complicated process. Adaptive Metropolis sampling was used as MCMC algorithm [9]. Being computationally expensive, we could not get enough samples to tune the MCMC algorithm. So, while some Markov chains did not achieve optimal sampling, the results nonetheless may provide insight into the posterior of parameters.

7.2 Finite Element Model

A finite element (FE) model in ABAQUS (version 6.14-2), shown in Fig. 7.1, is used to model the primary physics of the miter gate. We use this model both to simulate our data and to perform Bayesian inference. A 4-node doubly curved shell element is used to model the miter gate, and an 8-node linear brick element is used to model the wall and contact block. The damage (gap) is simulated by removing a part of the contact block which represents the connection in between gate and wall as shown in Fig. 7.2. The height of the removed part shows the intensity of the damage (gap length).

Figure 7.3 shows that the hydrostatic loading on the structure from both downstream and upstream. Fifty-eight strain gauges are considered in the FE model similar to those sensors used in the real asset monitoring program. These data are used in our Bayesian model to find the gap and loadings parameters.

In this work, we have four parameters which three of them (i.e. gap length, upstream and downstream levels of water) should be changed during MCMC in the FE model. An ABAQUS Macro is designed to handle the change in these parameters. It should be noted that the MCMC method is applied using PyMC3 package in python, while the mathematical model is a FE model in ABAQUS. Theano package is used to link PyMC3 to ABAQUS by using a Theano OP. In an inverse analysis, the output is 58 strain values at specified locations.

Running the MCMC on the ABAQUS model, the interaction between ABAQUS and PyMC3 creates a computationally expensive problem. In the result section of this paper, we will report the details about the computational cost of this problem.

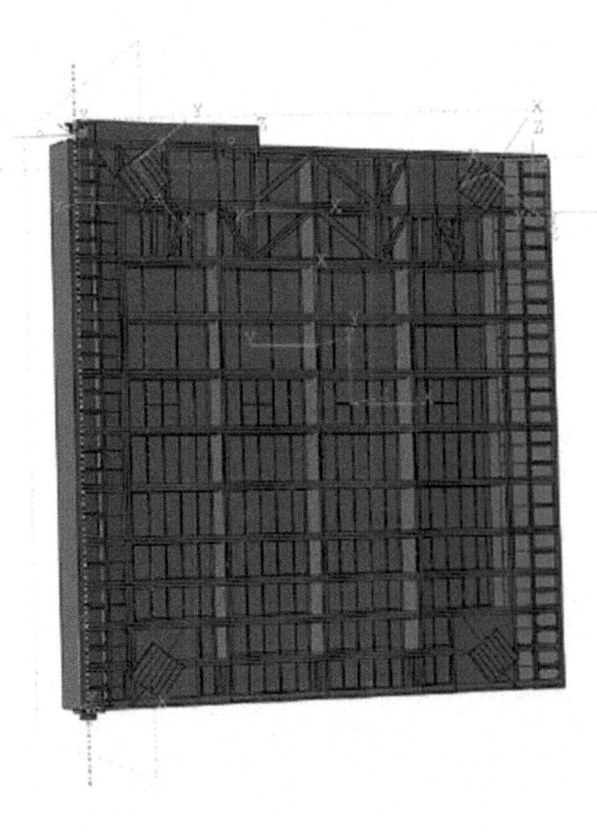

Fig. 7.1 Miter gate

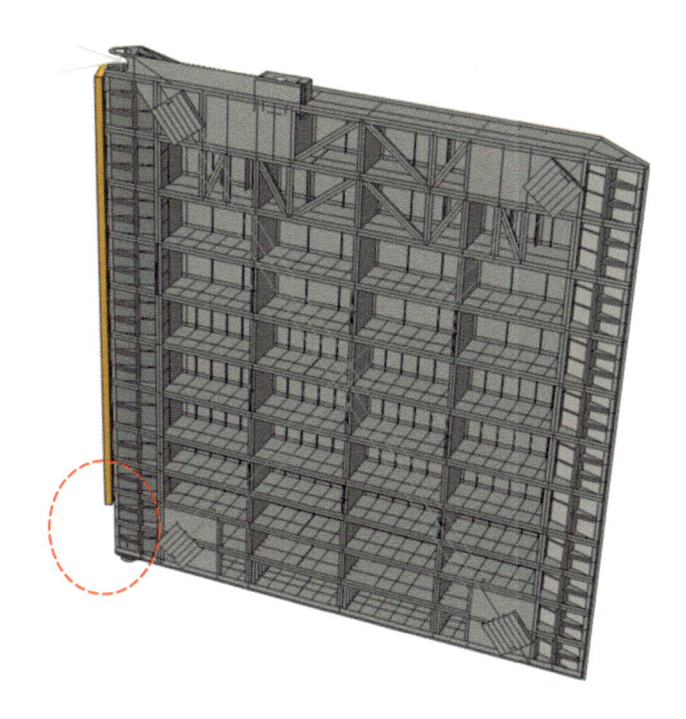

7 Bayesian Damage Identification Using Strain Data from Lock Gates

Fig. 7.2 Damage of miter gate

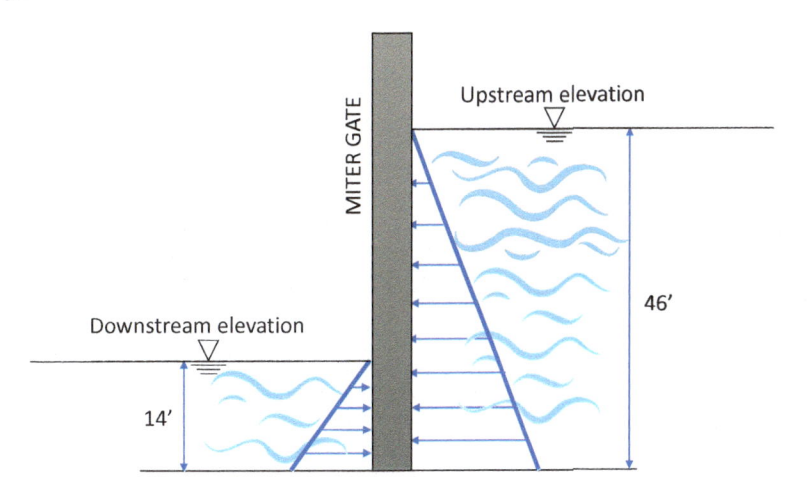

Fig. 7.3 Hydrostatic loadings

7.3 Bayesian Inference

In the structural health monitoring (SHM) field, in order for damage detection estimates to be meaningful to owners or operators, concern must be given to the degree of confidence in those estimations. Bayesian parameter estimation does not provide a single value as its best estimate; instead, it considers the belief about the true value of parameters and updates that belief by observing new data. One advantage of performing Bayesian parameter estimation is that the experience of the modeler or expert in the field is incorporated into priors and affects the posterior prediction estimations.

Equation (7.1) shows the familiar Bayesian formula. Here, θ is a vector of parameters and $p(\theta)$ is our priors for them. The observed data is denoted by D, and $p(D|\theta)$ denotes the likelihood. It should be noted that in this work we used a normal distribution for the likelihood, and we assume the standard deviation of this distribution as one of our unknown parameters (σ_{lik}). The posterior is shown by $p(\theta|D)$ in Eq. (7.1).

$$p(\theta|D) \propto p(D|\theta) \cdot p(\theta) \tag{7.1}$$

Table 7.1 Prior Distributions of Parameters and their "true" values

Parameters of interest	Prior distributions	"True" value	Units
Gap length	$Uniform(0, 180)$	75	in.
H_{down}	$Normal(\mu = 168, \sigma = 20)$	168	in.
H_{up}	$Normal(\mu = 552, \sigma = 10)$	552	in.
σ_{lik}	$Halfnormal1(\sigma = 5e-6)$	$1e-6$	–

7.3.1 Simulated Data

A single run of the model uses the previously specified parameters with their "true" values are shown in Table 7.1. The simulated data is then obtained by adding a noise to the results of the single run of the model. We assumed $1e-6$ as the true value for the standard deviation of the likelihood since it was the standard deviation of the noise we added to the strain gauges output.

7.3.2 Priors

Before observing the data, we built on existing experts' experiences on the parameters of interest which compiled and formed the prior distributions of these parameters. For example in a previously published work about damage detection in miter gates, a large gap was introduced as a gap with a length of 180 inches [6]. We used this to consider an upper bound for possible gap size. Also, we know the upstream and downstream levels of water are around 552 inches and 168 inches, respectively. We assumed a normal distribution for each of them with standard deviations of 10 inches and 20 inches, respectively. These standard deviations illustrate our uncertainty about these values. Table 7.1 shows prior distributions of these four parameters that we are interested in exploring.

7.3.3 Adaptive Metropolis-Hastings Sampling

In this work, an Adaptive Metropolis sampling algorithm was used to perform MCMC analysis. The Algorithm 1 shows a general logic behind Metropolis-Hastings sampling method. The proposal distribution is updated after taking several samples in the Adaptive version of the algorithm. Generally, enough numbers of samples should be considered for proper tuning of this algorithm. However, in this case due to the considerable computational cost in running the problem, it was not feasible to assign enough numbers of samples for tuning.

Algorithm 1: Adaptive Metropolis-Hastings sampling

Initialize $x^0 \sim q(x)$
for $i = 1, i \leq n, i++$ do
    $x^* \sim q(x^i | x^{i-1})$
    $\alpha(x^* | x^{i-1}) = \min\{1, \frac{q(x^{i-1}|x^*)p(x^*)}{q(x^*|x^{i-1})p(x^{i-1})}\}$
    $u \sim Uniform(0,1)$
    if $u < \alpha$ then
        Accept: $x^i \leftarrow x^*$
    else
        Reject: $x^i \leftarrow x^{i-1}$

7.4 Results

7.4.1 Computational Cost

The runtime of an ABAQUS model, and its interaction with PyMC3 and implementation of Metropolis-Hastings algorithm created a computationally expensive problem. MCMC sampling required several runs of FE model to obtain each sample. We decided to take 2000 samples when performing the MCMC sampling. On a 4 GB RAM computer, it takes on average 2576 seconds per sample. This number of samples is not enough to tune the proposal distribution of MCMC sampling. However, our results still showed that MCMC sampling can provide us with information about the posterior of our parameters. We also discarded the first 500 samples to remove samples that are not in the stationary state. These are called the burning samples.

7.4.2 Bayesian Inference for Gap Length

The MCMC chain for the gap length is shown in Fig. 7.4. The starting point was not close to the "true" value of the gap. When the starting point is not close to the "true" value, MCMC needs a number of iterations to reach to the stationarity. We deliberately selected a starting point far from the "true" value of the gap length to study how fast we converge to the stable state in MCMC chains. Considering the chain shown in Fig. 7.4, after few numbers of iterations, MCMC chain approaches to the "true" value around 75 inches. Here, we conservatively considered 500 samples for burn-in samples (Shown by red color in the Fig. 7.4).

Convergence performance of the chain is examined using Geweke scores [11]. This method compares the mean of a segment from beginning of the chain with another segment from last of the chain. Here, 20 segments from the first half of the chain, x_a, are compared with the second half of the chain, x_b, after burn-in samples are discarded. Geweke scores, obtained as shown in Eq. (7.2), should oscillate between -1 and 1 if the chain is converged. Figure 7.5 shows Geweke scores for the MCMC chain of the gap length. In Eq. (7.2), $E(*)$ and $V(*)$ are expected value and variance of the samples, respectively.

$$Geweke\ score = \frac{E(x_a) - E(x_b)}{\sqrt{V(x_a) + V(x_b)}} \tag{7.2}$$

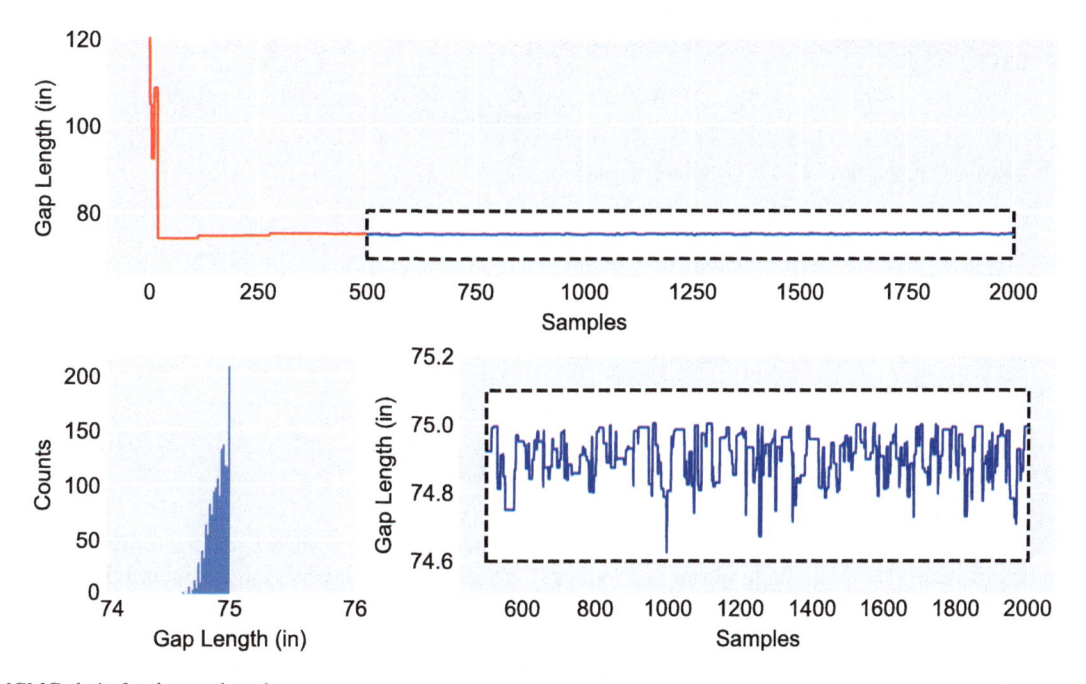

Fig. 7.4 MCMC chain for the gap length

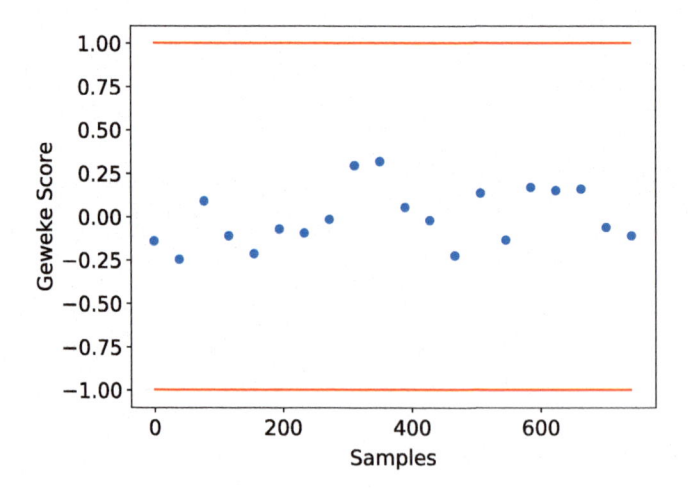

Fig. 7.5 Convergence diagnostics of the MCMC chain of the gap length

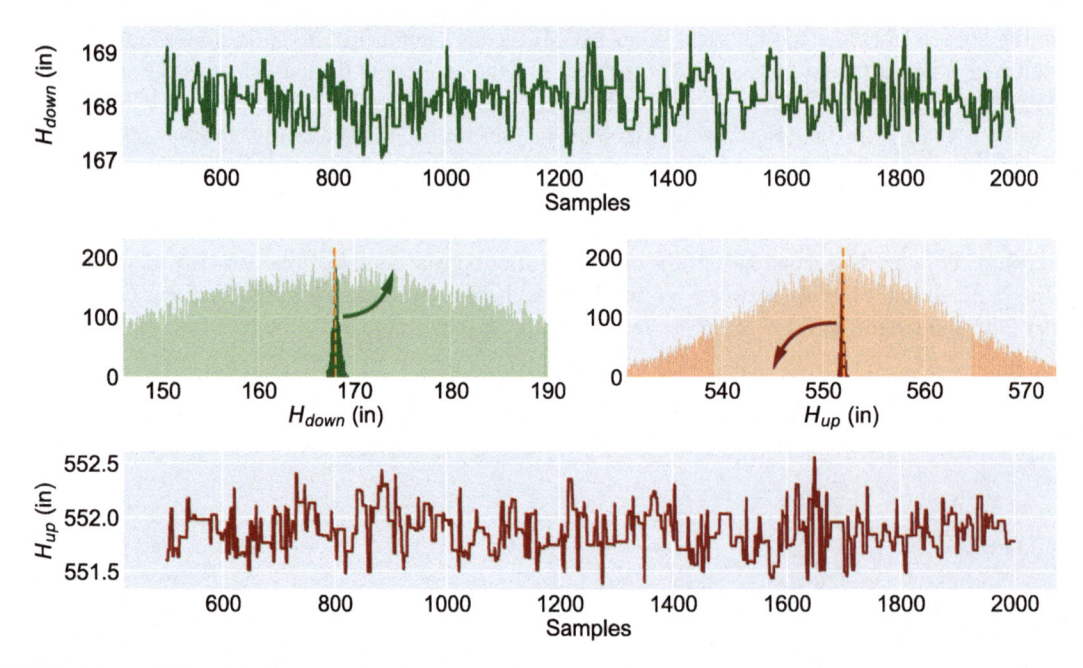

Fig. 7.6 MCMC chains and histograms for water elevation parameters

Although Geweke scores confirm the convergence in the chain (shown in Fig. 7.5), the shape of the chain implies many rejected samples. Probably reducing the proposal width could help obtain a denser chain. Since the problem was very computationally expensive, we could not tune the proposal distribution in MCMC algorithm to reach a dense distribution; however, the samples are very close to the "true" value.

7.4.3 Bayesian Inference for Hydrostatic Loading Parameters

Water levels are two parameters we used for hydrostatic loading. Figure 7.6 shows MCMC chains for the water elevation at upstream and downstream. We think the proposal distributions for these chains also had large standard deviations; however, the posteriors contain the "true" values of these parameters, shown by the dashed lines.

In Fig. 7.6, the priors are shown in the lighter colors, and their posteriors are shown on top of the prior distributions with darker colors. It is indicated clearly that our level of uncertainty about "true" values of these parameters is reduced after

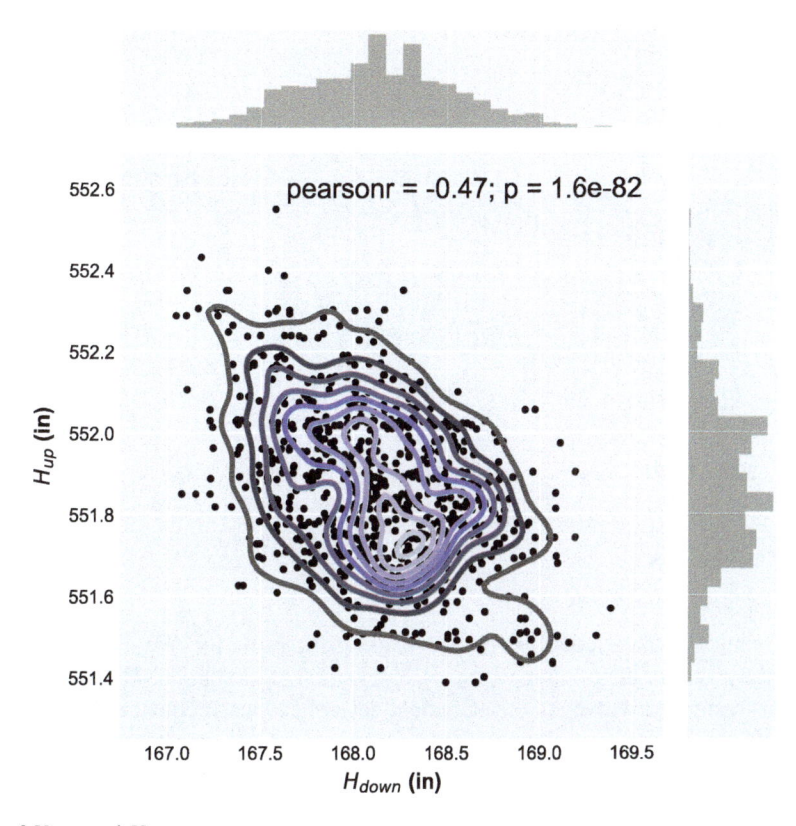

Fig. 7.7 Joint distribution of H_{down} and H_{up}

observing the strain gauges data. We also observed a negative correlation between these elevation parameters. Figure 7.7 shows joint distribution of these two parameters with Pearson correlation coefficient of -0.47.

7.4.4 Bayesian Inference for Standard Deviation of Likelihood

The standard deviation of the likelihood is another parameter which is considered in this work. It could be assumed as a parameter which shows the accuracy of input data. Figure 7.8 shows the MCMC chain and its corresponding histogram after discarding the burn-in samples. The time series shows a stationary and dense chain. Inference about this parameter is done using No-U-turn sampler (NUTS) instead of Metropolis Hastings algorithm [12]. We could not use NUTS for other parameters since the gradient field of those parameters is not available. Figure 7.8 also shows the posterior distribution in a darker color, which contains the value used for defining the noise in simulated data, on top of the prior distribution that is shown in a lighter color.

7.5 Concluding Remarks

In this work, we performed an inverse Bayesian analysis on a finite element model by using the strain gauge data for damage identification. We used MCMC methods to obtain the posterior for the damage parameter while the loading is assumed unknown. We successfully used a finite element model in ABAQUS integrated with the Bayesian script in Python. In this preliminary work, we showed, although the computational cost of this problem is considerable and tuning the adaptive MCMC algorithm is not feasible, the results are useful to describe the posterior of parameters. We showed that our initial uncertainty in the parameters, reflected in the prior, is reduced when the posterior for the parameters is obtained. Moreover, a moderate negative correlation was observed in loading parameters, while in prior information we did not consider any correlation between parameters. The results showed that the true parameter of damage, including the loading scenario

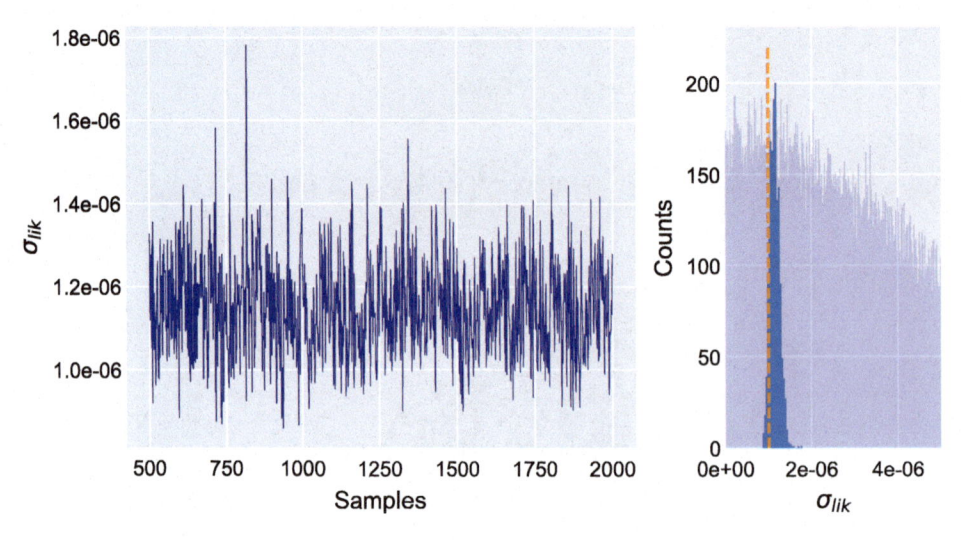

Fig. 7.8 Standard deviation of the likelihood

which is used to generate simulated data, could be predicted using the posterior. We also showed the standard deviation of input noise falls within the posterior distribution of standard deviation of likelihood. This means the standard deviation of likelihood can be used to estimate the accuracy of measurement data when measurement error follows a normal distribution.

Acknowledgement This work was supported by the US Army Engineer Research and Development Center (ERDC) under cooperative agreement W912HZ-17-2-0024.

References

1. Daniel, R.A.: Mitre gates in some recent lock projects in the Netherlands (Stemmtore in einigen neuen Schleusenanlagen in den Niederlanden). Stahlbau **69**(12), 952–964 (2000)
2. Richardson, G.C.: Navigation locks: navigation lock gates and valves. J. Waterways Harbors Div. **90**(1), 79–102 (1964)
3. Schwieterman, J.P., Field, S., Fischer, L., Pizzano, A.: An analysis of the economic effects of terminating operations at the Chicago river controlling works and O'brien locks on the Chicago area waterway system. In: System. Chicago, IL: Chaddick Institute for Metropolitan Development, DePaul University. http://www.unlockourjobs.org/wpcontent/themes/unlockourjobs/pdf/DePaul_University_Study.pdf, vol. 24, p. 2010 (2010). Accessed September 2010
4. Greimann, L.F., Stecker, J.H., Kao, A.M., Rens, K.L.: Inspection and rating of miter lock gates. J. Perform. Constr. Facil. **5**(4), 226–238 (1991)
5. Greimann, L.F., Stecker, J.H., Kao, A.M.: Inspection and rating of steel sheet pile. J. Perform. Constr. Facil. **4**(3), 186–201 (1990)
6. Eick, B.A., Treece, Z.R., Spencer Jr, B.F., Smith, M.D., Sweeney, S.C., Alexander, Q.G., Foltz, S.D.: Automated damage detection in miter gates of navigation locks. Struct. Control. Health Monit. **25**(1), e2053 (2018)
7. Wilkins, E.: Cumulative damage in fatigue. In: Colloquium on Fatigue/Colloque de Fatigue/Kolloquium über Ermüdungsfestigkeit, pp. 321–332. Springer, Berlin (1956)
8. Estes, A.C., Frangopol, D.M., Foltz, S.D.: Updating reliability of steel miter gates on locks and dams using visual inspection results. Eng. Struct. **26**(3), 319–333 (2004)
9. J. Salvatier, T. V. Wiecki, and C. Fonnesbeck: Probabilistic programming in python using PyMC$_3$. Peer J. Comput. Sci. **2**, e55 (2016)
10. Al-Rfou, R., Alain, G., Almahairi, A., Angermueller, C., Bahdanau, D., Ballas, N., Bastien, F., Bayer, J., Belikov, A., Belopolsky, A., Bengio, Y.: Theano: a python framework for fast computation of mathematical expressions. arXiv preprint (2016)
11. Cowles, M.K. and Carlin, B.P.: Markov chain Monte Carlo convergence diagnostics: a comparative review. J. Am. Stat. Assoc. **91**(434), 883–904 (1996)
12. Hoffman, M.D., Gelman, A.: The no-U-turn sampler: adaptively setting path lengths in Hamiltonian Monte Carlo. J. Mach. Learn. Res. **15**(1), 1593–1623 (2014)

Chapter 8
Dynamic Tests and Technical Monitoring of a Novel Sandwich Footbridge

Jacek Chroscielewski, Mikolaj Miskiewicz, Lukasz Pyrzowski, Magdalena Rucka, Bartosz Sobczyk, Krzysztof Wilde, and Blazej Meronk

Abstract A novel sandwich composite footbridge is described in this paper, for the first time after it has been put into operation over the Radunia River in the Pruszcz Gdański municipality. This paper presents results of dynamic tests and describes technical monitoring of the footbridge. The dynamic tests were conducted to estimate pedestrian comfort and were compared with the ones from numerical simulations made in the environment of Finite Element Method. A discussion of the obtained results is made then. The characteristics and capabilities of the network of sensors spread over the bridge to monitor its long-term behavior are described.

Keywords Fiber Reinforced Plastics (FRP) · Laminated shells · Technical monitoring · Footbridge

8.1 Introduction

A large number of bridges are made of traditional material like steel, concrete or wood. In recent years, there is a growing interest in the application of novel materials like composites in bridges construction industry. Moreover, many new procedures and technologies have been recently introduced into the design frameworks. Hence, high-tech materials having some profits and benefits compared to standard ones can be used by civil engineers.

This paper describes a novel composite footbridge. The bridge was a final product of the FOBRIDGE project implemented in the years 2013–2015, in the consortium associating Gdańsk University of Technology (GUT), Warsaw Military University of Technology and ROMA Co. Ltd. The single element, sandwich composite footbridge of a length of 14 m was constructed by means of the vacuum infusion technology.

The superstructure is completely made of hi-tech composites so it meets the conditions set in current trends in mechanics of materials (see e.g. [1–6]). The simply supported, shell-like bridge has a U-shape cross section. It was performed as sandwich structure with two skins and a core. The core was made of a Polyethylene terephthalate (PET) foam of a thickness of 100 mm and a density 100 kg/m^3. Fiber reinforced plastic (FRP) skins were made of stitched and balanced BAT [0/90] and GBX [45/−45] E-glass and vinylester resin. To protect the bridge from negative environmental and exploitation effects and to improve the bridge overall look, the load-bearing structural layers were additionally covered with external coating layer consisting of: topcoat, gelcoat and anti-skidding layer. The total weight of the footbridge superstructure is 3200 kg.

For the purpose of design of the footbridge, the Eurocodes were used, especially to establish loading conditions and combinations [7]. Since there are no procedures formulated exclusively for the purpose of the estimation of load capacity of FRP skins, other approaches were utilized, as the codes for the design of tanks [8]. This aspect was also supported by the use of literature containing a thorough review of hypothesis and criterions aiming to estimate the moment of failure initiation in FRP (see for instance [9, 10]). In consequence the bridge is capable of carrying 5 kN/m^2 dense crowd of pedestrians.

After the formal project ending, the project leader—GUT, initiated some discussions and negotiations and made efforts in order to put the bridge into operation under real pedestrian-cyclists traffic. Local government of the Pruszcz Gdański municipality expressed their interest in this field. Therefore, the substructure was designed and the bridge was finally installed on abutments in 2018. Now, it enables pedestrians and cyclists to cross the Radunia river, as it is shown in Fig. 8.1.

The footbridge was tested to establish its static and dynamic performance during the project (refer to [11, 12]) as well as before it was opened to traffic (6th of July 2018), see Fig. 8.2, which is required by the governmental authorities. Since the

---

J. Chroscielewski · M. Miskiewicz (✉) · L. Pyrzowski · M. Rucka · B. Sobczyk · K. Wilde · B. Meronk
Department of Mechanics of Materials and Structures, Faculty of Civil and Environmental Engineering, Gdansk University of Technology, Gdańsk, Poland
e-mail: mikolaj.miskiewicz@pg.edu.pl

Fig. 8.1 The footbridge: side view (**a**), pedestrian perspective view (**b**)

Fig. 8.2 The footbridge testing: geodetic measurement (**a**), fiber optic strain measurement (**b**), laser scanning (**c**) [15, 16]

tests revealed proper behaviour of the structure, it has been opened for traffic. It is worth to mention that it is the next FRP bridge to be opened in Poland recently (see [13, 14]).

During the tests on 6th of July 2018 its static and dynamic properties were evaluated. Additionally, technical monitoring of the bridge was installed and put into operation to observe and report its long-term performance. An attention is focused on the modal properties of the bridge, as it is a lightweight structure and thus may be vulnerable to dynamic loads causing excessive vibrations. The dynamic testing and the monitoring system are described in the next chapter.

8.2 Dynamic Testing

The dynamic testing of a footbridge is very important in the context of further operation. Footbridges are often lightweight structures; therefore, they may be vulnerable to vibrations. Thus, their appropriate design in terms of meeting users comfort conditions is very crucial. Nevertheless, before the dynamic test were conducted, the static load was applied to check whether the static response is in agreement with the assumptions made during the design. The comparison of the measured and calculated deflections in the mid-span section of the footbridge is shown in Fig. 8.3. The numerical calculations were made with the use of the Finite Element Method and were described in details in [11]. In general, the numerical and in situ values are in good agreement. The structure appeared to be a little bit stiffer than it was assumed during the design.

The dynamic tests were launched after the static tests. At first, the natural frequencies and mode shapes were identified experimentally. Then vibrations of the bridge were measured. The vibrations were induced by: groups of moving crowd (different types of march and run of groups of 3, 6 or 12 persons), jumps (groups of 3, 6 or 12 persons) and vandalism acts in the form of forced lateral pulling and pushing of handrails. The different stages of tests are depicted in Fig. 8.4.

The vibrations of the bridge were registered by the multichannel recorder LMS SCADAS SCR202 and seismic, ceramic-shear accelerometers PCB 393A03. Accelerometers were attached to the handrails at six measurement points located at the quarter-spans and mid-span (see Fig. 8.5 for their location). Vibrations of the bridge were measured at six measurement

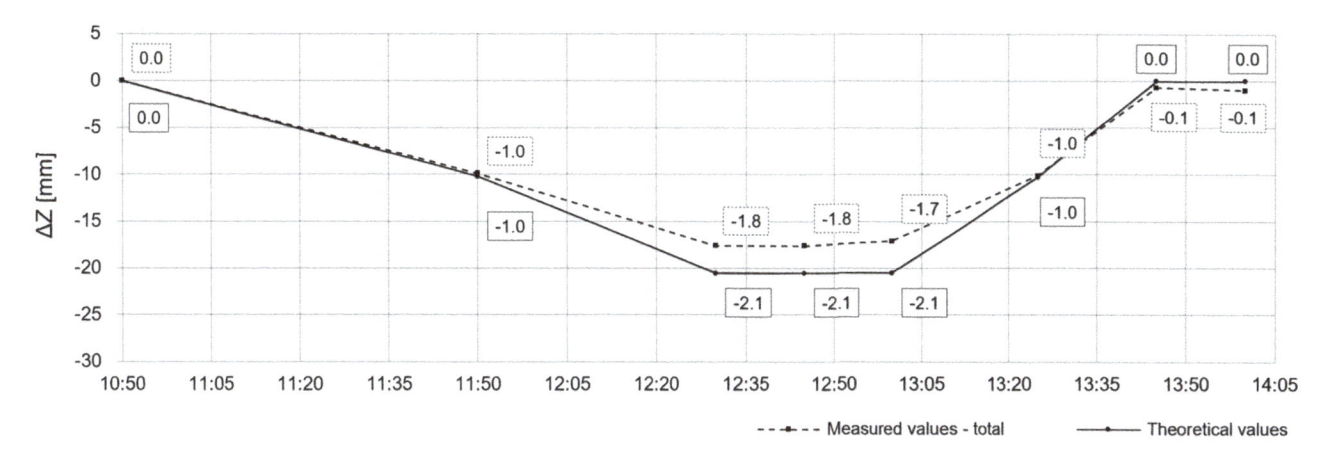

Fig. 8.3 Comparison of calculated and measured deflection of the footbridge under the static test load

Fig. 8.4 Dynamic tests of the footbridge: march of 12 people (**a**), forced vibrations of handrails (**b**), jumps of 3 people (**c**), impulse modal hammer test (**d**)

points. In each of measurement points, vibrations in two directions were recorded (along y and z axis). Large-sledge impulse hammer PCB Piezotronics 086D50 was utilised to for performing the experimental modal analysis resulting in determination of natural frequencies, corresponding mode shapes and damping ratios.

The experimentally identified natural frequencies and damping ratios are collected in Table 8.1 and compared with numerically calculated natural frequencies. The experimentally identified mode shapes are: 1st—torsion of the whole bridge, 2nd—vertical vibration of the whole bridge. 3rd–5th local oscillations of the handrails. The modes 1st–3rd are shown in Fig. 8.6.

The results from Table 8.1 show that the numerical model is well correlated with the real response of the structure. What is more the 1st measured eigenfrequency is very high and thus according to some footbridge design guidelines (see [17, 18])

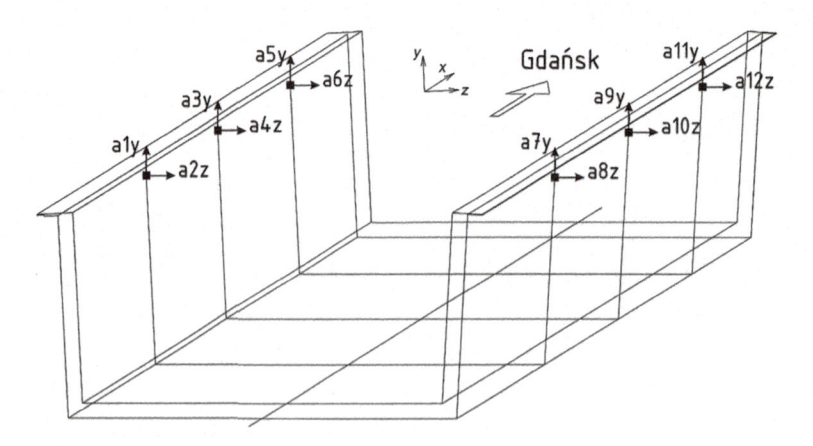

Fig. 8.5 Location of acceleration measurement points

Table 8.1 Experimentally identified natural frequencies and damping ratios compared to numerically calculated natural frequencies

Number of shape	Experimentally identified natural frequency [Hz]	Numerically calculated natural frequency [Hz]	Experimentally identified damping ratios [%]
1	8.4	7.8	2.7
2	9.5	9.8	1.6
3	13.3	12.5	1.1
4	15.4	14.8	1.5
5	19.6	18.4	1.8

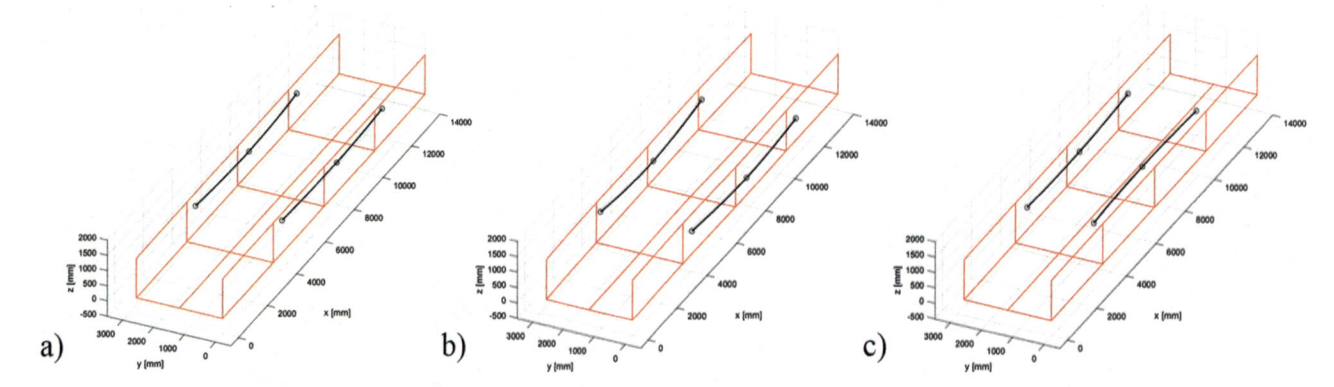

Fig. 8.6 Experimentally identified mode shapes 1st (**a**), 2nd (**b**) and 3rd (**c**)

the risk of resonance vibrations caused by pedestrians is negligible. In consequence, very good comfort for future users is achieved. This was also confirmed during the induced vibrations tests as march or jumps.

Finally, it can be concluded that the bridge has good dynamic properties and gives high comfort of exploitation.

8.3 Technical Monitoring

Although the bridge properties were tested during the project and just before it was put into operation, an additional system of technical monitoring was installed over the bridge [19–22]. It enables measurements of displacements, strains and accelerations. The technical monitoring system is composed of the following components: Fiber Bragg Gratings (FBG) optic fibre sensors and accelerometers and gyroscopes. These sensors are attached to the bottom of the deck, as shown in Fig. 8.7 and under the handrail.

The proposed system of health monitoring allows reporting and observing behaviour of the bridge in long-term conditions, which is important, as it is a novel structure.

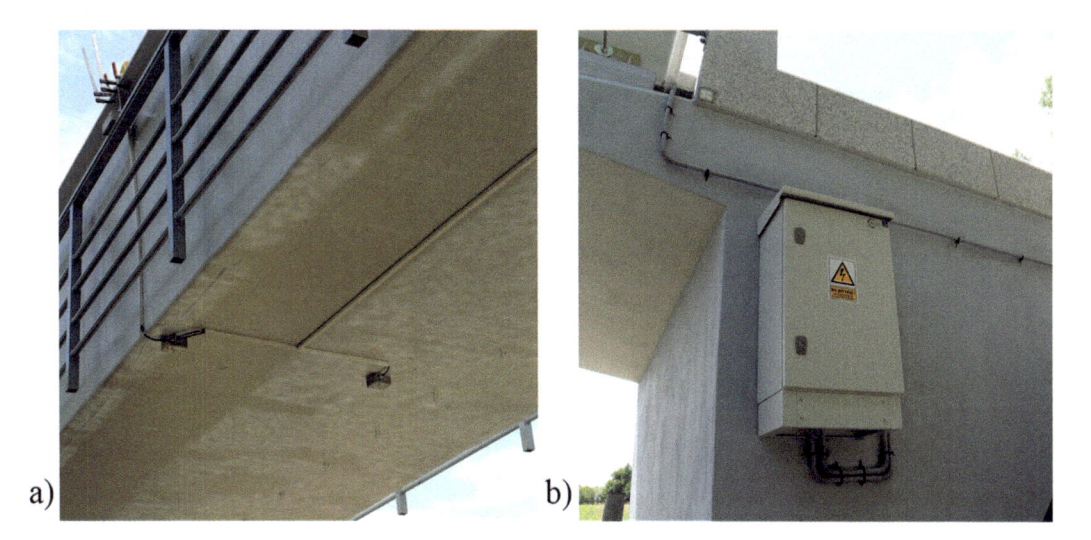

Fig. 8.7 Elements of the footbridge technical monitoring: sensors (**a**), central station (**b**)

8.4 Final Remarks

The footbridge described in this paper is a good alternative to a one made of regular materials (e.g. steel or concrete), because it is cheaper (including all the maintenance) assuming that its life cycle is 50 years. The experimental results obtained before it was put into operation and presented in this paper show that it has very good static and dynamic properties. Therefore, it gives very good comfort to the future users. Additionally, in order to check its long-term behaviour a system of technical monitoring was installed.

Acknowledgements The study was supported by the National Centre for Research and Development, Poland, grant no. PBS1/B2/6/2013.

References

1. Rozylo, P., Debski, H., Kubiak, T.: A model of low-velocity impact damage of composite plates subjected to Compression-After-Impact (CAI) testing. Compos. Struct. **181**, 158–170 (2017)
2. Wang, F.S., Yu, X.S., Jia, S.Q., Li, P.: Experimental and numerical study on residual strength of aircraft carbon/epoxy composite after lightning strike. Aerosp. Sci. Technol. **75**, 304–314 (2018)
3. Burzyński, S., Chroscielewski, J., Daszkiewicz, K., Witkowski, W.: Elastoplastic nonlinear FEM analysis of FGM shells of Cosserat type. Compos. Part B Eng. **154**, 478–491 (2018)
4. Sabik, A.: Progressive failure analysis of laminates in the framework of 6-field non-linear shell theory. Compos. Struct. **200**, 195–203 (2018)
5. Perrella, M., Berardi, V.P., Cricrì, G.: A novel methodology for shear cohesive law identification of bonded reinforcements. Compos. Part B Eng. **144**, 126–133 (2018)
6. Baranowski, P., Damaziak, K., Malachowski, J., Mazurkiewicz, L., Muszyński, A.: A child seat numerical model validation in the static and dynamic work conditions. Arch. Civ. Mech. Eng. **15**, 361–375 (2015)
7. PN-EN 1990:2002+A1: Eurocode—basis of structural design
8. PN-EN 13121-3+A1:2010E: Ground containers made of plastics reinforced with glass fibre. Part 3. Design and production control
9. Chroscielewski, J., Klasztorny, M., Nycz, D., Sobczyk, B.: Load capacity and serviceability conditions for footbridges made of fibre-reinforced polymer laminates. Roads Bridg. Drog. i Most. **13**, 189–202 (2014)
10. Soden, P., Kaddour, A., Hinton, M.: Recommendations for designers and researchers resulting from the world-wide failure exercise. Compos. Sci. Technol. **64**, 589–604 (2004)
11. Chroscielewski, J., Miskiewicz, M., Pyrzowski, L., Sobczyk, B., Wilde, K.: A novel sandwich footbridge—practical application of laminated composites in bridge design and in situ measurements of static response. Compos. Part B Eng. **126**, 153–161 (2017)
12. Chroscielewski, J., Miskiewicz, M., Pyrzowski, L., Rucka, M., Sobczyk, B., Wilde, K.: Modal properties identification of a novel sandwich footbridge—comparison of measured dynamic response and FEA. Compos. Part B Eng. **151**, 245–255 (2018)
13. Siwowski, T., Rajchel, M., Kaleta, D., Wlasak, L.: The first polish road bridge made of FRP composites. Struct. Eng. Int. **27**, 308–314 (2017)
14. Siwowski, T., Kaleta, D., Rajchel, M.: Structural behaviour of an all-composite road bridge. Compos. Struct. **192**, 555–567 (2018)
15. Tysiac, P.: Laser scanning of a soil-shell bridge structure. In: 2018 Baltic Geodetic Congress (BGC Geomatics), pp. 61–66 (2018)
16. Szulwic, J., Tysiac, P.: Searching for road deformations using mobile laser scanning. In: MATEC Web of Conferences, vol. 122, p. 04004 (2017)

17. Sètra. Technical guide—assessment of vibrational behaviour of footbridges under pedestrian loading. Service d'Etudes techniques des routes et autoroutes (2006)
18. Directorate-General for Research and Innovation (European Commission). Human-induced vibration of steel structures (Hivoss). Design of footbridges. Guideline (2010)
19. Miskiewicz, M., Pyrzowski, L., Chroscielewski, J., Wilde, K.: Structural health monitoring of composite shell footbridge for its design validation. In: Proceedings 2016 Baltic Geodetic Congress (Geomatics), pp. 228–233 (2016)
20. Hou, J., Jankowski, L., Ou, J.: An online substructure identification method for local structural health monitoring. Smart Mater. Struct. **22**, 095017 (2013)
21. Clemente, P., De Stefano, A.: Novel methods in SHM and monitoring of bridges: foreword. J. Civ. Struct. Heal. Monit. **6**(3), 317–318 (2016)
22. Michalcova, L., Belsky, P., Petrusova, L.: Composite panel structural health monitoring and failure analysis under compression using acoustic emission. J. Civ. Struct. Health Monit. **8**(4), 607–615 (2018)

Chapter 9
Assessment and Control of Structural Vibration in Gyms and Sports Facilities

Aliz Fischer, Rob Harrison, Mark Nelson, François Lancelot, and James Hargreaves

Abstract Structures of various function can be affected by human induced vibration in many ways. Depending on the type of the structure, the load, the structural response, and the serviceability criteria to be met vary and may not be clearly defined. There are several guidance documents covering the evaluation of the footfall aspect of human induced vibration in buildings and structures, but this does not cover all types of buildings and structures, and in particular, does not cover gyms and sports facilities.

This paper outlines Arup's holistic approach to assess gyms and sports facilities in mixed-use structures for human induced vibration. It typically incorporates on-site testing, finite element modeling, performance criteria selection, and design and validation of potential structural mitigation solutions. The methodology is presented throughout several case-studies highlighting the value of each element of the process.

Keywords Vibration · Assessment · Criteria · Mitigation · Gym · Sports facility · Mixed-use building · Structure · Modeling · Experimental methods · Passive control · Active control · Analysis · Football · Testing · Vibration testing · Motion platform

9.1 Introduction

The design of gym, sport, and leisure facilities involves consideration of dynamic loads from human activities including aerobics, weight training, and treadmills. These dynamic loads can lead to perceptible vibration or noise in the structure which needs to be assessed at the design stage to enable vibration allowance to be made within the structural design. Also, to identify the need for and extent of any countermeasures needed to control vibration; such countermeasures are also known as mitigation.

It is not uncommon for a gym facility to be included in the design of a new office building or introduced as part of an office refurbishment. In both cases, the transmission of structural vibration into adjacent offices needs to be assessed. Sometimes vibration concerns arise during commissioning and possibly in operational service.

The assessment methodology, appropriate criteria, and mitigation are not fully defined in the current literature. Case studies are presented in this paper which describe our experiences with gym vibration assessment and includes criteria selection, modelling and analysis, mitigation options, and verification of countermeasures after commissioning. For each case, the paper summarizes the currently available information in industry and illustrates how mitigation options have been implemented in practice.

A. Fischer (✉) · F. Lancelot
Advanced Technology and Research, Arup, San Francisco, CA, USA
e-mail: aliz.fischer@arup.com; francois.lancelot@arup.com

R. Harrison
Advanced Technology and Research, Arup, Manchester, UK
e-mail: rob.harrison@arup.com

M. Nelson
Advanced Technology and Research, Arup, New York, NY, USA
e-mail: mark.nelson@arup.com

J. Hargreaves
Advanced Technology and Research, Arup, Solihull, UK
e-mail: james.hargreaves@arup.com

9.2 Assessment Workflow

The typical assessment workflow adopted in these case studies to evaluate and control structural vibration is shown in Fig. 9.1. Where there is a concern with an existing facility then vibration testing on site would be the priority. This would involve measurement of the structural dynamic properties and the vibration response at receiver locations of interest. For both assessment of existing structures and assessment at design stage, another priority is to define what is the vibration for a given situation. This involves a review of the facility in question and the selection of appropriate criteria, to establish the limit for structural vibration along with a metric which can be evaluated for measured or predicted data. The gym, and in some cases adjacent office, structure, or site of interest is typically modelled and analyzed using finite element software. Such computational models enable vibration to be evaluated at the design stage but are also a useful tool for the resolution of operational vibration concerns; they can be used to evaluate and help design mitigation. Mitigation may take the form of stiffness, mass, damping, or isolation depending on the nature of the problem. Mitigation can be included in the computational model so that its performance can be evaluated against the criteria.

9.3 Vibration Testing

For situations where there is an existing facility or during commissioning, vibration testing of building floorplate structures is a well-established method. The dynamic properties of the floorplate structure can be estimated by applying a heel-drop (CCIP-016 App. B [1], SCI P354 [2]), where an impulse is applied to the floor by a person dropping their heels onto the floorplate (Fig. 9.2, left). Vibration response is measured at locations of interest but would typically include the most mobile positions such as mid-span. The vibration response can be analyzed to estimate the floorplate natural frequencies and structural damping. Similarly, a vibration shaker can be used to generate applied loading from which response can be measured using one or more vibration transducers (accelerometers).

The measurement of floorplate vibration response due to specific activities such as aerobics, falling weights, and treadmills will enable the vibration magnitude and frequency content to be determined and compared with the criteria (Fig. 9.2, right). Critical structural dynamic frequencies can be identified which can help determine the type of mitigation that might be needed.

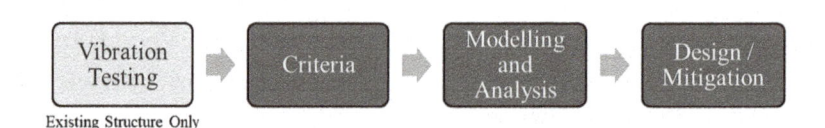

Fig. 9.1 Process overview

Fig. 9.2 Heel drop test during construction (left) and Rhythmic jumping testing during construction of Sky Health & Fitness Centre (right)

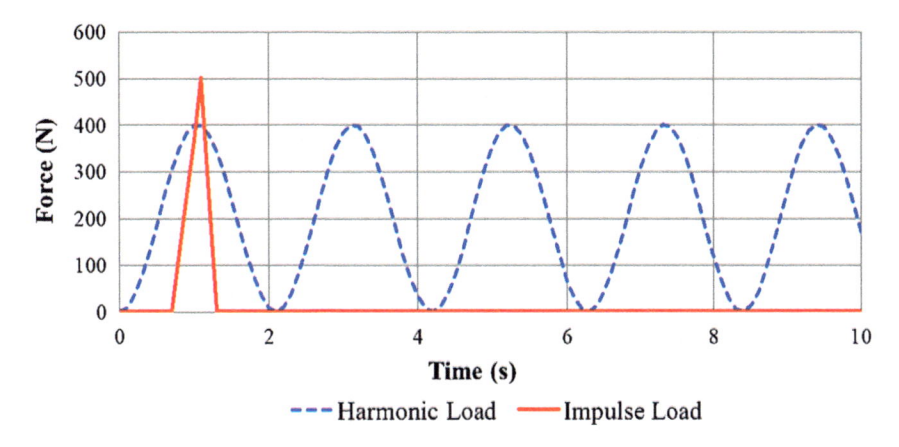

Fig. 9.3 Illustration of different loading types

9.4 Criteria

Acceptable vibration for office, residential, and other floor plate structures are well defined in a number of national and international published sources (CCIP, SCI P354, AISC [3], ATC [4], ISO [5]). In the context of this paper there are two main situations to consider: (1) The effect of vibration transmission from a gym on a nearby sensitive occupancy, such as an office or treatment area (e.g. physio room). (2) The vibration response of the gym floor itself.

Gym activities with harmonic loading which may result in resonant response, such as use of treadmill or aerobic activities, should be handled differently from impulsive activities, such as weight-drops, which result in a sudden response that dies down relatively quickly. Impulsive acitivities often result in structure-borne noise as well, that can also cause complaints. Harmonic and impulse type loading are illustrated in Fig. 9.3.

The number of occurrences and duration of vibration is important; people are less likely to complain about excessive vibration if it only happens a few times throughout a workday, for a short duration.

9.4.1 General Criteria

Human perception of sensitivity to vibration is influenced by the magnitude, frequency, direction, and duration of the vibration, as well as the general expectation of users for the type of structure or building. Furthermore, human sensitivity is frequency dependent. In this paper, while we are only concerned with the effects of floor vibration in the vertical direction, it may be felt by office occupants who are seated or standing, or by treatment area occupants who may be lying down.

Criteria require two main elements to be defined:

- The vibration magnitude, that depends on the function of the space (active vs. passive)
- The vibration metric, that depends on the type of the floor response (resonant vs. impulse)

Response factors (RF) are the most common metric used in assessing human response to floor vibration performance (CCIP-016, BS 6472:1992 [6]). A RF of 1.0 corresponds to the average threshold of human perception with other levels of vibration perception being expressed as multipliers of RF 1. For example, for a premium quality office space, RF 4 (i.e. 4 times the level of perception) would be specified. For a typical office space, a criterion of RF 8 (i.e. 8 times the level of perception) would be often allowed.

A metric also in use is the Vibration Dose Value (VDV) (ISO 10137:2007, BS 6472-1:2008 [7]). This takes into account the number and duration of the vibration events in a given assessment time period, for example daytime. The VDV metric lends itself well to the assessment of vibration from sources which are well defined events of known durations such as trains. VDV can also be applied to other scenarios, for example a basketball game with known duration. Limits for VDV are published in terms of typical human responses such as "adverse comment possible", "adverse comment probable"; this indicates the fuzzy nature of these criteria where precise statements about human response to a given vibration level cannot be made with certainty.

9.4.2 Gym Criteria

While criteria for application to general gym spaces, e.g. treadmills and weights floors, are not fully-defined or published, criteria applicable to rhythmic activities are published in several guidance documents. (NBC [8], ATC [4], AISC [3]).

It is often the case that people in adjacent occupancies to a gym, e.g. offices and dining areas, will be more sensitive to vibration than the users of the gym themselves. In other words, active participants in say, an aerobics class, will tolerate much higher vibration levels than those passively working at an office desk.

For activities resulting in resonant response, NBC 2010 states that adverse comment within an office space due to adjacent gym activities becomes probable when peak accelerations exceed approximately 0.5% of gravity. For areas where people are participating in rhythmic activities though, the code suggests that peak accelerations over 10% of gravity can be acceptable. An example of chosen acceleration limits for the various rooms in the Arup engineered Sky Health & Fitness Centre is shown in Fig. 9.4.

The most common, and often the most onerous, example would be jumping in an aerobics class context. Such activity would often involve several people jumping in a coordinated fashion at specific frequencies (i.e. to the beat of the music). This results in significant periodic loads being applied to the floor plate and can lead to high levels of vibration because the periodic loads are efficient in driving structural resonances.

Impulse loading in general results in large magnitude, short duration response, and is often associated with high frequency structures less prone to excitation from low frequency sources, such as jumping. This differs from resonant response for which the converse is true. Human sensitivity to single impulse loading is less researched, there is less guidance available. ISO 10137:2007 suggests that the tolerance is 30–60 times higher for impulsive vibration excitation than continuous or intermittent vibration. Although, the recommended values are applicable for events occurring only few times a day, currently, there is no determined correlation between tolerance and number of events per workday. For purely impact loading, the criteria should be selected case-by-case and the structure-borne noise level specified by the building code or building-specific lease agreement should also be taken into account.

Specific criteria for treadmill scenarios are not available in the published literature for gym floor areas in the vicinity of the treadmills. For adjacent occupancies, the relevant criteria for continuous vibration can be used.

A summary of potential applicable criteria for various gym activities are presented in Table 9.1.

For scenarios where criteria are not well defined, the expected performance can be simulated and experienced using a facility such as the Arup Motion Platform (Fig. 9.5). A motion platform would take a measured or calculated vibration or

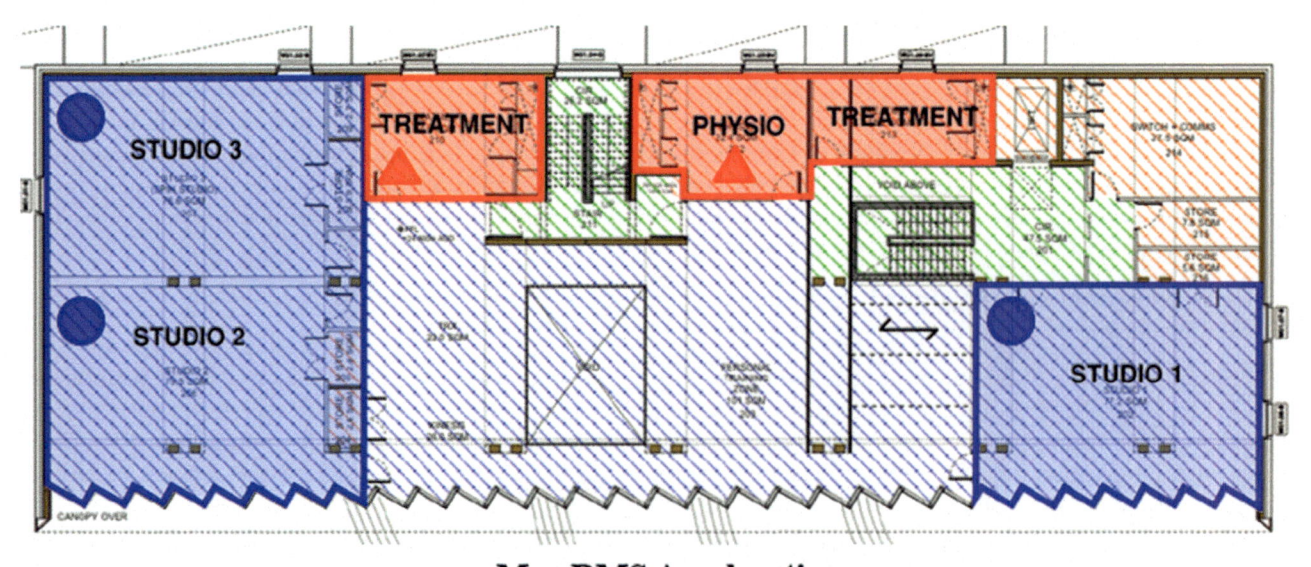

Fig. 9.4 Example of chosen peak RMS acceleration limits in a fitness center, based on NBC 2010

Table 9.1 Suggested metrics for various gym activities

Type of response	Example activity	Suggested metric	Standards
Resonant	Aerobics	Resonant maximum RMS acceleration, structure-borne noise	CCIP-016 [1], NBC (2010) [8], ISO 10137 (2007) [5], BS 6472 (1992) [6]
	Treadmill		
Impulse	Sports hall activity	Peak acceleration, VDV, structure-borne noise	ISO 10137 (2007) [5] Local building code or lease agreement
	Weight drop	Peak acceleration VDV Structure-borne noise	

Fig. 9.5 Motion Platform demonstration of floor response with and without damping mitigation

motion as an input, and reproduce it in real life as an output. By standing or sitting on the motion platform, subjects can experience the perceived performance and inform the decision making. It is recommended to pair the experience with virtual reality and a sound representation, given the perception of vibration can significantly vary with the environment the vibration is experienced in.

9.5 Modelling and Analysis

Modelling and analysis are significant in both the design stage and post-operation assessment. At design stage, the main purpose of modelling and analysis is to evaluate the vibration performance of the structure and to define the design allowance for vibration, over and above the strength design and such that acceptable vibration performance is demonstrated. For both design and post-operation assessment, this type of modelling and analysis can be used for the efficient design and optimization of any mitigation that might be needed. The ultimate goal is to demonstrate that the structure will meet its vibration criteria, if possible, already at the design stage.

The computational model, typically a finite element model, will be configured to address one of two main scenarios: (1) Transmission from a gym to an occupancy on the same floor plate (2) transmission to a different floor plate. The most common scenario is (1) above. For this scenario, the model would typically comprise the entire floorplate structure at the gym floor level as well as connected structural columns. The dynamic loads can be applied to the relevant gym area and response evaluated in the designated occupancy area.

The finite element model will be based on an implementation of standard methods published within industry guidance. The overall dynamic response of a structure is a function of its mass, stiffness, and damping. Whilst the stiffness is a function of the material properties and sizes of elements used, additional permanent mass (i.e. non-structural elements such as finishes, services, and partitions) should be modelled as accurately as possible. The structural damping and dynamic loads are discussed below.

9.5.1 Structural Damping

The structural damping of a floor plate is a function of hysteresis within the construction materials themselves and friction in the structural and non-structural connections. It is difficult to quantify with confidence at the design stage for a specific structure. However, values for typical structures can be found in the industry guidance and standards. Damping varies for different structural forms which might be entirely reinforced concrete (RC), or composite RC and steel. Damping tends to increase according to the level of fit-out present in a building and CCIP-016 in particular gives estimated ranges of structural damping, for application at the design stage, for a number of floorplate structural forms and fit-out levels. Estimates of structural damping are also available in SCI P354 and ISO 10137. Design damping values are typically expressed as a proportion of critical damping and are intended to apply to all structural modes predicted in the finite element model.

Damping is a critical design parameter for floor plates which exhibit resonant response, the response magnitude increases in inverse proportion to damping. The sensitivity of such a structure to the range of possible damping must therefore be considered at the design stage. A high frequency floor situation which exhibits non-resonant or impulsive/transient response is less sensitive to damping.

9.5.2 Dynamic Loads

While much of the industry guidance on the evaluation of footfall induced vibration is relevant to gym situations, the footfall dynamic loads have limited application to general non-walking gym load cases. For the footfall dynamic load, there is extensive research available that has led to design loads that can be used for both resonant and transient analysis cases. These design loads have been statistically determined from experimental research and in CCIP-016, for example, have a 25% probability of exceedance. For gym dynamic loads other than footfall the available literature is discussed, where dynamic load cases have not been published, methods have been proposed and discussed.

Aerobics and Similar Activities

For the aerobics class type of gym activity, which tends to include coordinated jumping, dynamic load information is available in published literature. In this case, a group of individuals may combine to generate higher dynamic loading than a single person with a musical beat serving to coordinate their movements.

NBC 2010 provides Dynamic Load Factors and suggest mass distributions for dancing and aerobic activities. Dancing in this context is believed to represent ballroom type dancing with two people and so loads are somewhat lower than for aerobics.

ISO 10137:2007 provides a description of the dynamic load impulse arising from a single person jumping from a defined height. Section A.1.1.2 of the standard provides a definition of a more general dynamic load for coordinated group jumping or rhythmic exercises. This includes discussion of static mass loading as well as extent of coordination.

AISC Design Guide 11 references both of the above sources in addition to a few more.

Treadmills

For treadmill loading published dynamic loads are not readily available. However, treadmills lead to dynamic loads that can conservatively be assumed to correspond to those arising from people running. Running produces vibration in the frequency range 1.6–4 Hz. This approach ignores the treadmill structure, mechanics, and floor mounting arrangement but this could be modelled if known.

By their nature treadmills generate vibration at the running speed and up to the first four harmonic frequencies and can therefore lead to the generation of resonant response. Treadmill usage durations vary but can be expected to be at least 5–10 min. Also, a gym will typically have several treadmills arranged in a row or an array of some kind, leading to situations where several treadmill vibration sources will be active at the same time, although the dynamic loads are not necessarily well correlated.

Sports Hall Activities

Sports hall activities, for example a basketball, squash, or soccer consist of non-rhythmic impulses imparted to the structure, including activities such as walking, running, jumping, changing directions (horizontal load), and balls bouncing on the floor. Basketball is particularly severe due to the high energy impulses generated by dribbling, both for vibration and structure-borne noise. Contrary to rhythmic activity, these do not impart a single frequency load, thus are less likely to cause resonant effects. They may however have a relatively predictable overall duration and frequency of occurrence. This type of loading is hard to quantify using available simplified models, and is best assessed using experimental methods. The example below illustrates the use of recorded accelerations of basketball activities as input loads for vibration analysis.

The aim of the project was to assess the expected vibration and structure borne noise in an office caused by a planned basketball court two floors above. The methodology used to determine the effects of the activity is presented in Fig. 9.6. The accelerations caused by basketball activities on a similar floor system were recorded. The recordings were then used as a reference event and the measured loading was applied to the numerical model of the building. A number of variables were examined, including the measurement of various numbers of players, various drills (shoot-around, pass drills, and fast breaks) and size/quality of players. These scenarios were then used to determine a range of input functions in the frequency domain, with standard deviations used to determine upper bound likelihood for design.

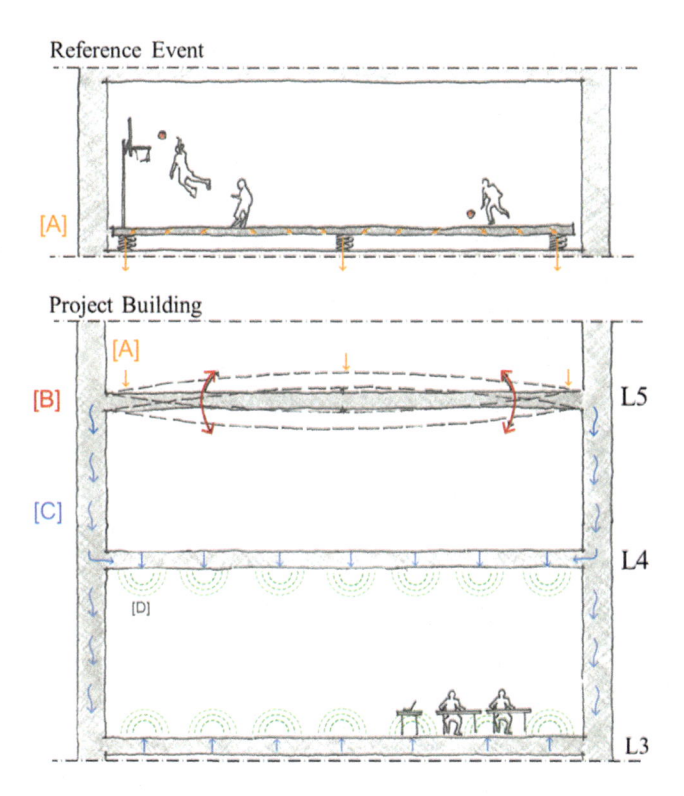

Fig. 9.6 Methodology to assess the effects of a planned basketball court above an office space

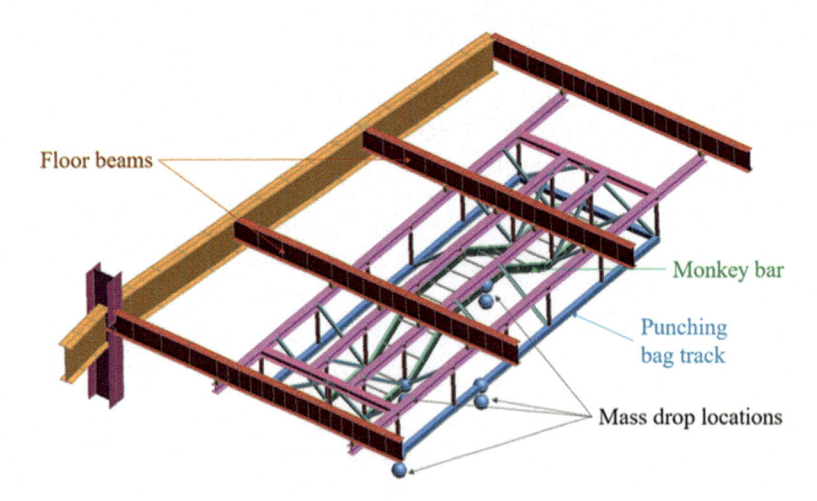

Fig. 9.7 Finite element model of monkey bars and support structure

Weights (Including Free Weights and Machines)

For weightlifting areas, published dynamic loads are not readily available. In this load case priority is normally given to the release of large weights following the lift. This leads to one or more impulses being applied to the floor structure as the weight falls, impacts, rebounds and falls again. These impulses can lead to both perceptible vibration and structure-borne noise. This dynamic load is impulsive by nature and therefore the corresponding floor vibration would be for a short duration. The mass of the weights will vary according to the type of gym, with performance gyms and elite sports facilities having much larger weights than a regular gym. Sometimes a mat is installed in weight areas to control the rebound of any falling weight. This mat will also influence the impulse transmitted to the floor structure beneath the mat. Where this load case is important for a project, then loading should be employed by explicitly modelled or experimentally measured values to evaluate the impulse loading for relevant weight and mat scenarios. An example case is shown below where boxing bags hanging from monkey bars were explicitly modelled and the loading was simulated. The example below illustrates the assessment of potential vibration issues induced by impulsive weight drops.

The project aimed to assess the vibration transmitted to the floor above by punching bags moved around and dropped on a suspended monkey-bar system. The goal was to get conservative estimates on structural response, forces in connection elements, and to evaluate the efficiency of hanging spring-dampers. The finite element model used for the analysis is shown in Fig. 9.7.

All frame elements are modelled with linear-elastic beam elements. The mass of the floor slab itself was incorporated by increasing the density of the steel floor beams. The recommended 2% system damping for steel structures was added to the model. Dropped masses were modeled at various locations, hanging from stiff (steel) cable elements using explicit time-history analysis. To model the drop, the mass elements were lifted 6 in. then released using prescribed movement. The nodal displacements, velocities, and accelerations of the floor beams were output.

The connection elements are shown in Fig. 9.8. Two types of connections were modelled to determine the effects of design interventions, a welded stiff connection and a hanging spring-damper connection. The maximum forces in these connections were also output allowing to size the corresponding structural elements. Our studies suggest that decreasing the connection stiffness is an efficient way of reducing the maximum acceleration in the flooring. The level of isolation should be calibrated keeping in mind the resulting maximum displacement in the monkey bar system as well, ensuring it is maintaining full functionality.

Other Activities (e.g. Spin Bikes, Cross Trainers etc.)

In addition to the above, other activities may take place in gyms such as spin bike classes and cross-trainer/stepper usage. While these activities will generate some dynamic loading, they are far less likely to govern the design of a space compared to the loadings described in the previous sections.

Guidance on loading from these types of activities is not readily available and so any assessment would need to consider these activities by either making first principles assumptions on loading or from measured data.

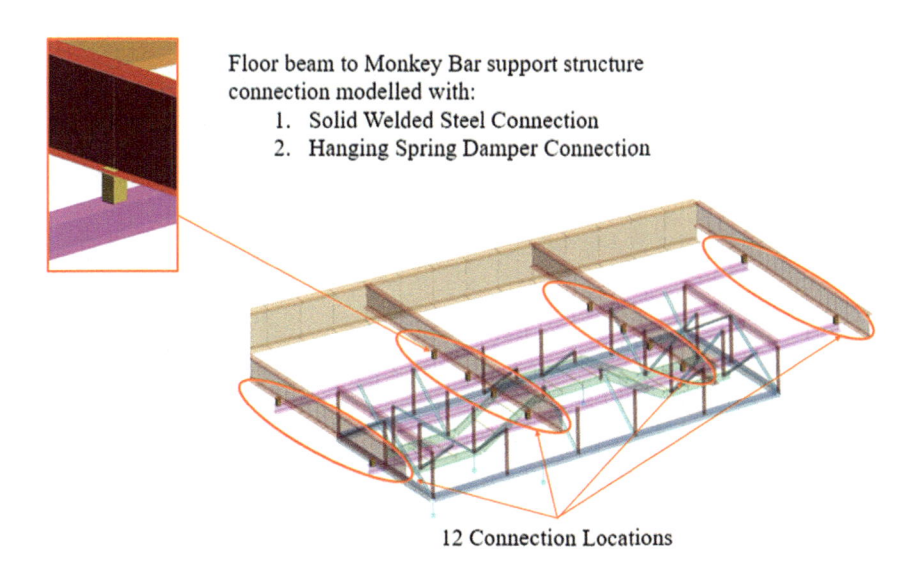

Fig. 9.8 Connection elements in the finite element model

Table 9.2 Gym activity types and their recommended analysis model

Activity	Modelling	
	Method	Standard
Aerobics	Coordinated jumping group	ISO 10137 (2007), NBC (2010), AISC (2016)
Treadmill	Individual running	
Sport hall activity	Measured	None
Weight drop	Modelled or measured	
Other activities (e.g. spin bikes, cross-trainers, steppers)	Modelled or measured	

9.5.3 Gym Dynamic Load Summary

A summary of the various load types in a gym facility is presented in Table 9.2. The table contains the general method a load type can be modeled, and standards, if available, that contain related guidance or design parameters.

9.6 Mitigation Options

Mitigation is a term used to describe countermeasures which can be included within or applied to a design to reduce or control the vibration level. Ideally the structure would be designed to meet the criteria for all of the dynamic loads it will undergo. This involves developing a floorplate that is designed to meet the static strength performance but also the vibration requirements. Where structural design alone cannot achieve this, or in existing building scenarios, mitigation needs to be considered.

There are a number of strategies available to mitigate gym floor vibration. A range of both common and innovative approaches are discussed below. The selection of appropriate mitigation options should always be done in coordination with the project architecture and engineering team, and considering possible impacts to other aspects of the project.

9.6.1 Stiffness

An increase in floor plate bending stiffness will lead to an increase in natural frequency for a given design assuming the increase in stiffness does not bring about the same relative increase in mass. This increase in frequency would have most benefit if it moved the vibration response from the resonant range to the transient frequency range. This could be achieved with an increase in the RC slab thickness and/or an increase in the steel beam depths. It could also be achieved with the introduction of steel beams for example tertiary beams. The use of a structural screed material would also lead to some increase in natural frequency. Reducing span lengths for a given structural framing layout is usually the most effective method for increasing stiffness.

9.6.2 Mass

Increasing the mass in the structure is not normally effective. While it increases the inertia of the floor plate it tends to lead to a reduction in natural frequency often making the floor more prone to resonant response. Reducing the mass in a structure could potentially lead to an increase in natural frequency providing stiffness was also not lost in the process; this might be difficult or impractical to achieve.

9.6.3 Damping

For resonant response cases, an increase in floorplate structural damping theoretically leads to a proportional decrease in vibration magnitude. However, when group activities are taking place (e.g. aerobics classes), consideration needs to be given to the fact that people can introduce damping themselves leading to an overall 'system' damping which is higher than the structure alone. This can be beneficial in reducing the structure response but needs careful consideration when assessing supplementary damping solutions. The relative increase in damping provided may not be as high as expected if only the structural damping has been considered as the base damping. NBC 2010 Section 11 recognizes this and suggests that for floors with many people on it, the damping ratio is approximately double that of the same floor with few people on it. Although not gym loading, UK IStructE guidance [9] for stadium loading takes account of this by modelling the crowd as spring mass damper units. Some potential supplementary damping solutions are covered in the following sections.

Tuned Mass Damper (TMD)

This method of damping involves the introduction of a spring mass damper system tuned to the frequency of the structural mode most relevant to the vibration concern. The vibration in the floorplate is transmitted into the TMD which then generates strain in the damper whereby vibration energy is dissipated as heat. TMDs are commonly used where there is just one structural mode to control, or several TMDs can be tuned to multiple modes.

TMDs can be installed on either the source or the receiver side. For gym applications, if the receiver occupancies have raised floor system, it is usually favorable to install it on the receiver side, because it is possible to locate the TMD system within the raised floor void therefore concealing it and avoiding obstruction at the same time. If this is not possible, an alternative is to hang the damping from the underside of the structure. An example for TMDs installed in the raised flooring of the receiver side is presented below.

A gym, located in on the seventh floor of an office building, caused unacceptable vibration in the office located a floor below. To mitigate the response in the office, a total of five TMDs were installed under the office's raised floor. The efficiency of the dampers is shown in Fig. 9.9, that summarizes the floor slab behavior at one of the TMD locations, before and after the TMD was installed. The measured structural damping doubled from 3 to 6%. The plot in the frequency domain (left) shows that this solution is only efficient in the target frequency range; FFT magnitudes further from the target 5 Hz were not significantly affected. The figure in the time domain (right) shows the decrease in the response factor due to the additional damping throughout resonant excitation.

There are various commercial products available. An Arup-designed Low Profile TMD installed in an office floor cavity is shown in Fig. 9.10. This product was designed for a high-end office building to mitigate footfall vibration, with the objective of low cost, and quick manufacture and installation. The damper would work similarly well to reduce the effects of gym activities.

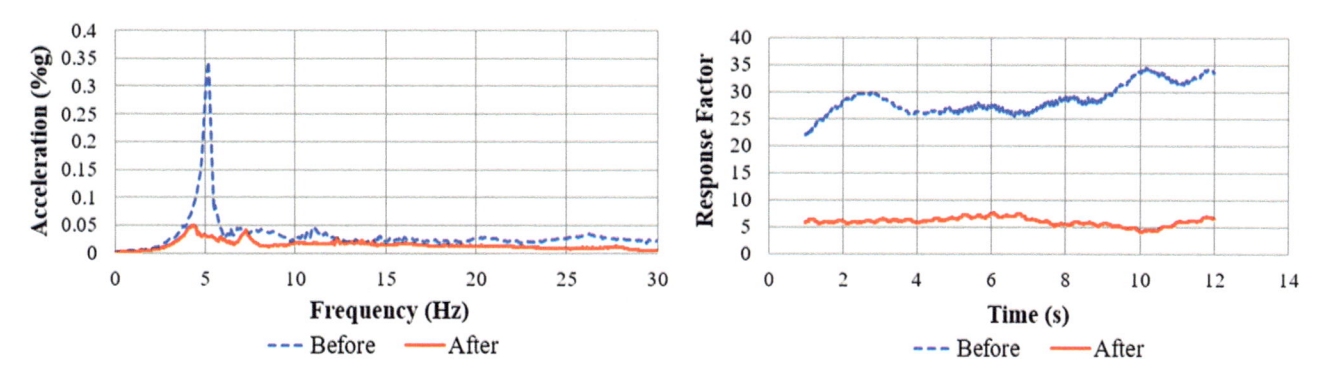

Fig. 9.9 FFT from heel drop at a TMD location (left) and RF (weighted) time history from bouncing at the same TMD location (right)

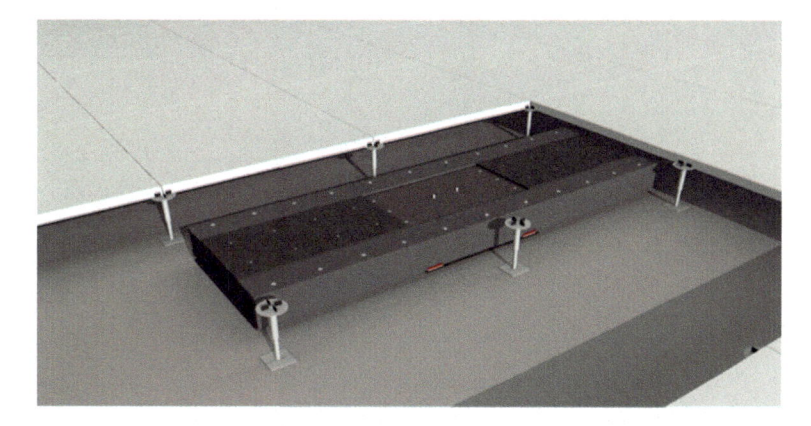

Fig. 9.10 Arup Low Profile TMD located in an office floor cavity

Constrained Layer Damping (CLD) Treatment

This method of damping involves the introduction of a thin layer of elastomer or polymer type material, with a backing plate, to the floor slab. The elastomer goes in between the concrete and steel of the composite slab, in the shear region. The bending of the floor slab with a given structural mode then induces shear in the damping material which then dissipates the vibration energy in the form of heat. For composite structural designs proprietary solutions such as Resotec can be incorporated between the slab soffit and the steel beam flange. This involves some loss of composite action and therefore strength, something which needs to be checked during design.

The effectiveness of this treatment varies with frequency as it depends on the structural mode shape and hence how much shear can be generated. This type of treatment will lead to modest enhancements of structural damping.

To evaluate the effectiveness of a given CLD design and demonstrate that the performance requirements have been met then a more detailed model of the floor structure and the CLD treatment is necessary. The shear in the elastomer depends on the mode shape detail, thus, the interaction between the CLD and the structural dynamics should be represented in the model in order to predict it accurately.

9.6.4 Active Vibration Control

A novel form of vibration control is an active vibration control device (AMD) which is a system which actively monitors floor vibration levels and provides a force based on this feedback to a moving mass to counteract the vibration. There are a number of benefits to this type of mitigation over TMDs, namely, the system can be used to target multiple modes and the mass of the moving mass can be much lower. The downside is that, as present, costs are high for such systems and they have therefore not been used extensively to solve structural dynamics issues. Additionally, since the vibration reduction results from force cancellation, not added damping, once the capacity of the system is exceeded, it becomes progressively less effective. Another major downside to these systems is that they are active and require power, which means they are less reliable than passive systems, such as isolation or TMDs. Considerable technical understanding is required in the specification, design, and manufacture of these systems.

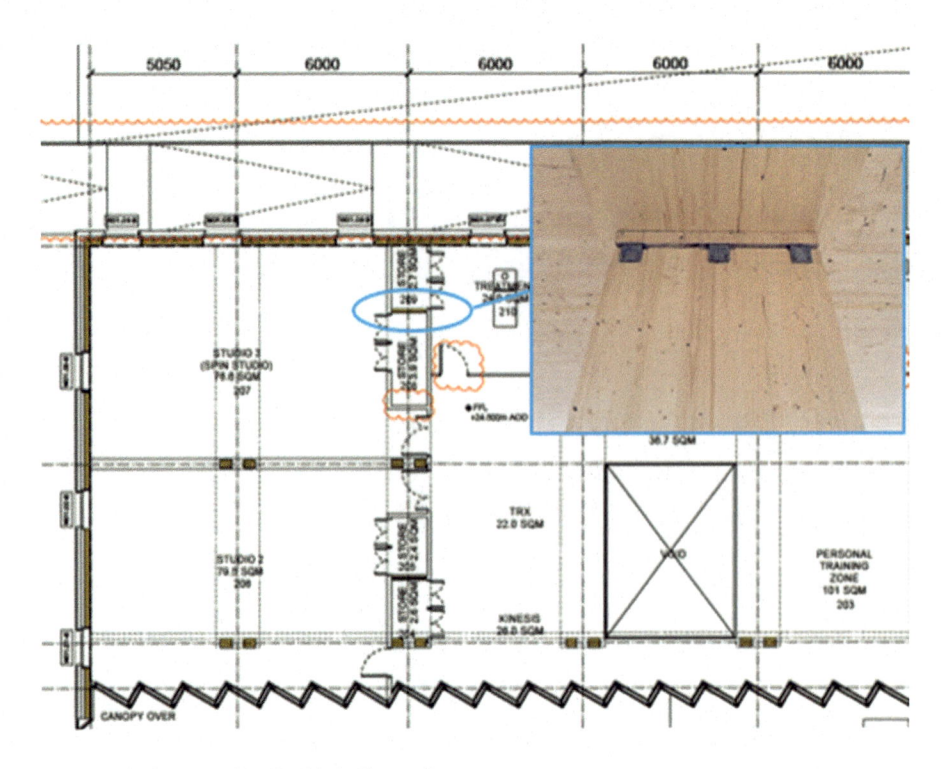

Fig. 9.11 CLT fins used to couple floors on Sky Health & Fitness Centre

9.6.5 Load Paths

For gym floors and occupancies at one floor level above the foundation slab it might be possible to introduce a load path between the gym floor and the foundation slab in the form of a column. The column location would be case specific but would generally target areas of high slab vibration, which would typically be at the anti-node of a structural mode.

Another example of load path mitigation would be in the form of tie-bars between two or more successive floor plate levels. Tie-bars couple the mass and stiffness of the two floorplates leading to a reduction in vibration for a given load. If tie-bars were selected they would normally be concealed within a partition where this is possible. It might not be practical to achieve this within a gym space and so would be considered more for a receiver occupancy.

An example of a structural load path to reduce vibration was used on Sky Health & Fitness Centre, as shown in Fig. 9.11. Here, cross laminated timber (CLT) fin wall elements were hidden within equipment cupboard walls to connect the gym floor to the rooftop plant deck. This increased the mass in the critical mode to reduce the response.

Contrary, if the issue is vibration at the floor above or below the floor of activity, the removal of the load path transferring the movement can be considered.

All of these forms of mitigation could readily be incorporated into a finite element model to evaluate their effectiveness and to develop the design.

9.6.6 Isolation

Vibration isolation involves the introduction of a low frequency isolation layer between source and receiver, either at source, in the structure, or at receiver. The isolation layer results in insertion loss at higher frequencies. This effectively reduces vibration transmission in a broad-brand of frequencies, including audible frequencies; however, it causes amplification at the isolation frequency. In a building context, isolation is often used to reduce vibration transmission into a building from an external source, such as railway. In a gym context, while isolation can be introduced, there are some important considerations.

Isolation of Gym Floors

General isolation of a gym floor plate structure is difficult because the isolation frequency is very likely to interact with the frequency of the dynamic loads. For example, the isolation frequencies of helical steel springs (2–6 Hz) and elastomer bearings (7 Hz and above) contain one or more harmonic frequencies of aerobic jumping and treadmills. The resulting interaction will lead to an increase in vibration on the gym floor at certain frequencies. NBC 2010 suggest that isolation could be used to reduced dynamic forces at the second and third harmonic frequencies. This may be helpful if vibration levels are being governed by floor modes at or close to second or third harmonic loading frequencies.

Isolation of Weight Floors

Local isolation of weight floors is an option to consider for the reduction of vibration transmission to occupancies adjacent to or below the weight floor. Such isolation might comprise a floating RC slab supported by isolation in the form of helical steel spring (HSS) or elastomer. Such isolation might also be required for the control of structure-borne noise. As noted above the isolation would act in series with the floating slab and any resilient matting used on the weight floor.

A finite element model of the floor plate structure, isolation, and floating slab can be used to evaluate the vibration performance. The dynamic loads for the falling weight scenarios however must be verified using experimental methods. The purpose of this analysis is to confirm the isolation frequency specification so that the vibration transmitted to the occupancies is acceptable in terms of perceptibility and structure-borne noise.

Some isolation systems vendors report considerable test data on their systems particularly in the audible frequency range, which can also be used for the specification.

9.6.7 Relocation of Equipment/Spaces

Relocation generally involves greater separation of the vibration source from the receiver occupancy. For example, treadmills could be positioned on a gym floor area that is more remote from a nearby office, or could be positioned nearer to columns or less mobile floor plate areas. It might also be possible to relocate weight areas to a more favorable location with respect to occupancies on the same floor plate or on the floor plate below. Relocation is often the easiest mitigation option from a design and construction perspective, but results in space programming consequences which are often unacceptable.

In terms of the location of a gym within a building, the foundation slab would always be favorable from a dynamic point of view given the reduced mobility and higher damping present when a slab is formed on soil/rock.

9.7 Summary

This paper summarized the vibration assessment and design of a gym or sports facility in a mixed-use building. It walked through the process, including the initial vibration assessment for existing buildings, and establishing the right criteria depending on structural response, and function of the space. It showed available methodologies to model loads resulting from various sports activities and provided a wide range of potential mitigation options with examples should the analysis show the vibration level is unacceptable.

For an existing building, the on-site vibration test is key in understanding its behavior. Besides the dynamic properties of the building, i.e. natural frequencies, mode shapes, and structural damping, it can serve to establish baseline loading. Testing best practices and guidelines are already relatively well-established for the building industry.

To identify the most applicable criteria metric and corresponding limiting value, the type of structural response (resonant or impulse) needs to be identified along with the intended use of the space (office, residential, gym, etc.). For resonant responses, there is a range of literature available, that gives guidance on acceptable response limits. This is not the case for impulse responses in a gym context. For impulse-type response in a gym or sports facility, the human tolerance is not very well researched. For either case, besides vibration limits, the noise requirements also need to be checked against the expected structure-borne noise.

Similarly to the vibration criteria, models are only widely available for loads resulting in resonant response, while there is limited data on other impulsive gym loadings. Until this data becomes available, the authors recommend the use of explicit or experimental models for these load cases. The paper gave an example for both options.

If structural vibration is determined to be an issue even with appropriate architectural layout, there are numerous mitigation options available. These vary in complexity, efficiency, and cost. The best and most common is the change of structural stiffness. Besides changes in element sizes, one should consider adding or removing loadpaths, that link multiple floors together. If the problem is resonant response, additional, active or passive dampers can be added to the structure, concealed, where possible. Also, the receiver can be isolated from the source.

9.8 Conclusion and Future Work

In conclusion, while both criteria and modelling is well researched for gym activities causing resonant response, there is currently not enough information available on modelling and assessing impulse loads, such as weight drops or basketball games. To model these load types, Arup currently uses experimental and numerical methods. To determine the acceptability of the behavior, Arup has been using the available guidance with the understanding that it is not gym and sport facility specific. To close this gap, Arup is planning to undertake broader studies in order to quantify the tolerance for these types of loads. Arup would be simulating and assessing different structural responses (both acceleration and noise) using motion platform, VR, and a sound system.

References

1. Willford, M., Young, P.: A Design Guide for Footfall Induced Vibration of Structures. Concrete Centre, Surrey (2006)
2. Smith, A.L., Hicks, S.J., Devine, P.J.: Design of Floors for Vibration: A New Approach. SCI P354, revised edn. The Steel Construction Institute, Berkshire (2009)
3. Murray, et al.: Structural design guide—Section 11. American Institute of Steel Constructors (1997)
4. Allen, D.E., Onysko, D.M., Murray, T.M.: ATC Design Guide 1 Minimizing Floor Vibration. Applied Technology Council, Redwood City (1999)
5. ISO (International Organisation for Standardisation): ISO 10137 Bases for Design of Structures—Serviceability of Buildings and Walkways Against Vibration. ISO, Geneva (2007)
6. British Standards Institution BS 6472:1992. Guide to Evaluation of Human Exposure to Vibration in Buildings (1 Hz to 80 Hz). BSI, London (1992)
7. British Standards Institution BS 6472-1:2008. Guide to Evaluation of Human Exposure to Vibration in Buildings. Vibration Sources Other than Blasting. BSI, London (2008)
8. NRC-IRC (National Research Council Canada—Institute for Research in Construction): Users Guide—NBC (National Building Code of Canada) 2010 Structural Commentaries (Part 4 of Division B): Commentary D—Deflection and Vibration Criteria for Serviceability and Fatigue Limit States. NRC-IRC, Ottawa (2010)
9. Institution of Structural Engineers (Great Britain): Dynamic Performance Requirements for Permanent Grandstands Subject to Crowd Action: Recommendations for Management, Design and Assessment. Institution of Structural Engineers, London (2008)

Chapter 10
A Large Scale SHM System: A Case Study on Pre-stressed Bridge and Cloud Architecture

Gabriele Bertagnoli, Francescantonio Lucà, Marzia Malavisi, Diego Melpignano, and Alfredo Cigada

Abstract In recent decades, external prestressing is increasingly being used especially in motorway and railway bridge structures due to the substantial savings in terms of construction time and costs. In such systems, internal and external steel tendons work together with concrete elements to withstand external actions. This means that the deterioration or failure of these elements reduces structural safety in a meaningful way. Real time monitoring of prestressing tendons can provide useful information on the health of the bridge under service loads, detecting possible fatigue, corrosion and damage/deterioration processes. However, most of the currently used structural monitoring systems are rather expensive and time consuming to install.

Although many papers address high density sensing as the proper solution thanks to the "internet of things" tool, both for hardware and software, there are not so many applications in which this approach is really put into service.

This paper describes the application of MEMS accelerometers in a high performance and cost-effective SHM system for bridge structures. In particular, data from a real time monitoring system installed in a box section composite highway bridge are presented. The external tendons of this bridge have been instrumented with a total number of 88 triaxial accelerometers. Changes in the dynamic characteristics of the monitored elements have been analyzed by detecting the shift in tendons' dynamic behavior. The main challenge was collecting a huge amount of data and find a way to properly process them, not requiring the operator's direct action, unless the observed situation is out of the "normal" scenario. For this purpose, simple but easy-to-implement specific data processing algorithms have been tested in order to check the real feasibility of such a SHM system first, and then to analyze the collected sensor data and provide an efficient real time damage detection.

Keywords SHM · Bridge · MEMS sensors · Prestressing tendons · Structural dynamics

10.1 Introduction

External prestressing is largely used in civil engineering and especially in bridges design to improve concrete performances by forcing the structure to be in a state of full compression through the use of multiwire steel tendons [1, 2]. As highlighted by some researches [3, 4], the damage or collapse of these elements could induce serious consequences for the integrity of the entire structure, which makes critically important the monitoring of tendons health condition. Thus, being able to assess the state of the tendons may grant higher safety levels for the users and a more efficient maintenance.

The research field that deals with the development of automatic strategies for damage detection is part of Structural Health Monitoring (SHM) [5]. Driven by the rapid development of more accurate and cost-efficient sensors, data acquisition systems and Internet of Things (IoT) architectures, an increasingly number of operating structures are being equipped with SHM systems [6–9], continuously getting data from the structures being monitored. Thus, the main challenge is extracting the useful information from a very big amount of data, synthesizing it in few indicators (or features) that could help assessing the integrity of a structure.

G. Bertagnoli · M. Malavisi (✉)
Department of Structural, Geotechnical and Building Engineering, Politecnico di Torino, Torino, Italy
e-mail: marzia.malavisi@polito.it

F. Lucà · A. Cigada
Department of Mechanical Engineering, Politecnico di Milano, Milan, Italy

D. Melpignano
STMicroelectronics, Agrate Brianza, Italy

Considering the specific case of pre-stressed bridge monitoring, a large number of SHM methodologies and systems have been proposed by many researchers over the years. Several examples of SHM implementation can be found in [10–17]. Among the possible strategies, a lot of interest is being paid to the vibration-based methods: the basic idea is that if a change in the structural properties occurs, it affects the modal parameters of the structure (natural frequencies, mode shapes and damping ratios). Thus, monitoring the trend in time of the natural frequencies may represent an effective strategy able to highlight changing in the tension of the tendons, caused by fatigue, corrosion and deterioration processes [18, 19]. Some specific sensors also offer the chance of measuring different quantities like vibration and rotation at the same time: in this case the information merge (sensor fusion) can be quite useful to have a cross check helping to assess data reliability.

In this work, an application of SHM to a real operating bridge is described. The structure under examination, is being monitored 24 h a day 7 days a week since September 2017. Vibration signals are acquired by a network of 88 MEMS tri-axial accelerometers placed on the tendons; these are g-sensitive and this property can be also exploited to use them as clinometer. After a pre-processing, data are sent, stored, and processed in an IoT Cloud that allows real time access to data and their management in a simple though effective form. As a feasibility study, attention has not been paid to the algorithms (the choice has been for those easiest to implement, with the literature known limits [20]), rather to the network management and to the quality of data through the measurement chain, from the accelerometer performances to those of the network up to the cloud.

A database of the statistical features and the natural frequencies for every cable has been therefore continuously updated since the system was put in operation, representing an interesting case study in how the issues of big data management in a SHM perspective can be overcome through the integration of IoT technologies. The results related to the first 8 months of operation are synthetically shown in this paper.

10.2 Bridge Description

The monitored structure is a highway concrete bridge located in Italy. The bridge, opened to traffic in 2006, is a composite box girder in which the concrete webs are replaced with corrugate steel plates to reduce the self-weight and simplify the construction. Mixed prestressing (internal/external) was used to strengthen the structure. Two abutments are supporting the bridge at the end points and five concrete piers clamped into the girder are holding up the six spans. The bridge is 580 m long and it is characterized by five equally spaced (120 m) hyperstatic spans and one isostatic span 43 m long. The main girder has a cross-section height varying from 6.0 m (at the bearings) to 3.0 m (on the centerline of each span). The structural details of the bridge are shown in Fig. 10.1.

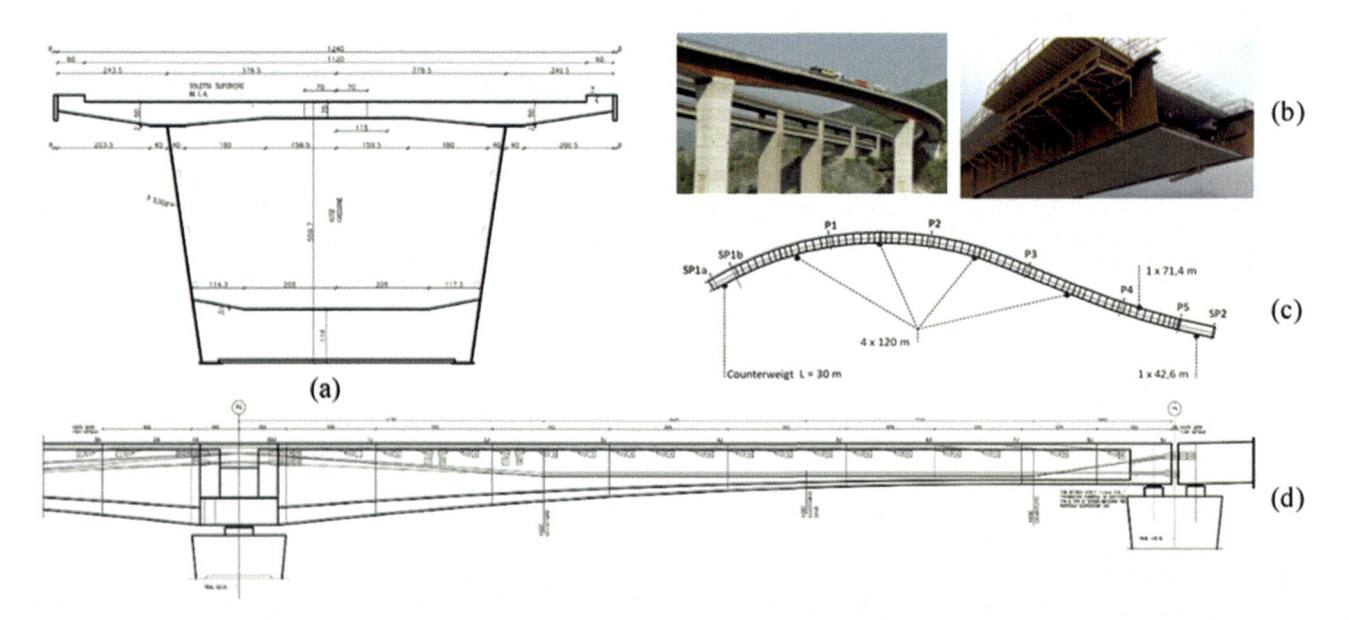

Fig. 10.1 Structural drawings of the bridge. (**a**) Bridge cross-section; (**b**) bridge images after and during construction; (**c**) plan view of the highway bridge; (**d**) longitudinal section of a span

The pre-stress of the structure is provided by means of bonded tendons arranged in the upper flanges of concrete slab and unbonded external tendons composed of 27 strands placed in the hollow section of the box girder.

The bridge was instrumented with a continuous monitoring system for the real-time detection of the tendon condition during the service life of the structure. The monitoring was instigated to check the behavior of external tendons after a failure of one tendon, probably due to an incorrect grout composition and in spite of checking the effects of the heavy daily traffic traveling on the bridge. Hence, in order to better understand the dynamic response of the bridge under operational conditions, external tendons were instrumented with 88 MEMS tri-axial accelerometers, 2 for each monitored tendon, between June and September 2017, the complete system being fully active since 20 September 2017.

Each tri-axial MEMS accelerometer provides data in the three orthogonal directions (x, y, z); in this way it is possible to capture the bridge vibration and deformation under traffic excitations and to obtain some dynamic parameters. Environmental data, such as air temperature and humidity, have been collected in order to remove the effect of these parameters (especially temperature) in vibration measurements and trace measured data back to a constant value of temperature.

10.3 Monitoring System

The accelerometers were mounted in ten different cross-sections, close to the steel protection screens in the upper part of the prestressing cables, as shown in Fig. 10.2.

A long lasting preliminary experimentation has allowed to assess that the accelerometer performances were fit for this kind of measurement, due to the high excitation especially produced by the heavy traffic travelling on the bridge day and night, with just a small reduction during the weekends. As MEMS accelerometers are g sensitive they can be used "twofold", both as accelerometers and clinometers: the fusion of these information can help detecting any ongoing damage: this could turn into both a change in the cable dynamic performances or in its attitude. Then the main problems to be faced were then the maximum data streaming rate, a real bottleneck when working with accelerometers, then, in turn, the maximum allowable sampling rate per channel.

The sensor alone was "updated" to a sensor node, including a microcontroller providing data sampling and some elementary management, ready for any further "edge" computing strategy. In addition, the node encodes the sampled data into a CAN BUS driven network, joining and sending data to a local gateway, which also can offer some storage and calculation capabilities.

The monitoring system is connected to the internet via a 5 GHz point-to-point Wi-Fi link between an access point located at one end of the bridge and an "Ubiquity Nano M5" station located at the P2 pier, halfway between the viaduct ends. An Ethernet cable connects the station to the two IoT gateways installed in the bridge: one for the right side and one for the left side. All the sensors are connected to these gateways, which, through the described path, send data to a specific cloud platform.

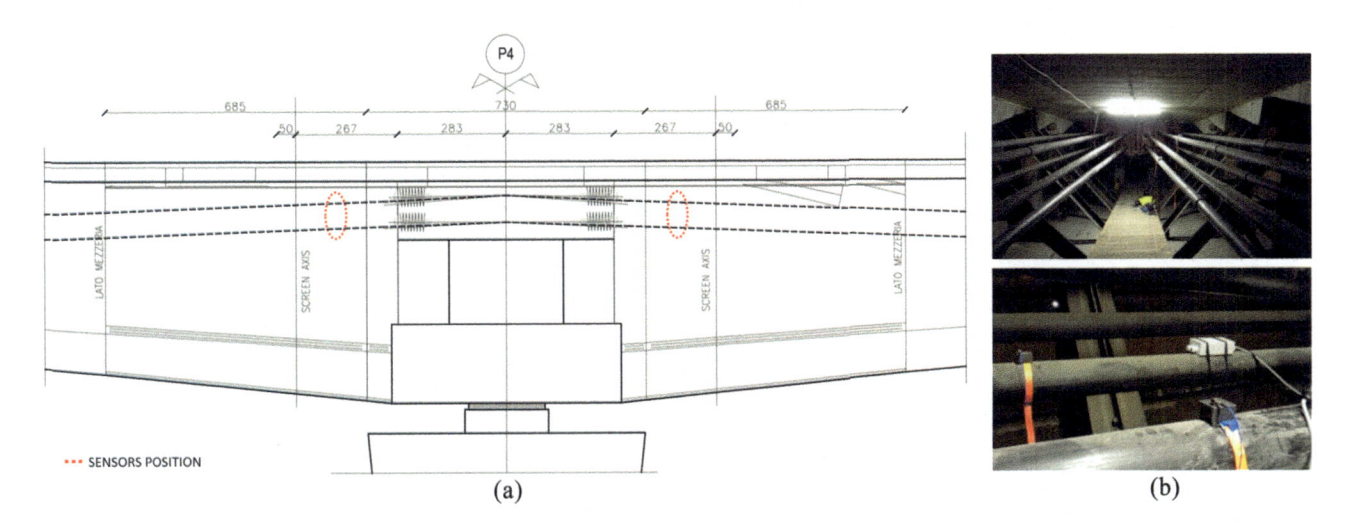

Fig. 10.2 Layout of the monitoring system. (**a**) Typical sensors position; (**b**) MEMS Accelerometers installed on the external steel tendons

The 88 acceleration time series as well as temperature and humidity data are stored in a cloud monitoring infrastructure which allows access to data in real time. The maximum sampling rate allowed by the encoding procedure has been 100 Hz. To prevent from aliasing data are sampled at the sensor level at 25.6 kHz, then filtered and down sampled at the sensor node, to make the data streaming manageable by the network. Some preliminary tests have shown that the final allowed bandwidth, in the order of 40 Hz, can be considered enough to get any eventual damage detection (Fig. 10.3).

Before performing any modal identification analysis, sensors data are pre-processed in order to make the monitoring architecture more effective in terms of performance and network use. In particular, the IoT gateway performs the pre-processing of information before they're sent to the data center. The software installed on the device is indeed dedicated to collect data from sensors, pre-process that data, and sending the results to the cloud.

Pre-processing is especially aimed at filtering acceleration data, in order to detect unusual patterns that do not conform to the expected behavior (outliers) and to generate anomaly alert messages when needed. In particular, data for a selected time span are analyzed by calculating the average (AVG), root-mean-square (RMS) and minimum/maximum (MIN/MAX) values on interval-by-interval basis. The obtained interval statistics are then compared with preset thresholds.

Data from sensors are then sent, stored, and processed in a IoT Cloud platform. The main challenge was to elaborate a huge amount of data in the shortest time by using cloud resources at best. Taking advantage of the high parallelism (up to 1000 simultaneous executions), data composed by time series of length T seconds can be subdivided in m slot of T_W seconds such that the algorithm can elaborate a slice of data with low time and memory consumption.

Although the parallelism provided by the IoT cloud is considerable, in order to save memory and time, the dataset is analyzed by a sequencing rule ordering the invocation of an action. Each performed action involves the steps of:

- download a dataset slice $D_{i,j}$ identified by some calculated input parameters;
- elaborate the dataset;
- upload elaboration results in a pre-determined location;
- invoke next action (with its input parameters) or terminate the sequence.

Figure 10.4 shows the time discretization scheme.

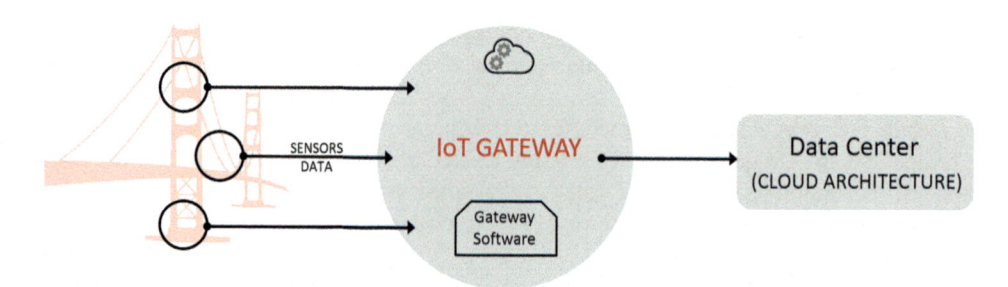

Fig. 10.3 Gateway architectural overview

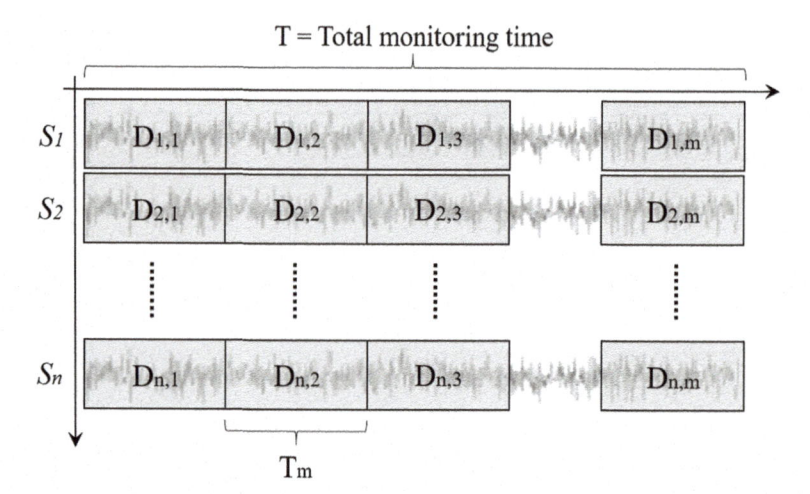

Fig. 10.4 Time discretization. $S_1 \ldots S_n$ represent readings from the n sensors over a time length T. Each column corresponds to one of the m time slot of length T_W. This subdivision create a grid of n × m dataset slices $D_{i,j}$ where $i \in \{1, \ldots, n\}$ and $j \in \{1, \ldots, m\}$

The static and dynamic parameter identification is thus performed for each interval time dataset slice $D_{i,j}$, lasting 30 min.

Even if some simple evaluations on the mean value, i.e. the actual cable position could be easily evaluated, the aim of this activity was to stress the system when dealing with a denser and intense data management, typical of dynamic analyses.

10.4 The Adopted Strategy

Since the signals are of random nature, the power spectral density (PSD) of the acceleration signals represents the starting point for the analysis in the frequency domain. In order to improve the accuracy of the parameter identification, the best balance between noise reduction and a good frequency resolution has been obtained by means of some preliminary tests carried out with lab instrumentation, made up of low noise piezo accelerometers, a data acquisition board with 24 bit and a sampling frequency of 2048 Hz. These tests have allowed to get that first of all measurements on the pre-stressed tendon sheaths were meaningful in the description of the cable behavior, then some fine tuning has allowed to fix the best choices for a series of parameters: sub-records of 200 s and an overlap of 66% (Hanning window) have proven to be the best compromise between the need to average data for a better signal to noise ratio, at the same time preserving almost constant temperature and insolation conditions, affecting the dynamic parameters. An example of the shape of the PSD is shown in Fig. 10.5, where it is in general possible to identify the peaks related to the modes of the considered tendon.

As this was a first approach, though being conscious about its limits [20–22], it has been decided to analyze damage in terms of natural frequency changes. Natural frequency changes have therefore been considered as the significant parameters to assess the damage of the pre-stressed cables: as stated attention was mainly paid to the possibility to apply this approach to a wide number of cables, exploiting the big amount of data both concerning the trend evolution of a single cable and a cross check among adjacent cables. In the case of cables injected with cement grout, only changes due to stiffness decrease are considered, since no mass losses are possible.

As well known the natural frequencies f_i for the cable vibration modes are derived as:

$$f_i = \frac{1}{2L_i}\sqrt{\frac{T_i}{m_i}}$$

where m_i, L_i and T_i are respectively the mass, length and axial force of the cable.

In the case of strand failure, the cross section of the tendon and consequently the corresponding axial force are reduced. As a consequence, a frequency shift should be observed.

Each frequency peak of the PSD can thus be used as a feature for damage detection. A peak identification algorithm has been developed in order to identify peaks in signal spectrum without prior knowledge on their number, shape or location. The variation of the frequency peaks over time has been then evaluated for each sensor. An experiment about a possible approach of edge computing has also been carried out by providing a synthesis over each considered window, including RMS, mean, max value and min value: this solution means a high data compression. As the traffic conditions over long time records are

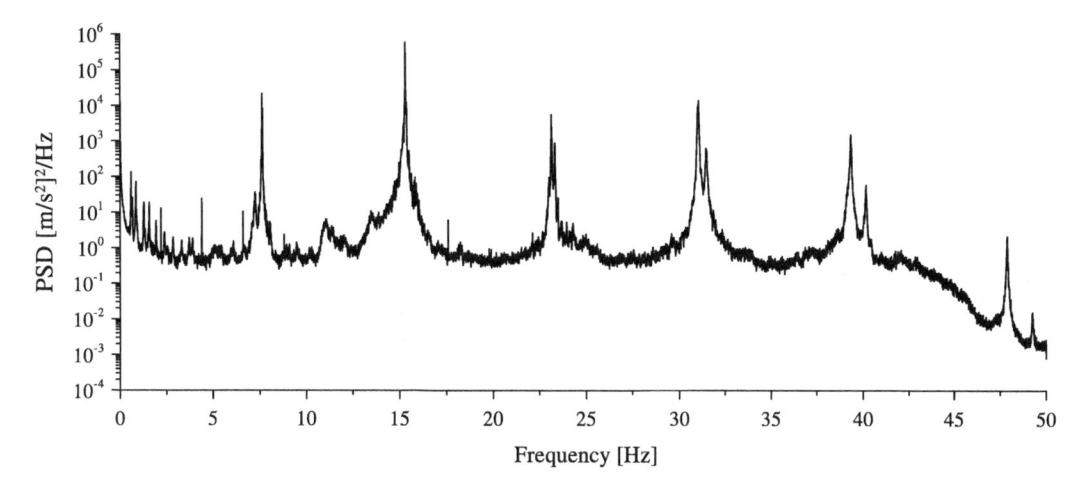

Fig. 10.5 Power spectral density (PSD) of the signal measured by one of the 88 accelerometers installed on the bridge

similar (at least considering working days), any trend in the RMS value can be considered a damage indicator, a change in the average is a change in the cable position, while this value, joined to the info from the max and min value are a sensor diagnostic indicator.

10.5 Results and Discussion

This section summarizes the results obtained from the large-scale monitoring system installed on the pre-stressed concrete bridge described in the previous paragraphs. In particular, experimental results obtained under normal traffic operation are presented here. The traffic is mainly composed of passenger cars as well as small and heavy trucks.

Statistical identification of standard trends is essential for handling large amounts of data while detecting changes and deviations over the monitoring time. The process of trend recognition has been carried out in the time domain first, by using mean and standard deviation values. In particular, for each measured direction (x, y, z), the signal has been pre-processed by applying a high pass filter. Mean (μ) and standard deviation (σ) values have been then obtained from the filtered data. Standard deviation has been considered as a good indicator of the average vibration activity induced by traffic loads, wind and/or other external agents under standard or exceptional conditions. In addition, maximum and minimum values have been considered as relevant for detecting possible anomalous behaviors of the tendons.

Figure 10.6 shows the variation of the aforementioned parameters (mean, 5% fractile which corresponds to $\mu \pm 1.64\sigma$, maximum and minimum acceleration values) for the three measured directions (x, y, z) of a sample sensor, named "sensor A", calculated over a time period of 24 h. A variable time window T_w of 1 min has been considered for the evaluation of daily trends.

From Fig. 10.6, it can be seen that the average acceleration value for the three directions is close to zero, resulting from an average of periodic and almost symmetrical oscillations of the cable to which the mean values has been subtracted.

The day-night cycle is clearly recognized: the maximum vibrations are mainly recorded during the day hours (from 6 a.m. to 10 p.m.) with an appreciable decrease at night (from 10 p.m. to 6 a.m.). This trend is also easily identifiable from the 5% fractile plot, which represents the 5% probability of the vibration data to exceed, in absolute terms, the $\mu \pm 1.64\sigma$ values.

Moreover, Fig. 10.6 shows markedly that the x and z axes are the most excited ones, with respect to the y axis. This is indicative of an elliptical vibration of the cable in the x-z plane, which is orthogonal to the longitudinal development of the

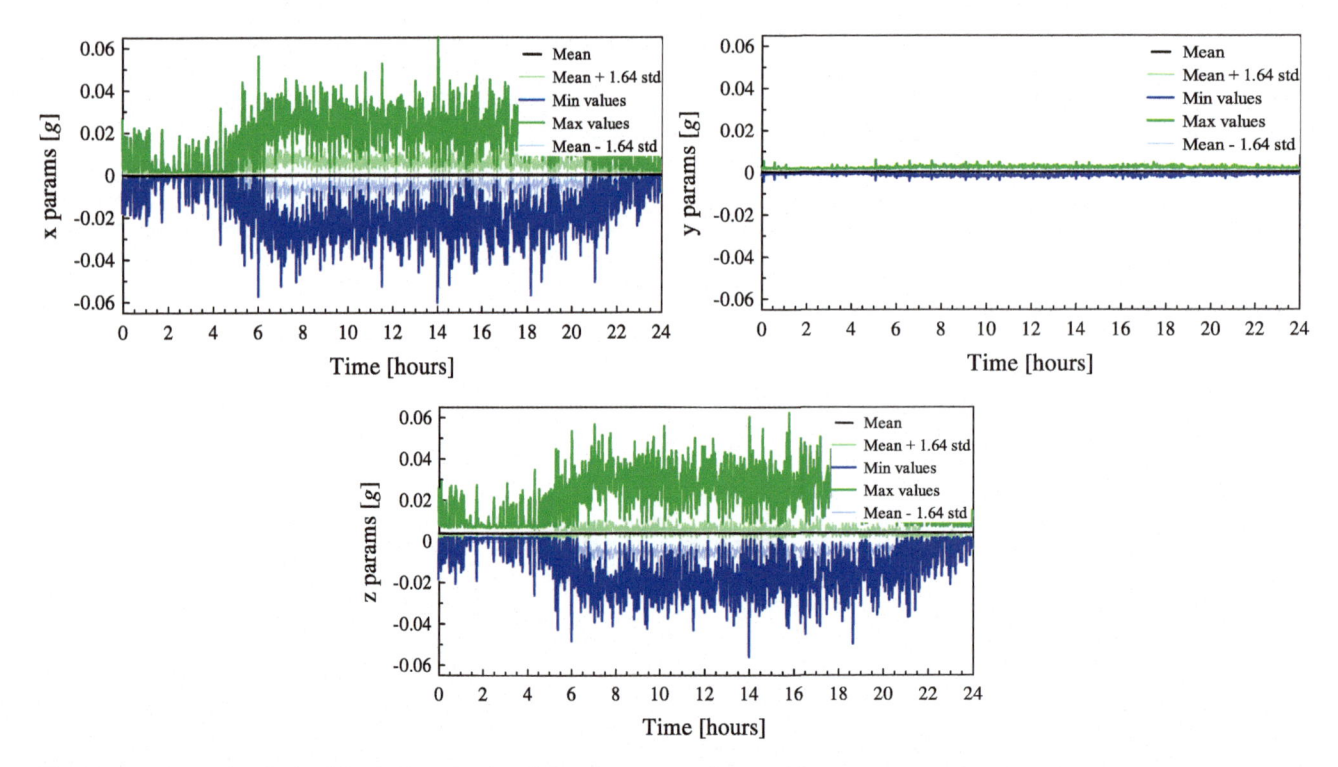

Fig. 10.6 Daily trends calculated for the three monitored directions x, y, z—"sensor A"

tendons. Y-direction has in fact reduced vibration levels (one order of magnitude smaller than x and z directions), meaning that it has a lower sensitivity with respect to traffic variations and environmental noise.

The trend of the vibration signals has been evaluated also over longer periods, for instance a month (a period suitable to appreciate the situation, longer periods make it more difficult to identify the situation). Figure 10.7 illustrates the variation of the considered parameters (mean, 5% fractile which corresponds to $\mu \pm 1.64\sigma$, maximum and minimum acceleration values), calculated over a time period of 2 months, for the x direction. A variable time window T_w of 30 min has been considered for the evaluation of monthly trends.

Figure 10.7 displays the tendon vibration response under different traffic conditions. Indeed, a higher acceleration variation can be observed from Monday to Friday with a significant reduction in traffic during the weekend. The trends observed during the monitored period allowed for the definition of a benchmark of measurements corresponding to the standard behavior of pre-stressing tendons subject to traffic conditions or non-exceptional external loads.

Together with the time domain outputs, in order to perform damage detection, the identification of standard trends has been performed by carrying out also analyses in frequency domain, meaning a heavier load for the gateway/cloud system.

In particular, the variation of frequency peaks over the entire monitoring period has been obtained. Since damage detection is accomplished by comparing frequency values that are obtained at different times, it is essential to remove any environmental influences affecting the considered parameters. In fact, natural frequencies are strongly dependent from temperature changes. Figure 10.8 represents a time frequency plot with the corresponding temperature readings for two sample sensor, named sensor "A" and sensor "B". All the detected peaks are symbolized by small black crosses and are tracked over a 8 months period.

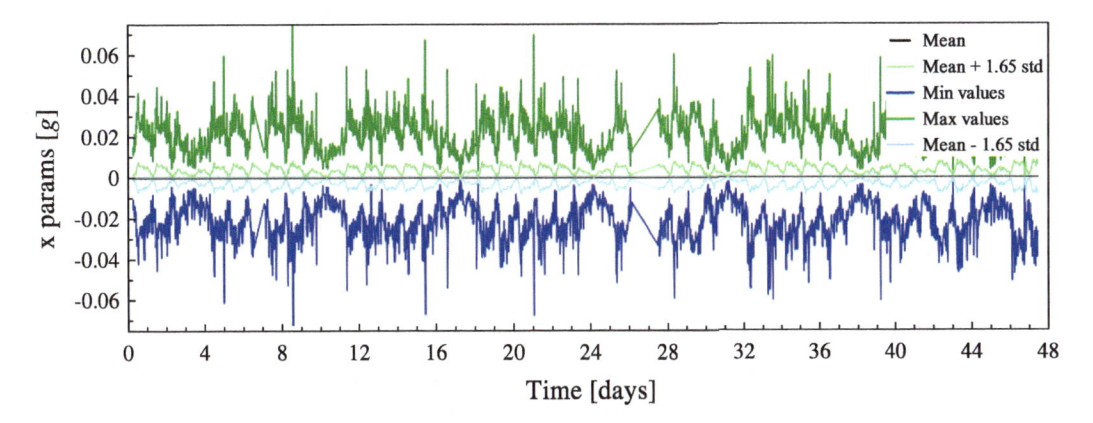

Fig. 10.7 Monthly trend calculated for x direction—"sensor A"

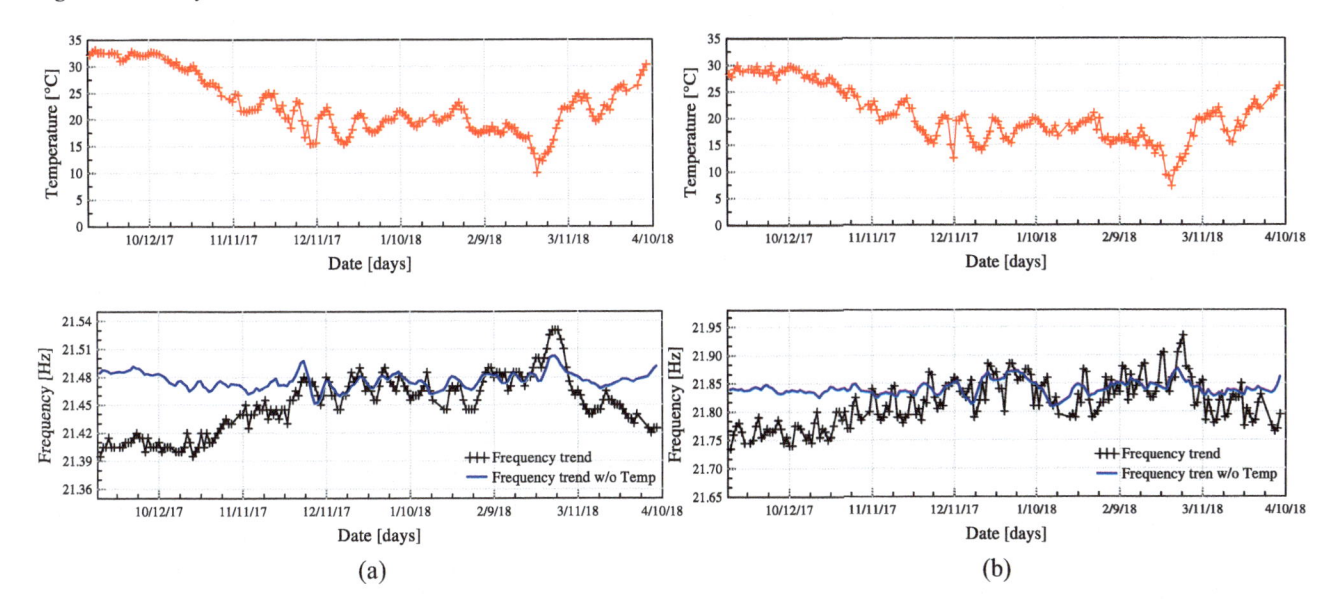

Fig. 10.8 Temperature and Frequency variation over time for two sample sensors. (**a**) Sensor A; (**b**) Sensor B

By comparing temperature and frequency trends, it is possible to observe that there is an explicit inverse correlation between them over the monitored period. Some noise on the peaks leads to the spread in the observed values, especially for sensor B. As this sometimes make it harder to identify the peak position, a smoothing procedure around the peaks has been carried out by means of a polynomial, to help keeping an automatic peak identification procedure. Anyway, for both sensors, frequency increases from September 2017 to February 2108, when temperature falls, and decreases from February 2018 to April 2018, when temperature starts rising. A standard regression analysis has been performed to assess the relationship between temperature and frequency evolution. The influence of temperature was therefore removed by the trend of frequency variation by applying the calculated regression coefficients.

The resulting trend, plotted in blue in Fig. 10.8, is almost constant for the entire monitored period. Moreover, it could be noted that no particular anomalies in peaks evolution have been observed over time. All the frequencies are included in a small variation range, without values clearly outside the normal oscillations around a constant mean value.

10.6 Conclusions

This paper presents results from a permanent dynamic monitoring for the real-time assessment of the health of a structure. In particular, a widespread, innovative and minimally invasive monitoring system has been installed on a concrete highway bridge in Italy. The peculiarity of the system lies in being composed of low-cost sensors based mainly on MEMS technology, capable of monitoring various physical quantities, and connected to each other with different technologies for data transfer and sensor power supply. A IoT cloud architecture has been developed to process in short time a huge amount of data in order to synthesize with few parameters the most relevant information about the behavior of the structure. Time domain analysis have been carried out to highlight some significant standard trends in order to identify threshold values for the generation of alerts in case of anomalies. Then, frequency domain approach has been implemented for determining a frequency domain model of the structure to use for damage identification. In particular, power spectrum peaks have been extracted from sensor signals and variations in frequency values have been evaluated over the monitored period. A strong and evident inverse correlation with temperature was observed, whose effect was eliminated through a linear regression analysis.

References

1. Combault, J.: Long span concrete bridges: influence of construction techniques. In Proc., IABSE Symp. Int. Assoc. Bridg. Struct. Eng. Zürich, Switz., vol. Proc., IAB (1991)
2. Miyamoto, A., Tei, K., Nakamura, H., Bull, J.W.: Behavior of prestressed beam strengthened with external tendons. J. Struct. Eng. **126**(9), 1033–1044 (2000)
3. Starossek, U.: Progressive collapse of bridges—aspects of analysis and design. In: Int. Symp. Sea-Crossing Long-Span Bridg., pp. 1–22 (2006)
4. Woodward, R.J.: Collapse of a segmental post-tensioned concrete bridge. Transp. Res. Rec. **1211**, 38–59 (1989)
5. Farrar, C.R., Worden, K.: An introduction to structural health monitoring. Philos. Trans. R. Soc. A Math. Phys. Eng. Sci. **365**(1851), 303–315 (2007)
6. Tokognon, C.J.A., Gao, B., Tian, G.Y., Yan, Y.: Structural health monitoring framework based on internet of things: a survey. IEEE Internet Things J. **7**(4), 4–7 (2017)
7. Lamonaca, F., Sciammarella, P.F., Scuro, C., Carni, D.L., Olivito, R.S.: Internet of Things for Structural Health Monitoring. In: 2018 Work. Metrol. Ind. 4.0 IoT, pp. 95–100 (2018)
8. Cremona, C.: Big data and structural health monitoring. In: IABSE Congr. Stock., no. October 2016
9. Li, H.-N., Ren, L., Jia, Z.-G., Yi, T.-H., Li, D.-S.: State-of-the-art in structural health monitoring of large and complex civil infrastructures. J. Civ. Struct. Heal. Monit. **6**(1), 3–16 (2016)
10. Robertson, I.N.: Prediction of vertical deflections for a long-span prestressed concrete bridge structure. Eng. Struct. **27**(12), 1820–1827 (2005)
11. Nair, A., Cai, C.S.: Acoustic emission monitoring of bridges: review and case studies. Eng. Struct. **32**(6), 1704–1714 (2010)
12. Casas, J.R., Cruz, P.J.S.: Fiber optic sensors for bridge monitoring. J. Bridg. Eng. **8**(6), 362–373 (2003)
13. Maeck, J., De Roeck, G.: Damage assessment using vibration analysis on the Z24-bridge. Mech. Syst. Signal Process. **17**(1), 133–142 (2003)
14. Yuyama, S., Yokoyama, K., Niitani, K., Ohtsu, M., Uomoto, T.: Detection and evaluation of failures in high-strength tendon of prestressed concrete bridges by acoustic emission. Constr. Build. Mater. **21**(3), 491–500 (2007)
15. Reda Taha, M.M., Lucero, J.: Damage identification for structural health monitoring using fuzzy pattern recognition. Eng. Struct. **27**(12), 1774–1783 (2005)
16. Webb, G.T., et al.: Analysis of structural health monitoring data from hammersmith flyover. J. Bridg. Eng. **19**(6), 1–11 (2014)
17. Cabboi, A., Magalhães, F., Gentile, C., Cunha, Á.: Automated modal identification and tracking: application to an iron arch bridge. Struct. Control Health Monit. **24**(1), e1854 (2017)
18. Saiidi, M., Douglas, B., Feng, S.: Prestress force effect on vibration frequency of concrete bridges. J. Struct. Eng. **120**(7), 2233–2241 (1994)
19. Kim, J., Ryu, Y., Yun, C.: Vibration-based method to detect prestress-loss in beam-type bridges. Smart Struct. Mater. **5057**, 559–568 (2003)

20. Doebling, S.W., Farrar, C.R., Prime, M.B.: A summary review of vibration-based damage identification methods. Shock Vib. Dig. **30**(2), 91–105 (1998)
21. Fan, W., Qiao, P.: Vibration-based damage identification methods: a review and comparative study. Struct. Health Monit. **10**(1), 83–111 (2011)
22. Farrar, C.R., Doebling, S.W., Nix, D.A.: Vibration-based structural damage identification. Philos. Trans. R. Soc. A Math. Phys. Eng. Sci. **359**(1778), 131–149 (2001)

Chapter 11
Vibration Serviceability Performance of an As-Built Floor Under Crowd Pedestrian Walking

Jinping Wang and Jun Chen

Abstract The vibration serviceability performance of an as-built floor was assessed under crowd pedestrian walking at prompted walking frequency and specified pedestrian density. The floor served in a laboratory building having 12 m × 12 m in size. First, modal parameters of the floor were measured by hammer tests and ambient vibration tests, the fundamental frequency was identified as 5.35 Hz. Then, controlled crowd walking test was conducted on the floor consisting of test cases at different pedestrian densities restricted by certain distances among pedestrians, i.e., 0.6, 0.8 1.0 1.2 1.5, and 2.0 m. Due to the difference in density, the crowd size varies from 30 to 74 pedestrians. Additionally, in each pedestrian density test case, several prompted walking frequency cases (resonance and non-resonance) and one free walking case was included. Various representative values of structural response for vibration assessment are used to make comparisons among responses of different test cases in terms of both pedestrian density and walking frequency. Results indicate that the crowd induced vibrations do relate to the square of number of pedestrians walking on the structure. Moreover, the structural responses are also compared with available guidelines for vibration serviceability assessment and show good satisfaction of the requirements.

Keywords Vibration · Floors · Serviceability assessment · Crowd pedestrian walking · Human induced load

11.1 Introduction

The issue of structural vibration serviceability has become a big hit these days with the development of slender structures and human's increasing attention on life quality [1–5]. Human's feeling is the main evaluation criterion of the vibration, while actually, human activities also act as main source of structural vibration, which is called human-induced vibration, being the dominant design issue for slender structures like floors and footbridges. The excitation causing this kind of vibration is called human-induced load, mainly inclusive of walking, jumping, bouncing, and running, among which crowd pedestrian walking is the most common one. On the other hand, floors in buildings play the role of the structural components that occupants primarily contact, and are the second most frequent source of complaints from building users, second only to roofs. When a large number of people walk on floors, especially when they walk close to certain frequencies (i.e., resonance to structure's natural frequencies), severe vibrations may occur and influence the feeling of both walking pedestrians themselves and people standing/sitting on the structures [1, 6, 7]. On account of the limited knowledge of crowd modelling and field measurements on real floors, there still demands a better understanding of the response of floors to human induced excitation and validation of the practicability of design theories. For these reasons, this paper conducts a study on vibration serviceability performance of an as-built floor in an office under crowd pedestrian walking. First, modal properties of the floor were tested through hammer test. Second, crowds of different pedestrian densities were asked to walk on the floor under the guidance of metronome. Next, testing results of structural responses expressed in the form of peak acceleration and 10 s peak root mean square (10 s RMS) are compared with comfort criteria in several guidelines, and then, accordingly, structural responses in different pedestrian density cases are normalized to the square of number of pedestrians walking on the floor to find the relationship between structural response and pedestrian density.

J. Wang
College of Civil Engineering, Tongji University, Shanghai, P. R. China
e-mail: wangjinping@tongji.edu.cn

J. Chen (✉)
College of Civil Engineering, Tongji University, Shanghai, P. R. China

State Key Laboratory of Disaster Reduction in Civil Engineering, Tongji University, Shanghai, P. R. China
e-mail: cejchen@tongji.edu.cn

11.2 Description of the Floor

The tested floor served in an office room of the second story of a two-story laboratory (Fig. 11.1a). The floor is rectangular of 60 m × 36 m in size; has columns with 12 m in between dividing it into 5 × 3 spans (the layout of the whole floor is shown in Fig. 11.1b); has steel main beams between columns and one-way secondary beams every 2.4 m. One of the mid spans was chosen for the crowd walking pedestrian testing as illustrated in Fig. 11.1a, b.

11.3 Modal Parameter Tests

To test the modal parameters of the slab of the chosen span, hammer tests were conducted. Acceleration responses of the slab at points A and B separately were recorded using Lance accelerometers (Model LC0132T) at a sampling frequency of 160 Hz. The accelerometers were placed on the slab at different locations shown in Fig. 11.2a. Structural natural frequencies, mode shapes, modal masses, and damping ratio of the first several modes were obtained from the vibration tests and testing

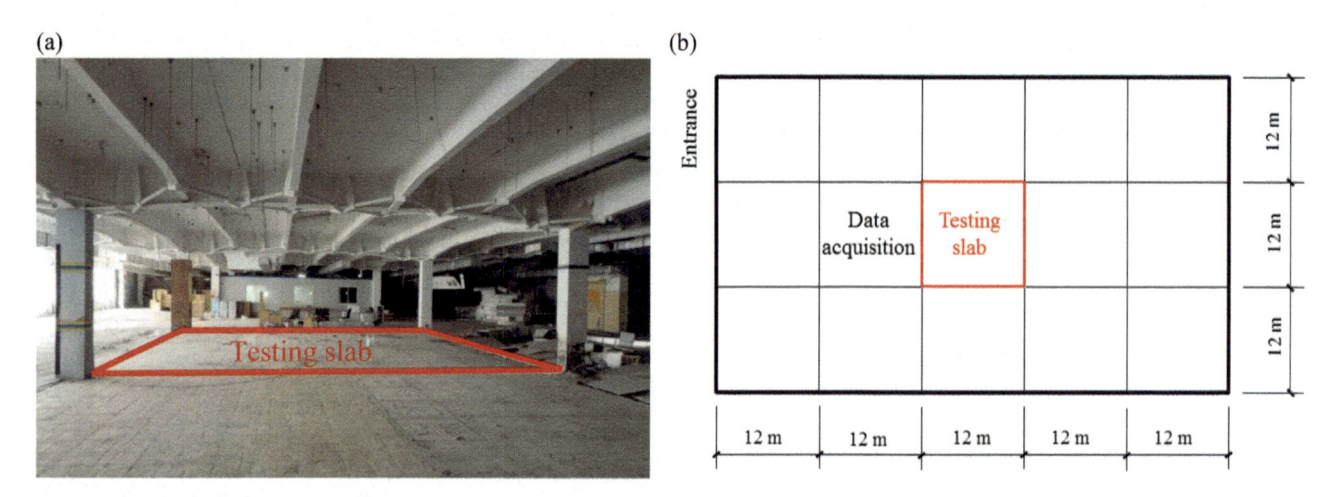

Fig. 11.1 (a) Photo of the testing slab (b) Overview of the whole floor

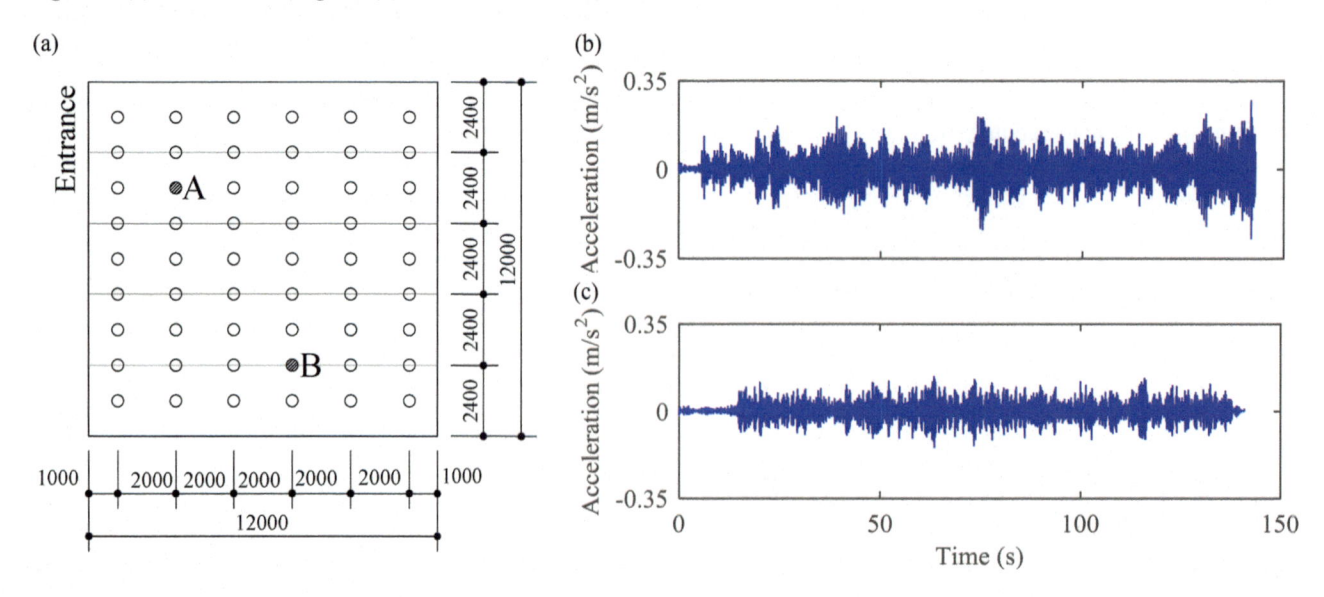

Fig. 11.2 (a) Locations of accelerometers and hammer impacting points in modal parameter tests (unit in mm). (b) Structural acceleration at the center of the slab under crowd walking at $f_p = 1.8$ Hz, $d = 1.0$ m. (c) Structural acceleration at the center of the slab under crowd walking at $f_p = 1.5$ Hz, $d = 1.0$ m

Table 11.1 Modal parameters of the testing floor

Mode		1	2	3	4
Natural frequency (Hz)	Impact hammer tests	5.351	8.170	9.291	10.538
	Ambient vibration tests	5.354	8.372	10.053	10.954
Damping ratio (%)		1.567	2.042	1.144	1.211

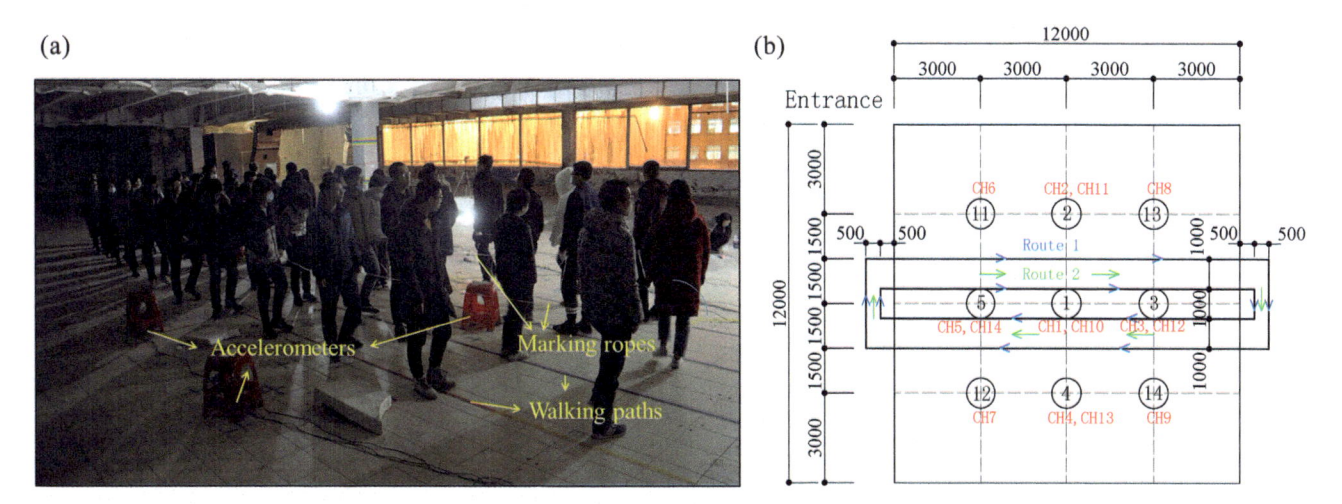

Fig. 11.3 Crowd walking pedestrian tests (**a**) Photo of the crowd walking testing (**b**) Locations of test points and walking routes for crowd pedestrian testing

results are listed in Table 11.1. Ambient vibration tests were also carried out on the slab to check the accuracy of the natural frequency: acceleration responses of the floor under ambient excitation without human occupants were recorded using the accelerometers, and the testing results were also listed in Table 11.1, showing that the first two modes coincide well, but the for the third and fourth modes, there exist some differences.

11.4 Crowd Walking Tests

As for crowd pedestrian walking, there are two main concerns: walking frequency and pedestrian density. To account for the two key parameters, the crowd walking pedestrian tests were so arranged: in each test case, test subjects were asked to walk at a specified walking frequency under the guidance of a metronome along the marked walking paths; spacing between every two test subjects was restricted by ropes with labels at certain points holding in their hands to control the crowd density (see Fig. 11.3a). The test cases consist of prompted walking frequencies $f_p = 1.5, 1.65, 1.8, 2.0, 2.2$ Hz and a free walking case, each with pedestrian spacing $d = 2.0, 1.5, 1.2, 1.0, 0.8$, and 0.6 m. It is worth mentioning that when $d = 0.8$ and 0.6 m, the spacing among neighboring pedestrians was too small to keep walking at high walking frequencies like 2.0 and 2.2 Hz, so these test cases were not included in the test.

Fourteen accelerometers were placed as shown in Fig. 11.3b to record the structural accelerations caused by crowd walking and the sampling frequency was 2048 Hz. Pedestrians walked in closed routes to keep a pedestrian flow and walked beyond the test slab when turning back. For most test cases, test subjects were asked to walk along Route 1 in Fig. 11.3b, and the only exception happened in the test case whose pedestrian spacing was 0.6 m, in which they turned to Route 2 due to the limitation of the number of test subjects. The mean body masses in test cases of pedestrian spacing $d = 2.0, 1.5, 1.2, 1.0, 0.8$, and 0.6 m are, respectively, 60.61, 62.69, 61.08, 61.93, 62.38, and 61.83 kg. Test protocol was approved by Tongji Medical Ethics Committee. When the commander commanded to start, the test subjects began to walk at the same time along the specified route with the guidance of metronome. Figure 11.2b, c are the acceleration time histories at the center of the slab (Point No. 1) at pedestrian spacing $d = 1.0$ m and walking frequencies of $f_p = 1.80$ Hz (resonance) and $f_p = 1.50$ Hz (non-resonance), respectively.

11.5 Vibration Serviceability Assessment

To analyze the vibration serviceability performance of the slab, assessment guidelines or design codes for this type of structure should be compared. In this paper, ISO 10137 [8] and AISC Design Guides [9] are involved, for the reason that these two guidelines are the most widely accepted ones and many subsequent guidelines were proposed on their basis. The structural response at the floor center was adopted in the assessment of vibration serviceability for the reason that it has the most violent vibration.

11.5.1 ISO 10137

The criterion in ISO 10137 [8] is expressed in form of curves obtained from a base curve (Fig. 11.4) with a multiplying factor, and an evaluation parameter of peak 10 s root mean square of acceleration (peak 10 s RMS) is recommended. The multiplying factor is chosen according to the function of the target building (e.g. critical working areas, residential, quiet office, general office or workshops), the time of the vibration exposure (day or night), and the characteristic of the vibration (whether it is continuous or impulsive). Different multiplying factors to the base curve are imposed for different situations. For walkways, a multiplying factor of 30 or 60 is imposed depending on whether there were people standing on the walkway. For stadia and floors of assembly halls, the multiplying factor is suggested 200. Seeing from Fig. 11.4a, measured values of peak 10 s RMS for the controlled walking cases satisfy the evaluation criteria on crowd cases easily.

11.5.2 AISC Design Guides

AISC Design Guides No. 11 [9] is another commonly used guide concerning the floor vibration serviceability. The criterion in this guide also benefits from the ISO baseline, but as mentioned above, the ISO Standard suggests multiples of the baseline curve targeted at RMS acceleration, while the multipliers recommended in AISC guides are targeted at peak acceleration. In Fig. 11.4b, measured peak values of acceleration in the crowd walking test cases fail to pass the criterion for office and residences; partly fail for indoor footbridges and shopping malls; and satisfy the criterion for outdoor footbridges.

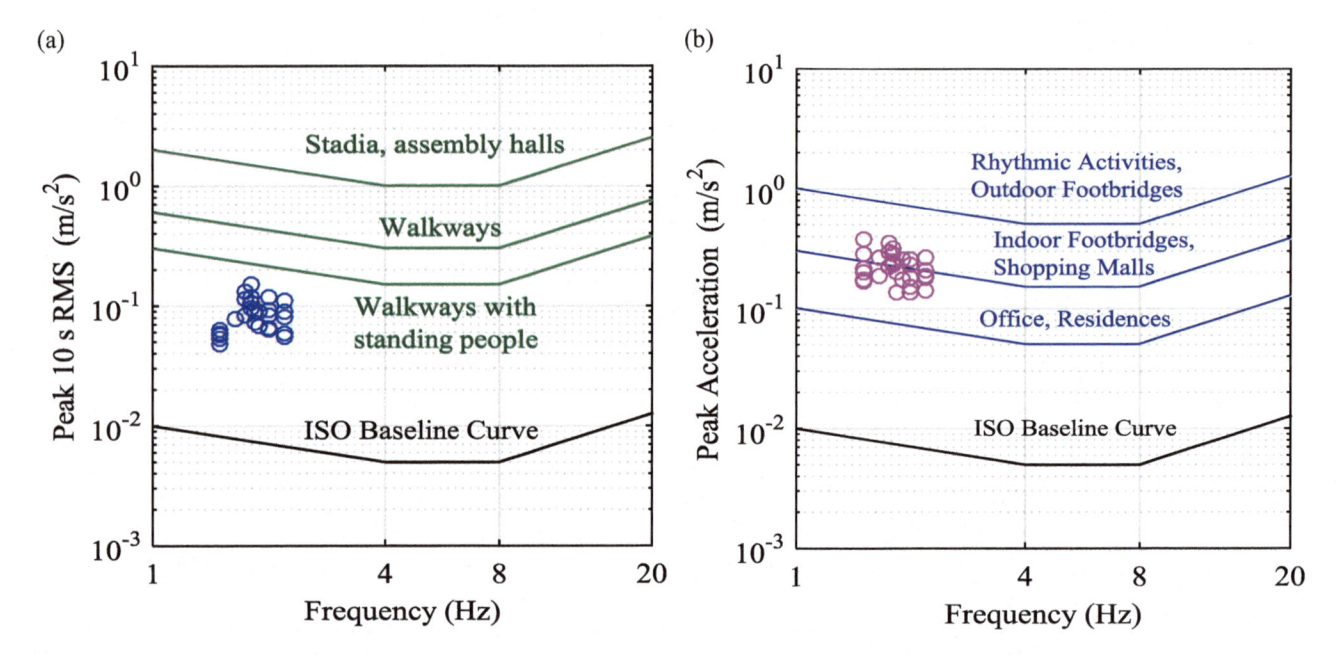

Fig. 11.4 Measured structural responses compared with guidelines (**a**) ISO 10137 (**b**) AISC Design Guides

11.6 The Influence of Pedestrian Density

Over the years, when calculating the vibration response to crowd traffic, crowd pedestrian induced vibration is always regarded to be the response to a single person excitation multiplied by a factor, and the factor is usually a function of the number of people walking on the structure at the same time. This approach is first proposed in the work of Matsumoto et al. [10], and still widely used due to its simplicity. The three design guidelines based on this approach are presented below, in which multiple factor should be paid special attention.

11.6.1 ISO 10137

In the ISO 10137 [8], dynamic load for a single pedestrian walking in the vertical direction is given as:

$$F_1(t) = W \left(1 + \sum_{n=1}^{k} \alpha_n \sin \left(2\pi f_p t + \phi_n \right) \right) \quad (11.1)$$

where W is the pedestrian's body weight, α_n is the dynamic loading factor, ϕ_n is the phase angle for the nth load harmonic, f_p is the pacing rate. For a group of N pedestrians, dynamic load for a single pedestrian is multiplied by $\sqrt{N}$ to obtain the total effective crowd pedestrian load.

11.6.2 French Sétra

In the design guidelines presented by the French road authorities [11], the dynamic crowd load per unit area is defined as

$$f_N(t) = 10.8 \frac{F_0}{A} \sqrt{N\zeta} \psi \cos \left(2\pi f_n t \right) \quad (11.2)$$

where A is the area of bridge deck, N is the number of people in the crowd, F_0 is the load amplitude of a single pedestrian; f_n is the natural frequency, ζ is the damping ratio of the structure; ψ falls in the range [0, 1] reflecting how far away the frequency of the load is from the average pacing rate.

11.6.3 Eurocode 1

U.K. National Annex to Eurocode 1 [12] suggests that the load for pedestrian load per unit area

$$f_N(t) = 1.8 \frac{F_0}{A} k \sqrt{\frac{\gamma N}{\lambda}} \sin \left(2\pi f_n t \right) \quad (11.3)$$

where A, F_0, f_n are defined the same with those in French Sétra, γ is for the lack of correlation between pedestrians in the crowd, k accounts for the excitation potential of the relevant forcing harmonic and probability of walking at the given resonant frequency, λ is used to adjust the number of effective pedestrians depending on their position related to mode shape ordinates.

The multipliers, i.e., the magnitudes of the crowd walking load normalized to single walking load, show great difference, which is originated from the different choices of the percentile value when modelling [13]. Yet, they all share the same idea that the crowd load is linear to $\sqrt{N}$, i.e., the square of number of pedestrians walking on the structures.

Since the dynamic load for crowd pedestrians walking is modelled linear to $\sqrt{N}$, the peak 10 s RMS values at test cases with different controlled walking frequencies and walking densities are normalized by the mean body mass W and $\sqrt{N}$. The normalized RMSs are classified by walking frequency and taken the mean values as plotted in Fig. 11.5. The standard deviations divided by mean values at walking frequencies 1.5, 1.8, 2.0, and 2.2 Hz are 0.080, 0.181, 0.075, and 0.097,

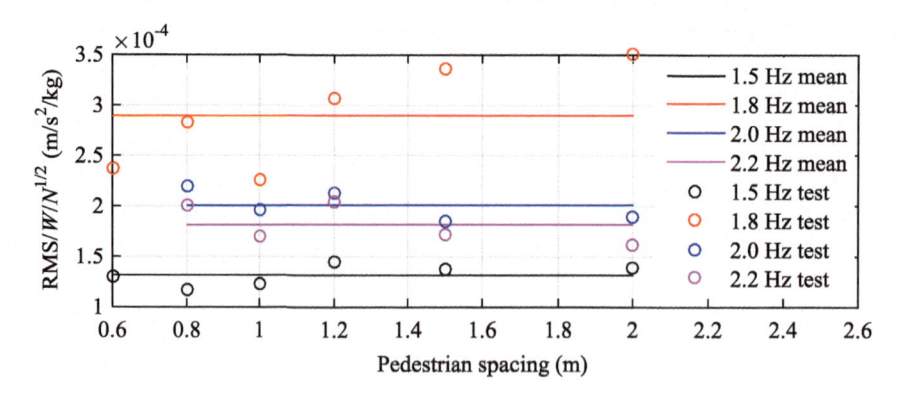

Fig. 11.5 RMSs normalized to $\sqrt{N}$ at different pedestrian spacing in each walking frequency test case

respectively. It illustrates that the structural response of resonance case (1.8 Hz) is not as good as those for non-resonance cases (1.5, 2.0, 2.2 Hz). But generally speaking, they follow the modelling strategy that crowd pedestrian is linear to $\sqrt{N}$.

11.7 Conclusions

The paper mainly introduced the vibration serviceability performance of an as-built floor under crowd pedestrian walking tests, in which the pedestrian density covered walking conditions in unrestricted traffic, restricted traffic, and extremely restricted traffic. For each walking spacing (pedestrian density), controlled walking frequencies in both resonance and non-resonance cases were included in the test, and one free walking case was also involved. In terms of vibration assessment parameters, peak 10 s RMS and peak accelerations are chosen. When the testing results compare with guidelines, peak 10 s RMS can easily satisfy the requirement from ISO 10137, while peak acceleration partly fail to meet the several demands form AISC design guides. Furthermore, by the virtue that in guidelines, the crowd walking excitation shows linear relationship with $\sqrt{N}$, the square of number of pedestrians walking on the structures at the same time, the testing results of RMS in each pedestrian spacing are then normalized to $\sqrt{N}$ for the aim of checking the suggested relationship. The results show a good match in the non-resonance cases, but cannot accord that well in the resonance case. All in all, there still calls for a better understanding on crowd pedestrian walking load.

Acknowledgements The authors would like to acknowledge the financial support provided by National Natural Science Foundation of China (51778465) and State Key Laboratory for Disaster Reduction of Civil Engineering (SLDRCE14-B-16). Moreover, the authors would like to thank all test subjects for participating in the project making possible the data collection.

References

1. Pavic, A., Reynolds, P.: Vibration serviceability of long-span concrete building floors. Part 1: review of background information. Shock. Vib. Dig. **34**, 191–211 (2002)
2. Racic, V., Pavic, A., Brownjohn, J.: Experimental identification and analytical modelling of human walking forces: literature review. J. Sound Vib. **328**, 1–49 (2009)
3. Wang, J., Chen, J.: A comparative study on different walking load models. Struct. Eng. Mech. **63**(6), 847–856 (2017)
4. Xiong, J., Chen, J.: Power spectral density function for individual jumping load. Int. J. Struct. Stab. Dyn. **3**, 1850023 (2018)
5. Wang, H., Chen, J., Brownjohn, J.: Parameter identification of pedestrian's spring-mass-damper model by ground reaction force records through a particle filter approach. J. Sound Vib. **411**(22), 409–421 (2017)
6. Chen, J., Zhang, M., Liu, W.: Vibration serviceability performance of an externally prestressed concrete floor during daily use and under controlled human activities. J. Perform. Constr. Facil. **30**(2), 04015007 (2016)
7. Varela, W., Battista, R.: Control of vibrations induced by people walking on large span composite floor decks. Eng. Struct. **33**, 485–2494 (2011)
8. ISO: Bases for Design of Structures—Serviceability of Buildings Against Vibrations. International Standardization Organization, Geneva (1992). ISO 10137
9. Murray, T., Allen, D., Ungar, E.: Floor Vibrations Due to Human Activity. AISC, Chicago (1997). AISC steel design guide #11
10. Matsumoto, Y., Nishioka, T., Shiojiri, H., Matsuzaki, K.: Dynamic design of footbridges. IABSE Period.-Proc., P17/78, pp. 1–15 (1978)

11. Sétra: Footbridges, Assessment of Vibrational Behaviour of Footbridges Under Pedestrian Loading, Technical Guide. Service d'Etudes Techniques des Routes et Autoroutes, Paris (2006)
12. British Standards Institution (BSI): UK National Annex to Eurocode 1: Actions on Structures—Part 2: Traffic Loads on Bridges. British Standards Institution, London (2008). NA to BS EN 1991–2: 2003
13. Živanović, S., Pavić, A., Ingólfsson, E.: Modeling spatially unrestricted pedestrian traffic on footbridges. J. Struct. Eng. **136**(10), 1296–1308 (2010)

Chapter 12
Identifying Traffic-Induced Vibrations of a Suspension Bridge: A Modelling Approach Based on Full-Scale Data

Etienne Cheynet, Jonas Snæbjörnsson, and Jasna Bogunović Jakobsen

Abstract The present paper introduces a procedure to identify traffic-induced vibrations from full-scale acceleration records from a long-span suspension bridge. First, an outlier detection algorithm coupled with a cluster analysis is applied to detect when a vehicle crosses the bridge. Then, the mass and average speed of each identified vehicle are estimated using a moving mass model. The current identification procedure requires a low wind speed and a low traffic density to isolate the background component of the displacement response from its resonant component. Eleven months of records from high-accuracy three-axial accelerometers were used to systematically identify traffic-induced vibrations and estimate the mass and speed of numerous vehicles. The computation of the combined effects of wind and traffic loading on the vertical bridge displacement response indicates that the study of the wind-induced vibrations of the Lysefjord bridge should account for traffic loading, even for wind speed above $10 \, \mathrm{m \, s^{-1}}$.

Keywords Full-scale; Traffic; Dynamic vibrations; Wind; Structural health monitoring

12.1 Introduction

The vibrations of a suspension bridge are mainly induced by wind and traffic. These two sources of excitation generally co-exist, and it can be difficult to identify from acceleration data what part of the bridge response is purely due to the wind and which is due to the traffic. Naturally, it can be assumed that the traffic dominates at low wind speed. However, it is not well known at what wind velocity level the traffic-induced load becomes negligible and what level of vibrations can be expected from heavy vehicles. This is primarily due to the fact that a limited number of full-scale studies have focused on this topic. Traffic-induced loading has mainly been studied through numerical models [1–5] and in some rare cases in laboratory experiments [6]. Traffic-induced vibrations are of key importance for the estimation of the serviceability limit state and often a significant source of discrepancies when modelling the overall bridge response.

Both wind- and traffic loading can be seen as random sources of excitation: the wind velocity fluctuations are generally modelled as a Gaussian stationary random process, whereas vehicle loads are responsible for a transient load with an unknown duration and starting time. The load applied by the vehicle is a function of the vehicle speed and mass, which are a priori unknown and can also be modelled as a random variable [1]. In the absence of surveillance cameras, additional sources of randomness are the number of vehicles moving simultaneously on the bridge as well as the direction of each of them.

Whereas the wind velocity characteristics can be assessed using sonic anemometers, there does not exist, to the authors' knowledge, efficient algorithms to automatically characterize traffic loading on a suspension bridge from full-scale acceleration records. By comparing the high-frequency and low-frequency acceleration bridge response Cheynet et al. [7] were able to establish whether traffic or wind was the dominant source of vibration of a long-span suspension bridge, based on the assumption that traffic loading produces a significantly larger acceleration response at higher-frequencies than the wind load. This approach is a simple tool to sort acceleration records, but it does not provide any information on the vehicles crossing the bridge.

E. Cheynet (✉) · J. B. Jakobsen
Department of Mechanical and Structural Engineering and Materials Science, University of Stavanger, Stavanger, Norway
e-mail: etienne.cheynet@uis.no

J. Snæbjörnsson
Department of Mechanical and Structural Engineering and Materials Science, University of Stavanger, Stavanger, Norway

School of Science and Engineering, Reykjavík University, Reykjavík, Iceland

In the present study, we aim to automatically identify the mass and speed of vehicles producing a noticeable bridge response. The collected information is then used to assess much more precisely under which conditions traffic-loading is dominating over wind-loading through a time-domain simulation. For that purpose, wind velocity and bridge acceleration records collected on the Lysefjord bridge (Norway) during a monitoring period of eleven months are used.

The bridge instrumentation and the identification algorithm are presented in Sect. 12.2, where the identification procedure is illustrated using one particular time series. In Sect. 12.3, the identification algorithm is applied to eleven months of data. The ability of the algorithm to provide relevant inputs for a moving-mass model is then assessed, followed by an analysis of the combined effect of traffic and wind loading on the vertical dynamic bridge displacement response.

12.2 Instrumentation and Method

12.2.1 The Lysefjord Bridge

The Lysefjord suspension bridge, located at the inlet of a narrow fjord in Norway, is surrounded by mountains and steep hills. It has a main span of 446 m and a 12.3 m-wide bridge deck with two traffic lanes and a walk-path. Although the traffic is on average low, it is also intermittent as the number of vehicles crossing the bridge is highly dependent on the schedules of a ferry, which lands only 2 km south of the bridge.

Since 2013, the Lysefjord bridge has been instrumented with a wind and structural health monitoring system consisting of nine sonic anemometers and four pairs of three-dimensional accelerometers. In Fig. 12.1, the position of the anemometers above the deck is defined using the hanger name HXY, where X is a digit between 08 and 24 indicating the hanger number, and Y denotes the west side (W) or east side (E) of the deck. Since two anemometers are mounted on the hanger west no. 08 (H08W), the notations H08Wb and H08Wt refer, in the following, to the sonic anemometer mounted 6 m (bottom) and 10 m (top) above the deck, respectively. Eight of the sonic anemometers are 3-D WindMaster Pro from Gill instruments (Lymington, UK), which can record the wind velocity and sonic temperature with a sampling frequency up to 32 Hz. The last sonic anemometer, mounted on H10W, is a Weather Transmitter WXT520 from Vaisala (Helsinki, Finland), which monitors the horizontal wind components, relative humidity, pressure and absolute temperature with a sampling frequency up to 4 Hz. The accelerometers are CUSP-3D from Canterbury Seismic Instruments which can operate with a sampling frequency up to 200 Hz. They are mounted inside the deck on each side of the girder to retrieve the horizontal, vertical and torsional bridge acceleration response (Fig. 12.1).

12.2.2 The Identification Algorithm

The identification of the parameters of a vehicle crossing a bridge has mostly been conducted for bridges with a main span shorter than 60 m [6, 8–11], where the modelling of the high-frequency bridge response is fundamental for the bridge-vehicle

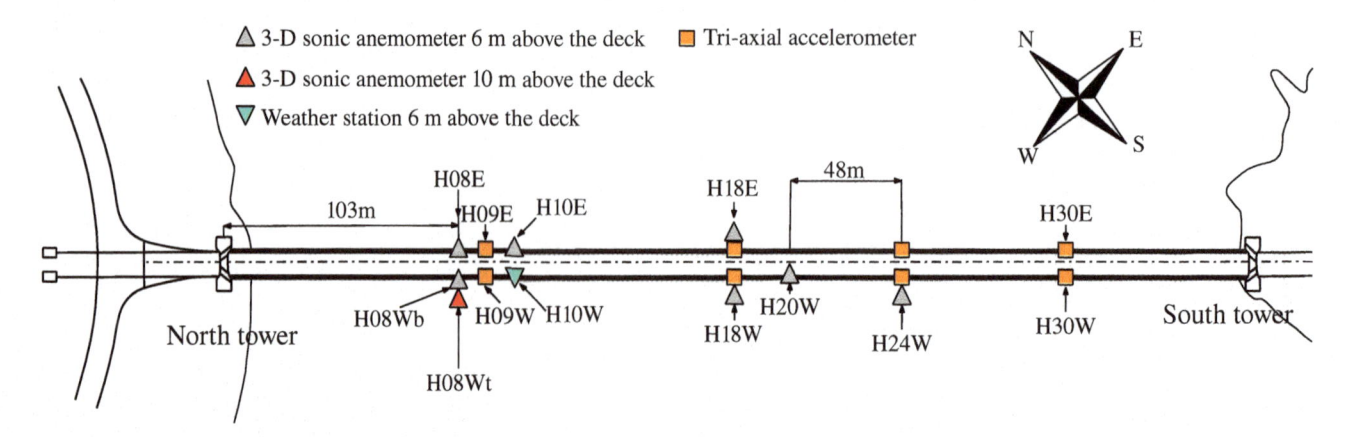

Fig. 12.1 Instrumentation of the Lysefjord bridge since July 2017

interaction. The situation described here is fundamentally different as vehicles are identified using the quasi-static bridge response only, i.e. the frequency range located below the first eigenfrequency. In addition, we focus only on the vertical displacement response of the Lysefjord bridge, which is dominated by only a few modes at frequencies below 1 Hz and allows the use of a moving-mass model to study the bridge-vehicle interaction. If the bridge acceleration response is studied, a more realistic model of the vehicles is required, which is out of the scope of the present study. This implies also that the approach used in Cheynet et al. [7] to study traffic-induced vibrations is not comparable with the one used here.

To accurately identify traffic-induced vibrations on a long-span suspension bridge, using this approach, the influence of the wind loading on the bridge structure needs to be negligible. Therefore, only acceleration bridge records associated with a mean wind speed below $5\,\text{m}\,\text{s}^{-1}$ and a turbulence intensity lower than 15% are here considered. At such low wind velocities, the majority of the wind records are non-stationary. Using 30 min-long time series, the first-order stationarity of each anemometer record is assessed using a moving mean with a centred half window of 5 min. Only samples with a moving mean that deviates by less than 20% of the static mean are kept as the other ones are assumed non-stationary. Under such conditions, it is observed that wind loading has a negligible contribution to the bridge response compared to traffic loading.

To identify if and when a vehicle crosses the bridge, the post-processing of the bridge acceleration data is done as follows:

(a) The vertical acceleration response at mid-span is transformed into the displacement response.
(b) A zero-phase digital low-pass and high-pass fifth order Butterworth filters are successively applied to isolate the background response of the bridge deck. Given the low frequency considered, a band-pass filter may be too unstable to be applied. The cut-off frequency for the high-pass filter is here 0.04 Hz. This limit is actually fixed by the accelerometers' performances, as measurement errors in the acceleration data gradually increase when the frequency decreases. The cut-off frequency for the low-pass filter is 0.15 Hz, which is lower than the first vertical eigenfrequency that has a value of 0.21 Hz. Not only does the use of fifth-order filter allow for a stable filter design but also help to enhance the displacement of the bridge due to traffic with respect to other sources of vibrations. To intensify even more the background response with respect to the resonant response, a fourth or even third-order filter may be reliably applied. For the sake of brevity, the study of the optimal filter order is not considered in the following and only the fifth-order filter is used.
(c) An outlier analysis, based on the generalized extreme Studentized deviate test for outliers [12], is applied to identify the peak response amplitudes associated with vehicles crossing the bridge. In particular, this approach works well when several outliers mask each other, i.e. when two vehicles cross the bridge in a short interval. Another possible outlier-detection algorithm relies on a moving median and seems to perform best when a high number of lightweight and heavy vehicles cross the bridge. It is, however, more sensitive to false-positive results. The outlier test based on a moving median is, therefore, not considered in the following.
(d) A clustering algorithm is applied to associate each group of outliers to a vehicle. A cluster analysis aims to partition scattered data into groups. For a single group, the data share the same properties, i.e. are homogeneous, whereas for multiple groups, a strong heterogeneity is visible. As the number of vehicles is a priori unknown, a hierarchical cluster tree is first built using the single linkage algorithm using Euclidian distance. Here, the term "Euclidian distance" has actually the dimension of time as two outliers are considered in the same cluster if they are separated by less than 30 s, which corresponds to the duration it takes to cross the bridge main span (446 m) at a speed slightly above $50\,\text{km}\,\text{h}^{-1}$. In the present case, up to nine clusters per 30-min records are detected. Clusters separated by less than 90 s are dismissed as they are assumed to correspond to vehicles too close to each other to be properly identified. Note that the duration of 90 s is arbitrarily chosen and is found appropriate for the case of the Lysefjord bridge. Clusters recorded at the very beginning or the end of the time series are also dismissed as they do not allow a reliable estimation of the arrival and departure time of the vehicle crossing the bridge. These four steps of the data processing are illustrated in Fig. 12.2.

Once each cluster is identified, the filtered displacement histories are cut into several segments, each of them centred around one cluster. For each segment, the bridge displacement response is normalized to minimize its dependency on the vehicle mass. The vehicle arrival time and average speed are then simultaneously estimated in the least square sense using a moving mass model (Fig. 12.3). For the sake of simplicity, only the average vehicle speed is estimated, although the vehicle speed is generally smaller on its north side than on its south side. The arrival time is defined here as the time at which the bridge response due to traffic loading is detected by the accelerometers at mid-span and corresponds to the time at which a vehicle arrives on either the north side or the south side of the bridge. The deck is modelled as a continuous single-span suspension bridge [13], the modal parameters of which have been estimated using the Galerkin method. Finally, the modal parameters have been corrected with respect to the full-scale structure using an operational modal analysis relying on an automated system identification algorithm [14, 15]. Table 12.1 shows the eigenfrequencies and modal damping ratios of the vertical symmetric modes, as only the vertical bridge response at mid-span is considered here.

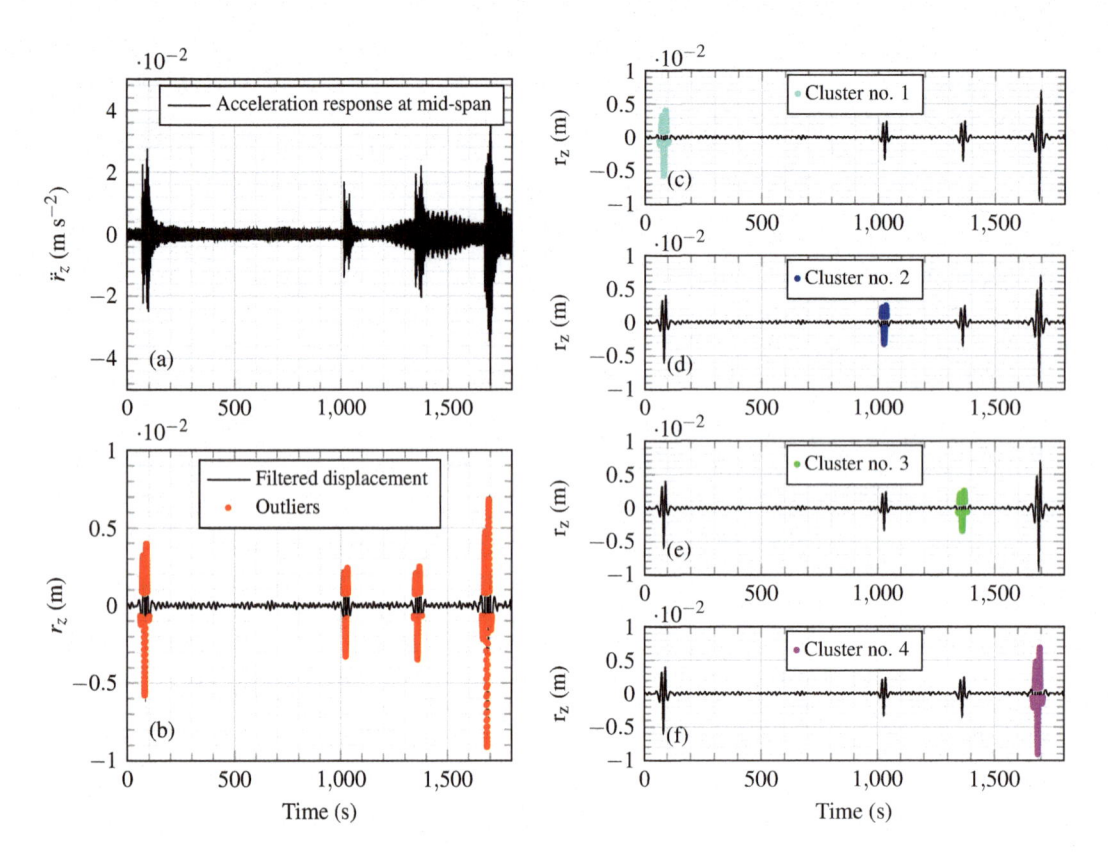

Fig. 12.2 Acceleration data (panel (**a**), top left) and associated filtered bridge displacement response at mid-span (panel (**b**), bottom left) recorded on 09-09-2017 at midnight. The Outlier detection algorithm leads to four clusters (panels (**c–f**), right) at around $t = 83$ s, $t = 1028$ s, $t = 1361$ s and $t = 1788$ s

Table 12.1 Eigenfrequency and modal damping ratios ζ of the vertical symmetric modes of the Lysefjord bridge computational model [15, 16]

Modes	V1S	V2S	V3S	V4S	V5S
Frequency (Hz)	0.294	0.408	0.853	1.520	2.380
ζ (%)	0.2	0.2	0.5	0.5	0.5

As the bridge computational model is symmetric, the direction of a single moving vehicle has no influence on the displacement response at mid-span. In full-scale, neglecting the vehicle direction is an acceptable approximation if the traffic density is low, which is generally observed for the Lysefjord bridge. Once the vehicle average speed and arrival time are known, the vehicle mass is estimated in the least-square sense using the non-normalized bridge displacement response and a moving-mass model.

Although a moving mass model [17] is one of the simplest approach available to describe a vehicle crossing a bridge, it is found suitable in the present study for the following reasons: (1) The high-frequency range of the displacement response of the bridge is negligible compared to its low-frequency response, as the structure acts as a low-pass filter; (2) In the present case, the use of a damped moving oscillator model leads to a similar bridge displacement response as the moving mass, which is attributed to the high damping ratio of the vehicle modelled (above 10%) and its relatively high eigenfrequency (above 1 Hz). Using a moving mass, each vehicle is defined by only three parameters: arrival time, equivalent mass and equivalent travelling speed. The adjective "equivalent" used here refers to the mass and speed estimated by the moving mass model at a constant speed that produces a computed response similar as observed in full-scale, even though the real mass and speed of the vehicle may be slightly different. For example, using the study case of Fadnes [18, p. 73], who observed a single 50-t truck crossing the Lysefjord bridge moving at an average speed of 45 km h^{-1}, the computational model found a mass of 58 t and a speed of 50 km h^{-1} for a first crossing and a mass of 45 t and a speed of 56 km h^{-1} for a second crossing, two hours later. The truck mass was not precisely known, its speed was not constant and its loading was likely different during the two trips, Nevertheless, the estimated parameters are consistent with those from the full-scale observations and lead to a good agreement between the computed and measured bridge displacement response.

The applicability of the moving mass model is illustrated in Fig. 12.4, where the computed and full-scale high-pass filtered bridge displacement response at mid-span are superposed. In this figure, the bridge response is computed based on the parameters identified in Fig. 12.2 and 12.3, for the four vehicles crossing the bridge. The good agreement between the two power spectral density estimates shows that for the wind conditions considered, the modelling of the turbulent wind load is not necessary and that the parameters used with the data post-processing are appropriate. Note that in the right panel of Fig. 12.4, the broad peak recorded below 0.1 Hz results from the application of the high-pass filter and has, therefore, no physical meaning contrary to those at higher frequencies, which correspond to the vertical eigenmodes.

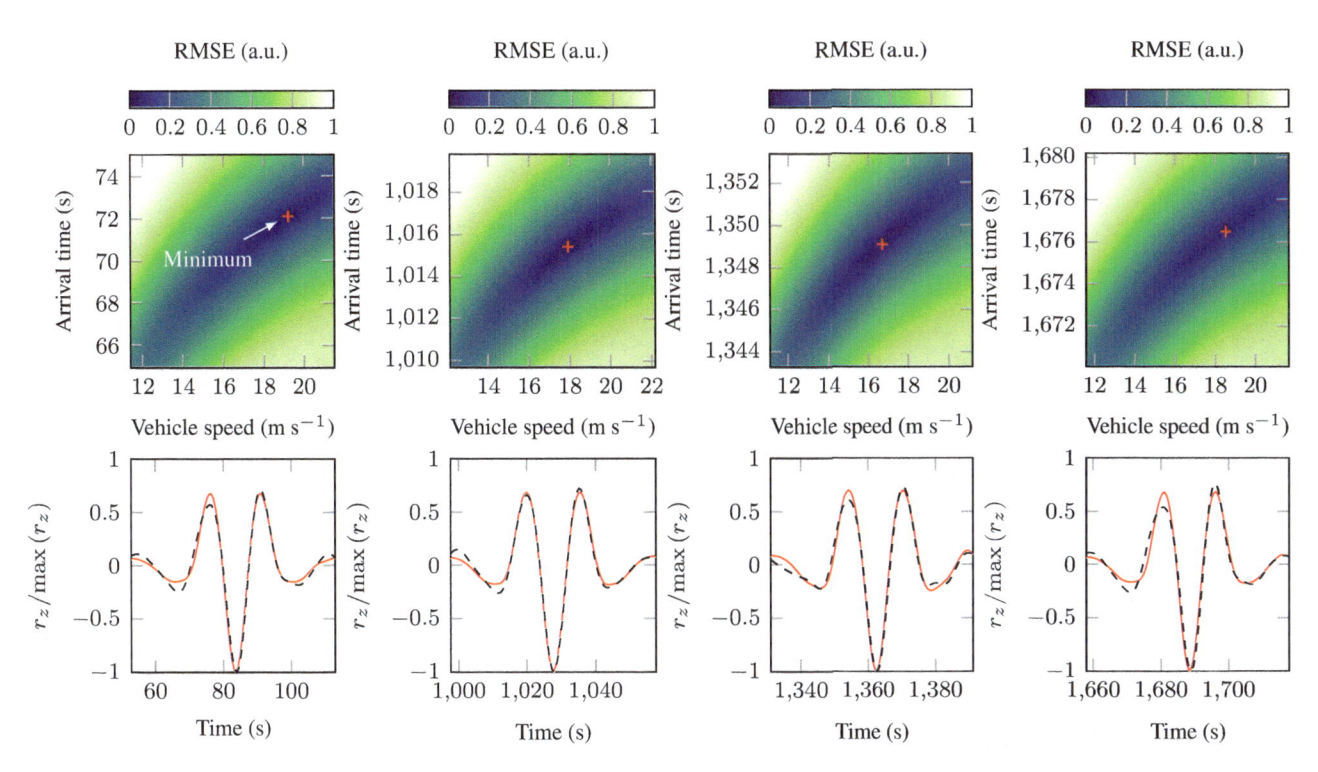

Fig. 12.3 Top panel: Root mean square error (RMSE) between the computed and measured displacement response for the four clusters identified and displayed in Fig. 12.2. Bottom panels: Measured (dashed line) displacement response superposed to the computed one (red line) after estimating the vehicle speed and arrival time

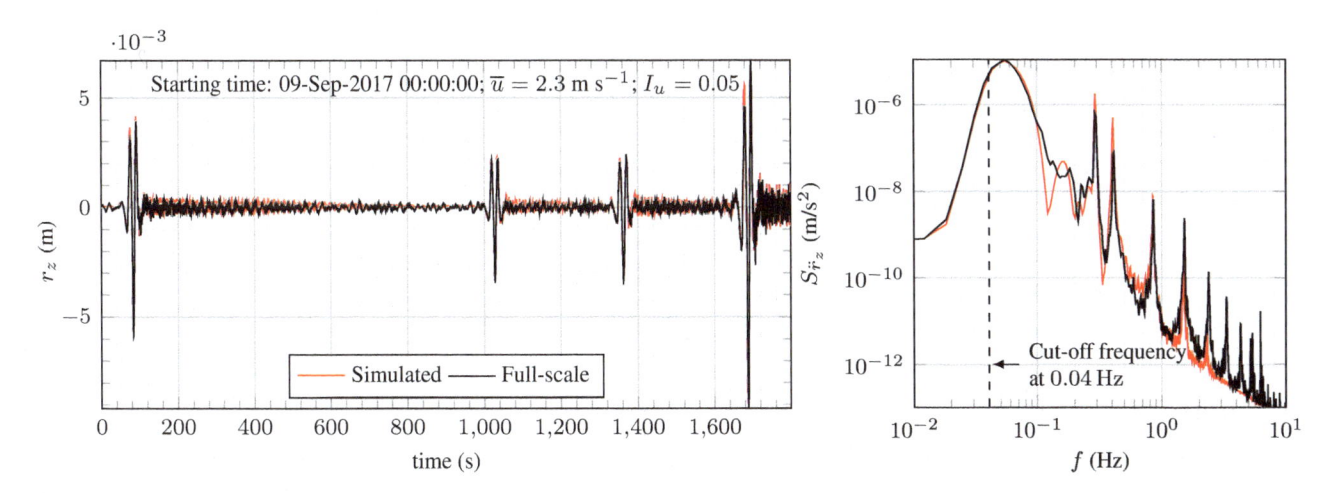

Fig. 12.4 Time series (left panel) and associated power spectral density estimate (right panel) corresponding to the measured and computed vertical bridge displacement response induced by the crossing of four vehicles, the parameters of which (arrival time, mass and average speed) have been automatically estimated

12.2.3 Wind Load Modelling

To evaluate the significance of traffic loading with respect to wind loading, their combined effect is briefly studied in Sect. 12.3.2 in the time domain. The bridge model is the same as in Sect. 12.2.2. The turbulent wind load is modelled using the Norwegian guideline for suspension bridge design [19], with a turbulence intensity at deck height $I_u = 0.15$ and $I_w = 0.7 I_u$, which is representative of the wind conditions from south-southwest at the bridge site. For $i = \{u, w\}$, the one-point velocity spectrum is modelled as,

$$\frac{f S_i}{\sigma_i^2} = \frac{A_i n_i}{(1 + 1.5 A_i n_i)^{5/3}}, \quad n_i = \frac{f L_i^x}{\bar{u}(z)} \tag{12.1}$$

where $A_u = 6.8$, $A_w = 9.4$ and L_i ($i = \{u, w\}$) is the integral length scale, defined as:

$$L_u^x = \begin{cases} L_1 (z/z_1)^{0.3}, & \text{if } z > z_{\min} \\ L_1 (z_{\min}/z_1)^{0.3}, & \text{if } 0 \leq z \leq z_{\min} \end{cases} \tag{12.2}$$

$$L_w^x = L_u^x / 12 \tag{12.3}$$

where L_1 is a reference length scale equal to 100 m and z_1 a reference height equal to 10 m, as prescribed in EN 1991-1-4 [20]. Finally, the lateral coherence is modelled using the Davenport model [21] and a decay coefficients $c_u^y = 10$ and $c_w^y = 6.5$ for the along-wind and vertical wind component, respectively. The stationary turbulent wind field is generated in the time domain using the traditional spectral representation approach [22]. The wind-induced response of the bridge is computed in the time-domain using the quasi-steady theory, in a similar fashion as by Wang et al. [16], except that the simpler continuous bridge model is here used instead of the finite element model.

12.3 Results

12.3.1 Automatic Identification of Vehicles

The identification procedure illustrated in Fig. 12.2 to 12.4 is repeated for each 30-min acceleration records between July 2017 and June 2018. The ability of the aforementioned algorithm to capture and model traffic-induced vibrations under low-wind conditions is displayed in Fig. 12.5, where the bridge response from the computational model is compared to the one estimated from the full-scale measurements. The bridge response is herein only studied in terms of the standard deviation and maxima of the bridge displacement response at mid-span. It should be noted that the load applied on the bridge is computed using a moving mass model and assumed to be entirely due to traffic since for the wind conditions considered the wind load is found negligible. The samples considered here are only those where the traffic density is low but at least one vehicle is identified. For each identified vehicle, the root-mean-square error (RMSE) between the computed and estimated displacement response quantifies the reliability of the estimated vehicle mass and speed, leading to a distribution of RMSE values. The 99th-percentile values of the RMSE are used to filter out vehicles that have not been successfully identified by the algorithm. A large RMSE may be due to a low signal-to-noise ratio or indicate that the outlier detection algorithm or the cluster analysis failed to detect multiple vehicles crossing the bridge close together.

In Fig. 12.5, the computed displacement response at mid-span shows a good agreement with the observed displacement response based on the recorded acceleration. The comparison is shown in terms of the peak value and standard deviation. However, the discrepancies are seen to be slightly increasing with increased amplitude of vibration. For $\sigma_{r_z} < 0.05$ m, the slight underestimation of the standard deviation of the simulated displacement response shown in the left panel of Fig. 12.5 is expected. Firstly because wind-induced vibrations are not modelled and secondly because we do not attempt to capture every vehicle crossing the bridge, especially when a mixture of multiple heavy and light vehicles are recorded in a short time span, such that the outlier detection algorithm ignores the lighter ones. This particular issue can be solved using an outlier detection algorithm relying on a local median, as described in Sect. 12.2.

The distribution of identified vehicle equivalent mass and speed is displayed in Fig. 12.6. The term "distribution" refers here to the joint distribution of mass and speed responsible for a noticeable bridge response as the outlier algorithm does not

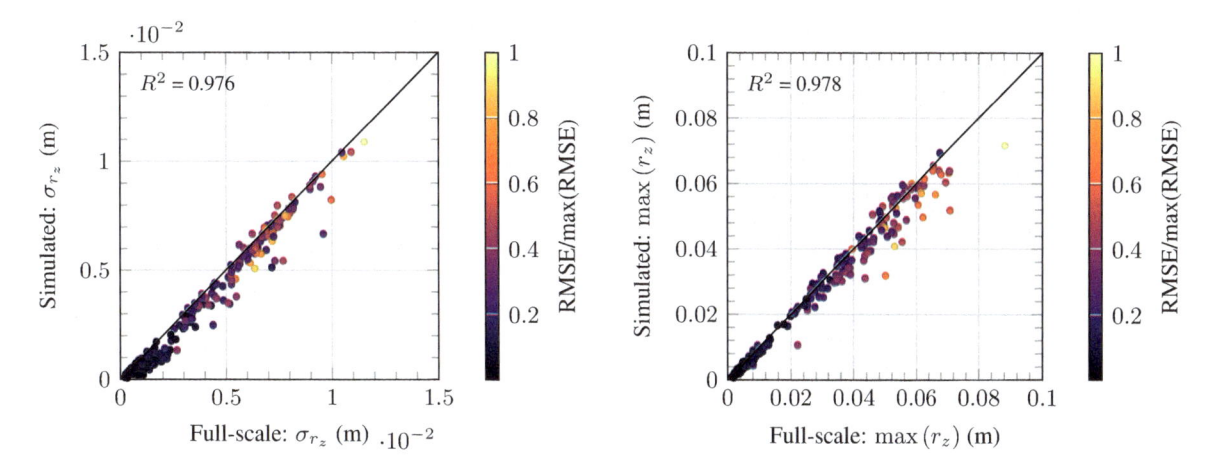

Fig. 12.5 Computed and estimated standard deviation (left panel) and maximal value (right panel) of the displacement response of the Lysefjord bridge at mid-span for each times series where at least one vehicle is successfully identified (245 samples of 30-min duration)

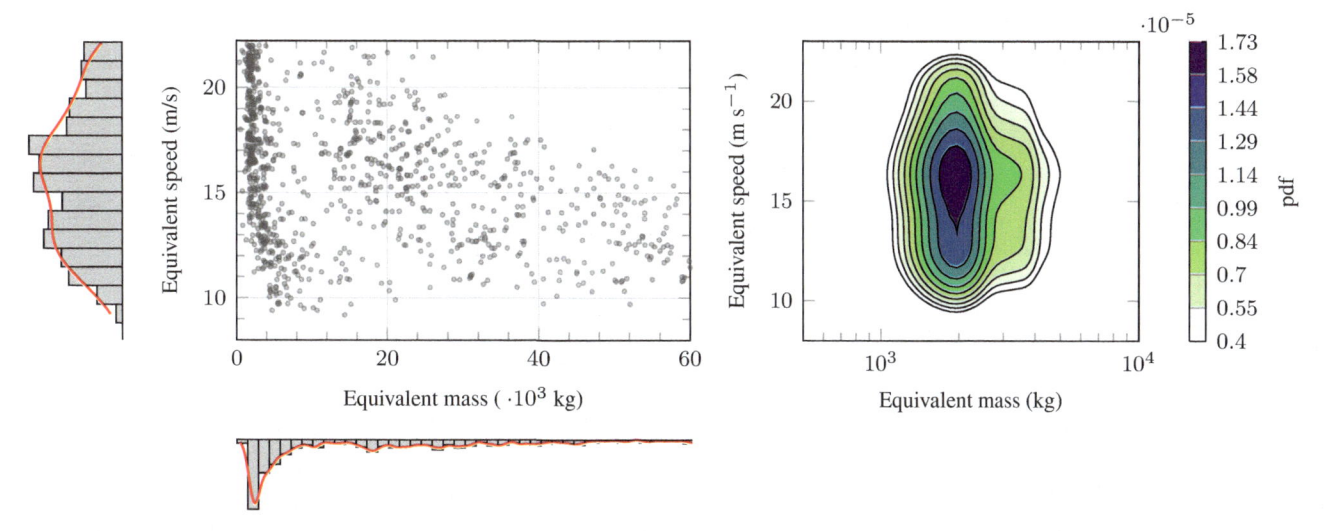

Fig. 12.6 Distribution of the equivalent mass and speed of each identified vehicle crossing the Lysefjord bridge and the associated kernel density estimate (left), which are combined into a joint-probability distribution (right)

detect every vehicle. In particular, the number of identified lightweight vehicles crossing the bridge at a low speed may be substantially lower than in reality as they produce a lower bridge response than heavyweight vehicles. Nevertheless, the range of vehicle speeds is consistent with those documented by Fadnes [18, p. 36] on 17-04-2017, although the lack of monitoring data from this particular day prevents a more detailed comparison.

12.3.2 Combined Effect of Wind and Traffic Loading

The combined effect of wind and traffic loading on the bridge response is here briefly assessed using three situations: (1) The case of a single 10-t vehicle crossing the bridge one time at a speed of $50 \, \text{km} \, \text{h}^{-1}$; (2) the case of a single 40-t vehicle crossing the bridge one time at a speed of $45 \, \text{km} \, \text{h}^{-1}$ and (3) the case of multiple vehicles crossing the bridge at different speeds and times, as observed on 23-Feb-2018 between 12:00 and 12:30. For the last case, the identified arrival time, equivalent mass and speed of the vehicles are given in Table 12.2. For the three situations considered, both the traffic and wind loads are simultaneously accounted for. To assess the relative importance of these two sources of vibrations, the computation of the bridge response is repeated for the same traffic load but for an increasing mean wind speed and thus an increasing dynamic wind load Eqs. (12.1) and (12.2).

Table 12.2 Parameters of the vehicles used in Fig. 12.7 for the case no. 3 "Wind + multiple vehicles"

Vehicle number	1	2	3	4	5	6	7
Arrival time (min)	12:02:39	12:08:04	12:12:26	12:20:04	12:21:54	12:25:02	12:28:46
Equivalent mass (10^3 kg)	15.5	51.8	56.7	24.2	17.0	15.5	26.4
Equivalent speed (km h^{-1})	66	53	49	61	66	66	57

The vehicle parameters are estimated from a 30-min acceleration records obtained on 23-Feb-2018 from 12:00, which was characterized by multiple heavy-weight vehicles crossing the bridge in less than 30 min

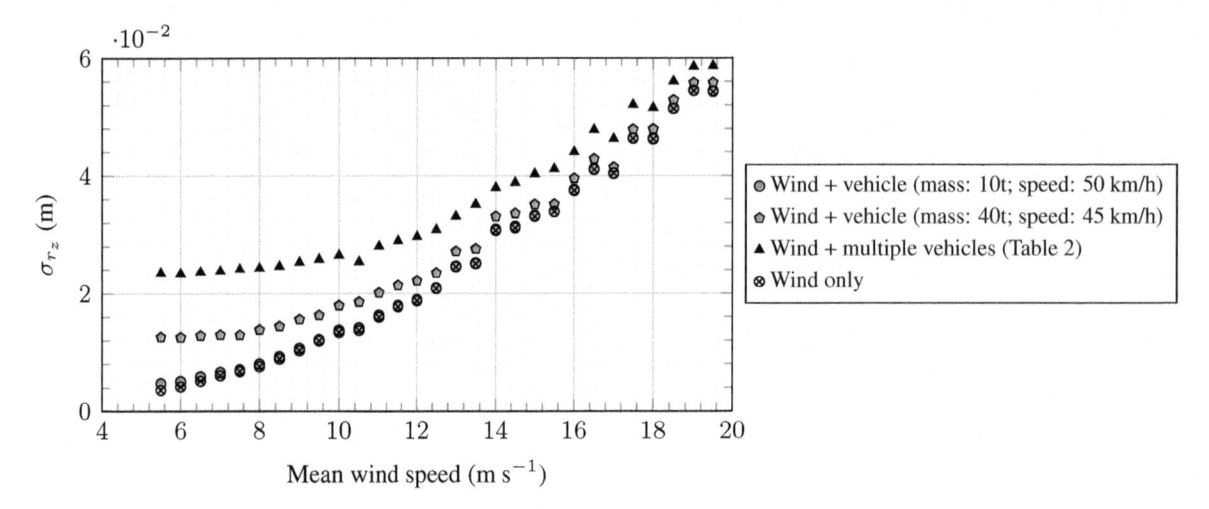

Fig. 12.7 Standard deviation of the computed vertical bridge displacement response at mid-span induced by wind turbulence and one or multiple vehicles crossing the main span

For the three situations considered, Fig. 12.7 shows that the standard deviation of the bridge displacement response is systematically underestimated if traffic is neglected, especially cases (2) and (3). The load induced by the 10-t vehicle contributes to less than 10% of the standard deviation of the displacement response for $\bar{u} > 7$ m s^{-1}, whereas the same contribution for the 40-t vehicle is obtained at $\bar{u} > 14$ m s^{-1}. In the case of multiple vehicles crossing the bridge, traffic is the dominant source of vibrations at $\bar{u} \leq 10$ m s^{-1} and its contribution remains above 10% of the vertical displacement response up to $\bar{u} = 18$ m s^{-1}. The results displayed in Fig. 12.7 are consistent with those from Macdonald [23], who noted that traffic loading on the Second Severn Crossing bridge, which has similar dimensions as the Lysefjord bridge, was dominant for a mean wind speed below 12 m s^{-1}. Said differently, Fig. 12.7 brings an additional proof that the study of the buffeting response of a suspension bridge in full-scale needs to account for traffic-loading. However, it should be noted that the comparisons shown in Fig. 12.7 depend on the bridge structure considered and the turbulence model used, in particular, the turbulence intensity, for which greater values imply a reduced influence of traffic loading on the bridge response compared to wind loading.

12.4 Conclusions

We present an algorithm that automatically identifies the equivalent mass and speed of a vehicle crossing a long-span suspension bridge using wind velocity and acceleration records at mid-span. The goal is to estimates vehicle characteristics that can be further used to gather information about the level of serviceability loading of the bridge, as well as providing relevant data for smart asset management.

The identification procedure relies on the analysis of the background component of the displacement response. First, traffic-induced vibrations are detected using an outlier detection algorithm and a cluster analysis. Then, a simple moving-mass model is applied to estimate in the least-square sense the time at which a vehicle arrives on the bridge as well as its speed and mass. As the latter two parameters are estimated knowing the bridge response only, they are referred to as "equivalent mass" and "equivalent speed". Despite its simplicity, the use of a moving-mass model is found appropriate to describe the vehicle-induced bridge quasi-static and dynamic displacement response. As the wind-induced vibrations are not modelled in the identification algorithm, it is primarily suitable for acceleration records associated with wind speed below 5 m s^{-1}. At present, the identification algorithm requires a low traffic density, as two vehicles crossing the bridge simultaneously or in a short amount of time lead to two mixed signal clusters that are not easily separated. As only vehicles inducing a noticeable bridge response are detected, the present algorithm is suitable to study vehicle-induced bridge vibrations rather than direct traffic monitoring.

The simulation of combined traffic and wind loading on the Lysefjord bridge shows that the buffeting response of the structure is in general underestimated if traffic loading is not accounted for. In some situations, as shown in Table 12.2, traffic can be the main source of excitation, even for a mean wind speed of $10\,\mathrm{m\,s^{-1}}$.

Acknowledgement The support of the Norwegian Public Roads Administration is gratefully acknowledged, as well as their assistance during the installation and maintenance of the monitoring system.

References

1. Bryja, D., Śniady, P.: Random vibration of a suspension bridge due to highway traffic. J. Sound Vib. **125**(2), 379–387 (1988)
2. Chatterjee, P., Datta, T., Surana, C.: Vibration of suspension bridges under vehicular movement. J. Struct. Eng. **120**(3), 681–703 (1994)
3. Yang, Y.B., Lin, B.H.. Vehicle-bridge interaction analysis by dynamic condensation method. J. Struct. Eng. **121**(11), 1636–1643 (1995)
4. Yang, Y.B., Yau, J.D.: Vehicle-bridge interaction element for dynamic analysis. J. Struct. Eng. **123**(11):1512–1518 (1997)
5. Bryja, D., Śniady, P.: Stochastic non-linear vibrations of highway suspension bridge under inertial sprung moving load. J. Sound Vib. **216**(3), 507–519 (1998)
6. Yu, L., Chan, T.H.: Recent research on identification of moving loads on bridges. J. Sound Vib. **305**(1–2), 3–21 (2007)
7. Cheynet, E., Jakobsen, J.B., Þór Snæbjörnsson, J.: Full scale monitoring of wind and traffic induced response of a suspension bridge. In: MATEC Web of Conferences, vol. 24, p. 04003. EDP Sciences (2015)
8. Deng, L., Cai, C.: Identification of parameters of vehicles moving on bridges. Eng. Struct. **31**(10), 2474–2485 (2009)
9. Yin, X., Fang, Z., Cai, C., Deng, L.: Non-stationary random vibration of bridges under vehicles with variable speed. Eng. Struct. **32**(8), 2166–2174 (2010)
10. Lalthlamuana, R., Talukdar, S.: Estimation of gross weight, suspension stiffness and damping of a loaded truck from bridge measurements. Struct. Infrastruct. Eng. **13**(11), 1497–1512 (2017)
11. Wang, H., Nagayama, T., Zhao, B., Su, D.: Identification of moving vehicle parameters using bridge responses and estimated bridge pavement roughness. Eng. Struct. **153**, 57–70 (2017)
12. Rosner, B.: Percentage points for a generalized ESD many-outlier procedure. Technometrics **25**(2), 165–172 (1983)
13. Strømmen, E.N.: Structural dynamics. In: Eigenvalue Calculations of Continuous Systems, pp. 89–159. Springer, Cham (2014). ISBN 978-3-319-01802-7
14. Magalhães, F., Cunha, A., Caetano, E.: Online automatic identification of the modal parameters of a long span arch bridge. Mech. Syst. Signal Process. **23**(2), 316–329 (2009)
15. Cheynet, E., Jakobsen, J.B., Snæbjörnsson, J.: Buffeting response of a suspension bridge in complex terrain. Eng. Struct. 128:474–487 (2016)
16. Wang, J., Cheynet, E., Jakobsen, J.B., Snæbjörnsson, J.: Time-domain analysis of wind-induced response of a suspension bridge in comparison with the full-scale measurements. In: ASME 2017 36th International Conference on Ocean, Offshore and Arctic Engineering. American Society of Mechanical Engineers, New York (2017)
17. Jeffcott, H.: Vi. on the vibration of beams under the action of moving loads. Lond. Edinb. Dublin Philos. Mag. J. Sci. **8**(48), 66–97 (1929)
18. Fadnes, T.O.: A full-scale study on traffic induced vibrations of a suspension bridge. Master's Thesis, University of Stavanger, Norway, 2017
19. NPRA: Bridge Projecting Handbook N400 (in Norwegian). The Norwegian Public Roads Administration, 2015
20. EN 1991-1-4: Eurocode 1: Actions on structures–part1-4: general actions-wind actions, 2005
21. Davenport, A.G.: The spectrum of horizontal gustiness near the ground in high winds. Q. J. R. Meteorol. Soc. **87**(372), 194–211 (1961)
22. Shinozuka, M., Deodatis, G.: Simulation of stochastic processes by spectral representation. Appl. Mech. Rev. **44**(4):191–204 (1991)
23. Macdonald, J.H.G.: Identification of the dynamic behaviour of a cable-stayed bridge from full scale testing during and after construction. Ph.D. Thesis, University of Bristol, 2000

Chapter 13
Floor Vibrations and Elevated Non-structural Masses

Christian Frier, Lars Pedersen, and Lars Vabbersgaard Andersen

Abstract Vibration performance of floors in buildings is an issue to consider if vibration sources (internal or external) can cause resonant action and behavior. Modal properties of floors are some of the parameters that determine how a floor reacts when exposed to vibration excitation. On floors in buildings, usually furniture will be present. The mass of furniture on a floor (an added non-structural mass) will influence the natural frequencies of the combined furniture–floor system, and hence how the floor will respond when exposed to vibration excitation. Employing a finite-element model and Monte Carlo simulation, the paper predicts the stochastic nature of floor frequencies for a concrete floor carrying different configurations of non-structural masses. Since in reality, centers of gravity of non-structural masses (furniture) are elevated above the horizontal floor plane, this elevation of mass is implemented in the finite-element model and hence in the predictions of floor frequencies. Other differences in the assumed configuration of the non-structural masses is considered in order to allow comparison between influences of different assumptions.

Keywords Modal parameters of floors · Floor dynamics · Numerical prediction · Serviceability-limit-state · Uncertainty

13.1 Introduction

Floors vibrations can be annoying to users or problematic because the floor carries vibration sensitive equipment. Problematic excitation sources range from external sources such as road or rail traffic [1–3] to internal excitation sources in the form of human activity or machinery operating inside the building [4]. The nature of excitation is different as is the frequencies at which the different sources add energy into the floor structure.

This paper investigates predictions of natural frequencies of a solid reinforced concrete floor carrying non-structural masses (masses not contributing with strength or stiffness to the floor). Non-structural masses would often be present on a floor in the form of furniture and it is this scenario which is considered. For the investigations, the floor is assumed to carry furniture at nine different positions (stations) on the floor at the same time. The presence, size and spatial distribution of these masses would influence the natural frequencies of the floor.

For design-stage predictions there would be uncertainties involved with predicting the total mass of furniture present on the floor when in service, how this mass will be distributed between the nine stations, etc. A central feature of the investigations and calculations of this paper is that they account for the fact that the centers of gravity of furniture and hence the masses of furniture are (in fact) elevated above the floor. This approach is taken because calculations reported in [5–7] have suggested that this may have a considerable impact on the natural frequencies of some eigenmodes of a floor (floor frequencies).

There would also be uncertainty related to the elevated height of the centers of gravity of furniture on the floor, and this uncertainty is also accounted for in the investigations by modelling this parameter as nine statistically independent random variables, one for each station, whereas the sizes of the non-structural masses are modelled as nine correlated random variables that together provide a total mass that is also addressed as a stochastic variable.

As uncertainty is modelled and accumulated, it is the stochastic nature of floor frequencies that is considered in this paper reported in the form of estimates of cumulative distribution functions for floor frequencies. The functions are computed under different assumptions: non-structural masses assumed placed at floor level versus accounting for the actual elevation of the masses, the size of non-structural masses equal or unequal, etc.

C. Frier · L. Pedersen (✉)
Department of Civil Engineering, Aalborg University, Aalborg, Denmark
e-mail: lp@civil.aau.dk

L. V. Andersen
Department of Engineering, Aarhus University, Aarhus, Denmark

The methodology is outlined in Sect. 13.2. The section describes characteristics of the floor and some of the features of the finite-element (FE) model developed for the empty floor and how the presence of non-structural masses is implemented in the FE model. The scenarios in terms of usage of the floor assumed for predicting floor frequencies are described as is the stochastic framework set up for computing floor frequencies and adjoining cumulative probability distribution functions. Section 13.3 presents results, and Sect. 13.4 summarizes the conclusions and discusses the findings.

13.2 Methodology

13.2.1 The Empty Floor

A solid reinforced pin supported two-way-spanning rectangular floor is considered. The thickness (t) of the 8 m by 9 m floor is 180 mm. The floor is assumed to be homogenous and isotropic, and linear elastic behavior is assumed. Young's modulus, E, is 30 GPa and Poisson's ratio, v, is 0.15. The mass density is 2400 kg/m^3.

Biquadratic quadrilateral shell elements [8] are employed for building an FE model of the floor, and a 12-by-12 element grid is used. More details about the FE model of the floor can be found in [5].

13.2.2 The Non-structural Masses on the Floor

For the calculations of this paper, the floor will carry nine lumped non-structural masses. These are assumed positioned at the predefined locations (stations) shown in Fig. 13.1. Each lumped mass, m_k, is allowed to be elevated above the floor with a height above the floor defined by the parameter z_k. For calculations, a variation in the size of the individual lumped masses is also allowed.

In the FE model accounting for the presence of non-structural masses, rigid massless elements are implemented in between the floor and the lumped masses. The lumped masses represent masses of furniture present on the floor. The parameter z_k defines the elevated height of the center of gravity of furniture above the floor.

13.2.3 Scenarios of Floor Usage

As mentioned, the floor area is assumed to carry the nine non-structural masses positioned in the horizontal plane as defined in Fig. 13.1. Having decided on this lay-out plan for the masses, there are still numerous possibilities related to establishing a calculation basis for computing floor frequencies. Decisions need to be made in terms of the vertical positioning of the masses (z_k) and the size of each mass (m_k). The scenarios considered for computations of this paper are outlined below.

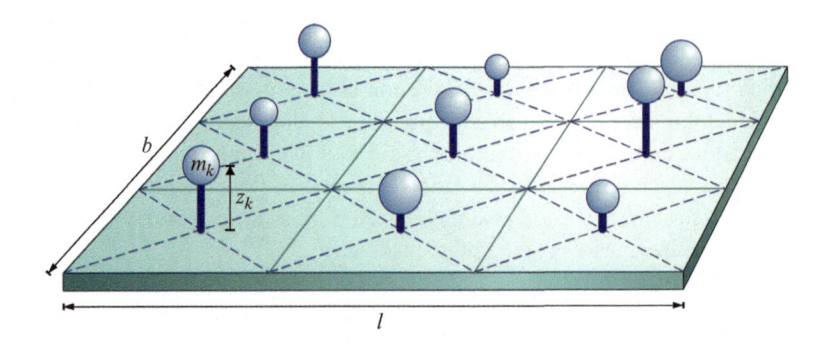

Fig. 13.1 Locations of lumped masses on the floor, and definition of the parameters m_k and z_k

Table 13.1 Assumptions about vertical positions and sizes of the nine non-structural masses

Scenario		Vertical position	Size of the masses
1	A	Not elevated	Equal
	B		Unequal
2	A	Elevated with equal elevation height	Equal
	B		Unequal
3	A	Elevated but unequal elevation height	Equal
	B		Unequal

Scenarios 1A and 1B

The simplest option is to assume that all non-structural masses are positioned at floor level (not elevated above the floor, i.e. $z_k = 0$ for all masses bonding them to the upper side of the floor) in combination with assuming that all masses are equal in size ($m_1 = m_2 = \ldots = m_9$). This scenario is referred to as Scenario 1A.

Scenario 1B accounts for the possibility that the size of the individual non-structural masses might be different, however, still assuming the masses to be positioned at floor level and adding up to the same total mass.

Scenarios 2A and 2B

Study assumptions with elevated positions of the masses are also considered. To this end scenarios in which all masses are elevated with equal elevation height are considered (Scenarios 2A and 2B). The difference between Scenarios 2A and 2B is that in Scenario 2A, the elevated masses are assumed equal in size, whereas in Scenario 2B the possibility of differences in the size of individual masses is considered.

Scenarios 3A and 3B

For completeness, scenarios in which masses are elevated, but not with equal elevation height, are considered as Scenarios 3A and 3B. Again, the scenario ending with an A represents the assumption that all masses are equal in size, whereas the scenario ending with a B considers situations in which the sizes of masses can differ.

The entire set of study assumptions is summarized in Table 13.1.

Scenarios 1A and 1B serve as reference scenarios in that they represent the most simple calculation basis as not at all accounting for the actual elevation of centers of gravity of furniture on the floor.

Scenario 2A represents situations in which a floor carries a set of identical furniture—for instance nine identical bookshelves. The scenario applies if there is an equal mass of books in the nine bookshelves and if this mass is placed in the same manner in all bookshelves (as hereby not only the elevated height of center of gravity but also the mass of the bookshelves including mass of books would be equal). Scenario 2B (instead of 2A) could apply if the nine bookshelves do not carry a similar amount of books or other scenarios where the centers of gravity of the nine non-structural masses are equal but differences exist in their masses.

Scenarios 3A and 3B serve as examples of applications of the floor with different types of furniture occupying the floor, the highest degree of diversity found in Scenario 3B where both the elevation heights of centers of gravity and sizes of masses of furniture changes from one station to the next.

There are different strategies that may be used when equipping a floor area with furniture. Scenario 2A represents the strategy sometimes seen for an office floor (using the same type of furniture at every working station). At the other end of the scale would be Scenario 3B in which elevation heights and sizes of masses are allowed to differ.

For the studies of this paper, a number of the parameters involved defining the calculation basis for determining floor frequencies are defined by random variables. Assumptions made in this context are described below.

13.2.4 Random Variables

The random variables are those addressed below.

The Elevated Height of Non-structural Masses

The parameters z_k, $k = 1, 2, \ldots, 9$, are in some scenarios (2 and 3) modelled as random variables. When this is the case, a uniform distribution is assumed with a lower limit of 0 and an upper limit of 1.5 m.

For Scenarios 2A and 3A, one value of z_k applies to all masses on the floor (i.e. a common value for all masses), whereas for Scenarios 2B and 3B uncorrelated values of z_k are generated for each individual mass on the floor.

The Size of Non-structural Masses

Values representing the size of individual masses, m_k, are obtained following the line of procedure described below.

First the size of the total occupancy mass, M, (corresponding to the sum of the nine non-structural masses) is determined. M is modelled as a random variable described by a Gumbel distribution. This is in line with [9] in which a Gumbel distribution with a variation coefficient of 0.2 is proposed for modelling the imposed load on an office floor. Eurocode 1 [10] suggests a characteristic value (98% quantile of the imposed load) of 2.5 kN/m^2 for an office. These assumptions (and values) are employed for setting up the stochastic model for M (only differing from the corresponding stochastic model for the distributed load (in kN/m^2) by a constant factor).

When a value of M has been sampled from the distribution, the values of m_k are determined by dividing this value with the number of stations (nine). This is the case for Scenarios 1A, 2A, and 3A, whereas for Scenarios 1B, 2B, and 3B (where variability in the sizes of the individual masses are aimed at) uniform distributions are employed for the relative distributions of the individual masses with the constraint that the sum of the nine masses is to equal the a priori sampled value of M.

Having defined the stochastic framework for the random variables involved with modelling the scenarios, Monte Carlo simulation has been performed to map the cumulative distributions functions for floor frequencies (100,000 simulation runs proved sufficient for obtaining converged results).

13.3 Results

In this section, results are presented in terms of the stochastic nature of floor frequencies. As mentioned, this will be done by displaying cumulative distribution functions computed under different study assumptions (different study scenarios). It is chosen to present these functions in the frequency range 0–200 Hz in order to monitor the behavior of a quite wide range of floor frequencies when non-structural mass is added to the floor. This is meaningful and of interest because excitation frequencies potentially causing problematic floor vibrations can encompass such wide frequency range. Attention is on the upper range of the cumulative distribution functions (CDF) (outcomes in the range 0.7–1). It is in this range that most realizations involving the lowest possible values of M are believed to occur, and (already) in this range noticeable diversity in the computed stochastic nature of floor frequencies (relatively between the studied scenarios) is observed and found relevant to discuss.

First, results obtained for Scenario 1 (the reference scenario encompassing Scenarios 1A and 1B) are presented.

13.3.1 Predicted Floor Frequencies with Masses Bonded to the Floor

Figure 13.2 shows CDFs for floor frequencies computed for the two scenarios in which the non-structural masses are assumed bonded to the floor.

First it is useful to mention that for a specific CDF, the highest quantile (the 100% quantile) corresponds to empty floor conditions. Hence, the 100% quantiles correspond to the floor frequencies of the empty floor. However, this the empty floor condition is never realized in simulations, and even the probability of encountering a small non-structural mass, close to the condition of the empty floor, is extremely low due to the modelled stochastic nature of the sizes of the masses. The Gumbel distribution leads to relatively few realizations of low total mass.

Hence, in numerical simulations, a non-structural mass ($m_k > 0$) will be present at all stations explaining why floor frequencies are generally predicted to be lower than the empty floor frequencies. As an example, the floor frequency of eigenmode 5 is 37 Hz but the 70% quantile for this mode is found at 27.5 Hz in Scenario 1A. In scenario 1B, the latter value is 28.0 Hz.

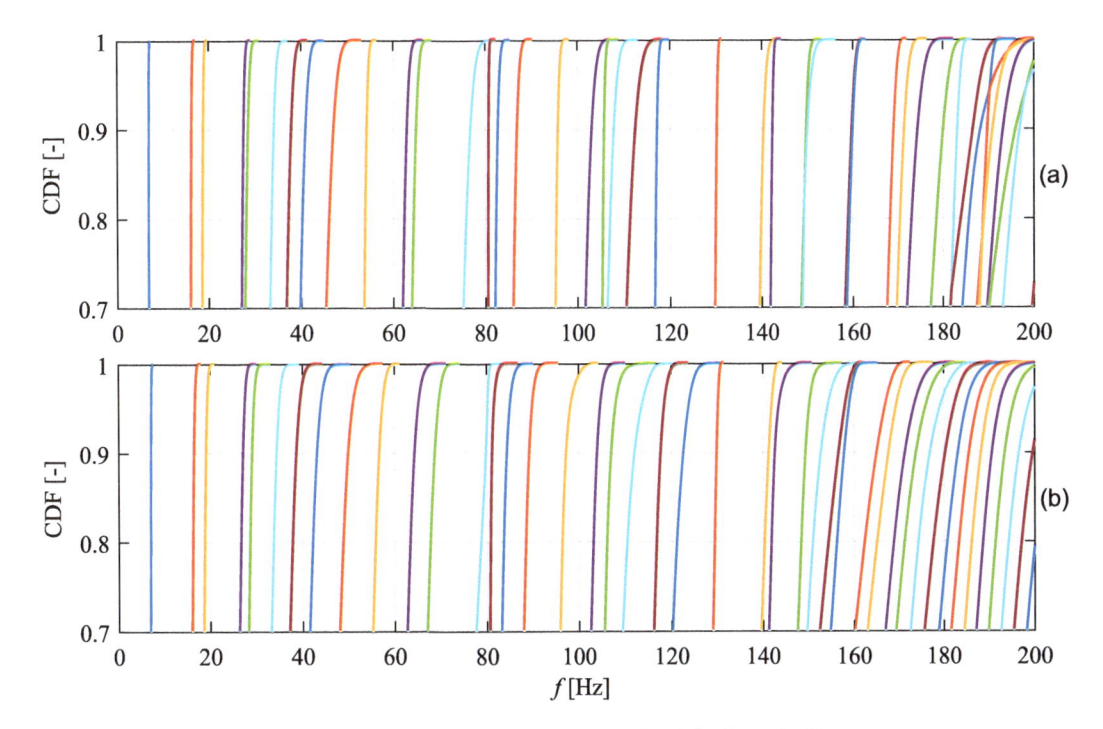

Fig. 13.2 Cumulative distribution functions for floor frequencies: (**a**) Scenario 1A and (**b**) Scenario 1B

In Fig. 13.2, it is difficult to see that those three values apply (37, 27.5 and 28.0 Hz) but they are taken from the data basis employed for drawing the two CDFs for this eigenmode of the floor. The difference between the two 70% is relatively small compared with the difference between the empty floor frequency of this mode (37 Hz) and the two 70% quantiles quantiles (27.5 Hz and 28.0 Hz). This is generally the case for other quantiles too, and for many of the eigenmodes for which CDFs are shown in Fig. 13.2.

This might suggest that for the study assumptions of this paper, the assumption about whether the sizes of the non-structural masses on the floor are equal or unequal have some but not a significant bearing on estimates of CDFs for floor frequencies.

The next section addresses results obtained accounting for elevated positions of non-structural masses allowing a more detailed perspective for a discussion of results.

13.3.2 Predicted Floor Frequencies with Masses Elevated Above the Floor

Figure 13.3 shows CDFs for floor frequencies computed for scenarios in which the non-structural masses are allowed to be elevated above floor level.

It would be obvious, by comparing CDFs presented in Fig. 13.2 with those in Fig. 13.3, that opening for the possibility of elevating the non-structural masses and by modelling the stochastic nature of elevation height as done in this paper, has an effect on the stochastic nature predicted for floor frequencies.

Scenario 2 (2A and 2B) represents situations in which all non-structural masses are assumed elevated in unisome ($z_1 = z_2 = \ldots = z_9$) and Scenario 3 (3A and 3B) represents situations where inequality between elevation height of masses is modelled.

Where scenario 1 suggested rather step declines in CDFs for almost every eigenmode in the region 0.7 to say 0.97, this is not the case especially for some of the eigenmodes in scenario 2. In this scenario, the inclination of the CDFs is quite low particularly for a number of the higher eigenmodes. As an example, the 100% quantile for eigenmode 19 is 126 Hz being the empty floor frequency for this mode regardless of whether scenario 1 or 2 is considered. In scenario 1, the 70% quantile is found to have a value close to 107 Hz whereas assuming scenario 2, the corresponding quantile assumes a value of in the vicinity of 48 Hz.

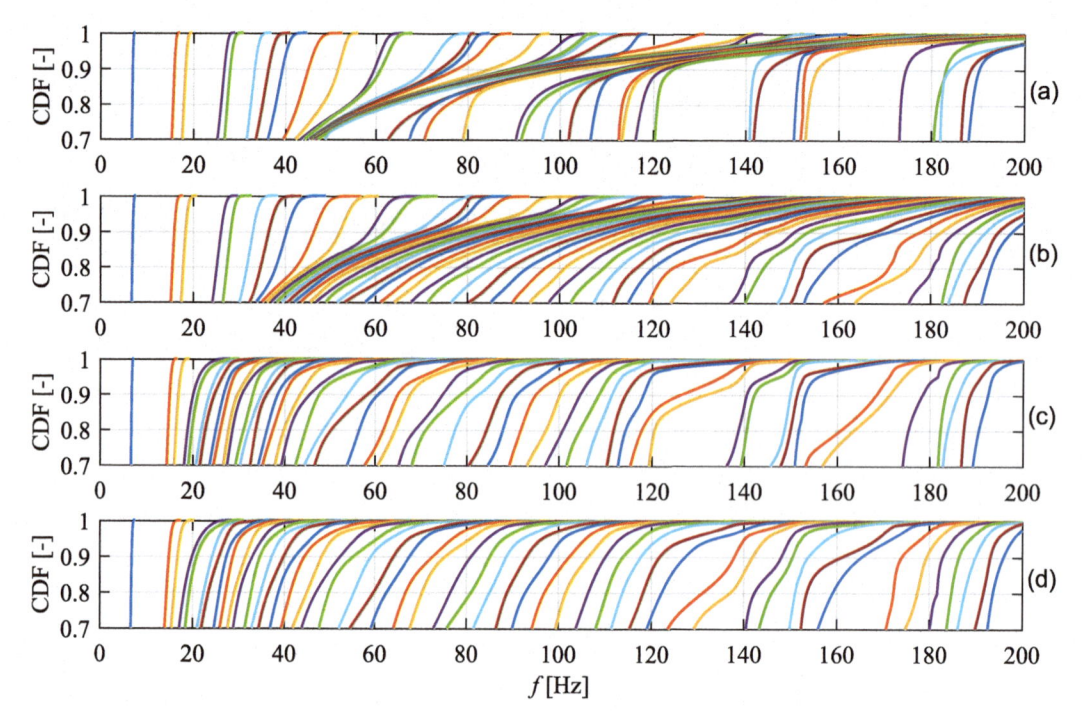

Fig. 13.3 Cumulative distribution functions for floor frequencies: (**a**) Scenarios 2A, (**b**) 2B, (**c**) 3A, and (**d**) 3B

A similar tendency is seen for many other modes and, hence, accounting for elevated positions of the individual masses is seen to have a considerable effect on frequencies of some eigenmodes. Interestingly, a number of CDFs is observed to follow a similar path of decline—at least for some part of their decline. For example, in Scenario 2A, the 70% quantile of the floor frequency is seen to be close to 45 Hz for a significant number of eigenmodes, and the tendency of clustering of CDFs is also seen at other quantiles.

The clustering of CDFs is less significant in Scenario 2B, and it is not observed in Scenario 3, or in Scenario 1 for that matter.

Differences between and similarities in CDFs are subjects for further investigation and discussion in the next section.

13.3.3 Predicted Floor Frequencies for Selected Eigenmodes

Figure 13.4 shows estimates of CDFs computed for the Scenarios 1, 2 and 3 for selected eigenmodes. In this presentation of results it is obvious that some quantiles of floor frequencies are overestimated by assuming the non-structural masses to be bonded to the floor (Scenarios 1A and 1B). As can be seen, there is not a significant difference between the CDFs computed for Scenarios 1A and 1B compared with the differences observed to exist between the functions computed for Scenarios 1, 2 and 3. A similar observation is made when comparing functions obtained for Scenario 2 (Scenario 2A versus Scenario 2B) and scenario 3 (Scenario 2A versus Scenario 2B).

As may be recalled, the difference between Scenarios 1, 2 and 3 is the assumptions made about the elevation height of the masses (from being bonded to the floor (1) over equal elevation height for all masses (2) to unequal elevation height for the masses (3)). Whether 1, 2 or 3 is assumed, is seen to have a higher bearing on predictions of CDF of floor frequencies than whether the individual masses are assumed identical or different (Scenarios A and B, respectively).

For the uppermost quantiles of floor frequency (those in the vicinity of unity regarding the CDFs), it can be observed that Scenario 3 (in which inequality between elevation height of masses is assumed) predicts values for the quantiles lower than those predicted for Scenario 2 (in which equality in elevation height is assumed). For instance, for f_{19} the 90% quantile is close to 50 Hz assuming Scenario 3, whereas it is close to 85 Hz assuming Scenario 2. For this specific quantile, Scenario 1 predicts a value of about 109 Hz. Hence, results obtained assuming Scenarios 2 and 3 differ, although they both model the allowance of non-structural masses to be elevated above the floor.

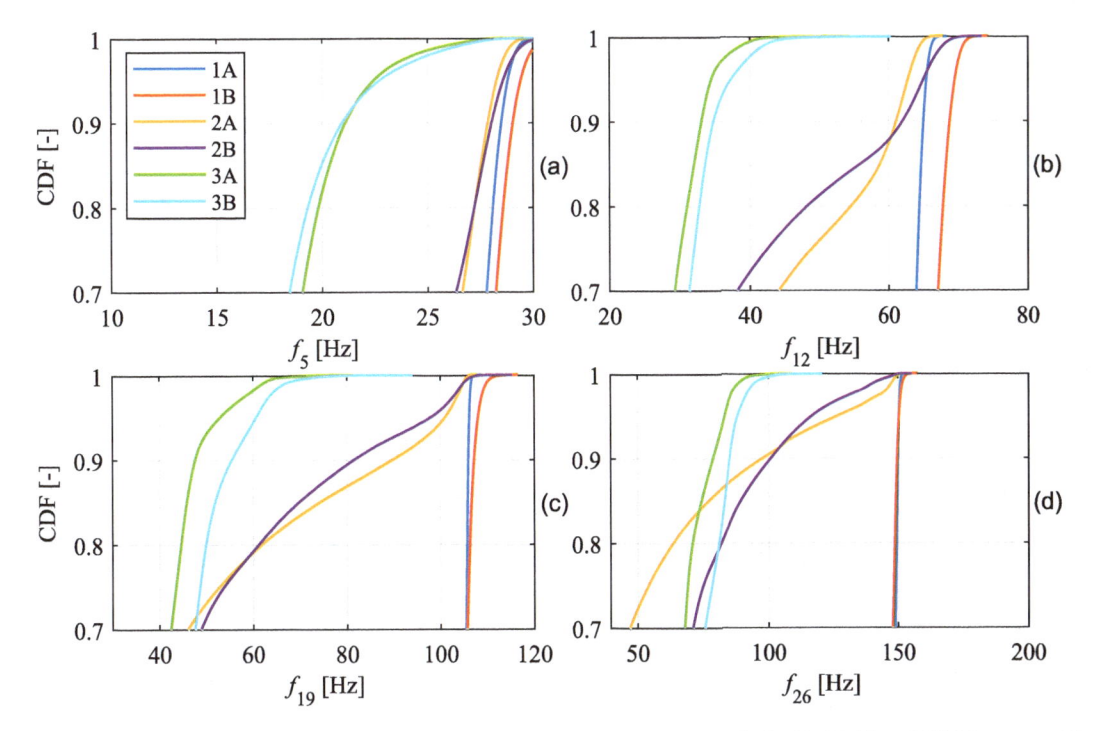

Fig. 13.4 Cumulative distribution functions for floor natural frequencies for eigenmodes: (**a**) 5, (**b**) 12, (**c**) 19, and (**d**) 26 computed for Scenarios 1A, 1B, 2A, 2B, 3A and 3B

13.4 Conclusion and Discussion

By FE calculations and Monte Carlo simulations, the paper has mapped the stochastic nature of floor frequencies in the range 0–200 Hz under six different sets of assumptions in terms of usage of the concrete floor. In all cases, the floor was equipped with furniture at nine stations on the floor. In order to be able to compute cumulative distribution functions (CDF) for floor frequencies, a stochastic model for the total occupancy mass on the floor was assumed. This model was the same for all six scenarios of floor usage considered in the paper.

Still, results showed markedly different CDFs for floor frequencies, especially between scenarios in which centers of gravity of furniture (artificially) were assumed bonded to the floor and scenarios in which these centers of gravity were allowed to be elevated above the floor (as they will be in practice).

A noticeable difference was also observed between CDFs for floor frequencies computed under the assumption that centers of gravity of furniture at all nine stations were equal and under the assumption that they were not.

This observation indicates that the strategy employed for equipping a floor area with furniture (even with predefined positions of furniture in the horizontal plane) can have a rather significant impact on floor frequencies and hence (potentially) on the vibrational behaviour of the floor. As an example, a high degree of equality in terms of elevated heights of centers of gravity of furniture on the floor may result in a clustering of floor frequencies in a quite narrow frequency range.

The results presented in the paper also suggests that how the size of masses of furniture is assumed distributed between stations does not have as high an influence on the stochastic nature of floor frequencies as other variations considered in this paper are found to have.

Conclusions drawn above cannot immediately be transformed to any other type of floor structure than the one studied in this paper.

Acknowledgements This research was carried out in the framework of the project "UrbanTranquility" under the Intereg V program with participation of Aalborg University as well as Aarhus University. The authors of this work gratefully acknowledge the European Regional Development Fund for the financial support.

References

1. Nagy, A.B., Fiala, P., Márki, F., Augusztinovicz, F., Degrande, G., Jacobs, S., Brassenx, D.: Prediction of interior noise in buildings generated by underground rail traffic. J. Sound Vib. **293**, 680–690 (2006)
2. Fiala, P., Degrande, G., Augusztinovicz, F.: Numerical modelling of ground-borne noise and vibration in buildings due to surface rail traffic. J. Sound Vib. **301**, 718–738 (2007)
3. Lombaert, G., Degrande, G., Cloutaeu, D.: Road traffic induced free field vibrations: Numerical modelling and in situ measurements. In: Chouw, N., Schmid, G. (eds.) Wave 2000: Wave Propagation—Moving Load—Vibration Reduction: Proceedings of the International Workshop, Ruhr-University, Bochum, Germany, 13–15 December 2000, pp. 195–207 (1999)
4. Bachmann, H., Ammann, W.: Vibrations in Structures—Induced by Man and Machines, Structural Engineering Documents, vol. 3e. IABSE, Zürich (1987)
5. Pedersen, L., Frier, C., Andersen, L.V.: Flooring systems and their interaction with usage of the floor. In: Caicedo, J., Pakzad, S. (eds.) Dynamics of Civil Structures, vol. 2: Proceedings of the 36th IMAC, A Conference and Exposition on Structural Dynamics 2017, pp. 205–211. Springer, Berlin (2017)
6. Frier, C., Pedersen, L., Andersen, L.V., Persson, P.: Flooring systems and their interaction with furniture and humans. Procedia Eng. **199**, 146–156 (2017)
7. Frier, C., Pedersen, L., Andersen, L.V.: Non-structural masses and their influence on floor natural frequencies. In: Pakzad, S. (ed.) Dynamics of Civil Structures, vol. 2: Conference Proceedings of the Society for Experimental Mechanics Series, pp. 59–65. Springer, Cham (2019)
8. Ahmad, S., Irons, B., Zienkiewicz, O.: Analysis of thick and thin shell structures by curved finite elements. Int. J. Num. Meth. Eng. **2**, 419–451 (1970)
9. NKB/SAKO: Basis of design of structures, proposal for modification of partial safety factors in Eurocodes. 1999:01 E, Nordic Council of Ministers, Oslo (1999)
10. EN 1991-1-1 Eurocode 1: Actions on structures. Part 1-1 General actions. Densities, selfweight, imposed loads for buildings, CEN (2002)

Chapter 14
Vibration Performance of a Lightweight FRP Footbridge Under Human Dynamic Excitation

Stana Živanović, Justin M. Russell, and Vitomir Racic

Abstract Fibre-reinforced polymer (FRP) composites are increasingly used as main load bearing materials in design of pedestrian bridges. The FRP footbridges are typically characterised by high strength, and relatively low mass and stiffness. These properties could lead to excessive vibration response under human-induced dynamic loading. This paper studies dynamic performance of a 19.8 m long, simply supported, FRP footbridge exposed to walking and jogging. Moreover, the vibration response of this bridge is compared and critically evaluated against the response of an equivalent, in terms of natural frequency and span length, composite steel-concrete structure. The main factors that drive the vibration performance of the FRP structure are discussed and some recommendations for vibration serviceability checks are made.

Keywords FRP composites · Footbridge · Walking · Jogging · Vibration

14.1 Introduction

Fiber-reinforced polymer (FRP) structures are known to be responsive to dynamic excitation due to their relatively low weight and stiffness. This paper investigates responsiveness of an FRP shallow-truss footbridge structure having 19.8 m × 2.1 m deck area. The bridge spans 16.8 m, and it has the first vertical flexural vibration mode at 2.53 Hz. The modal mass for this mode is 650 kg and damping ratio is 1.69% [1]. The performance of this bridge, referred to FRPB hereafter, will be compared against a classical composite steel-concrete bridge, referred to as CSCB, of similar size and natural frequency. The CSCB has a deck area of 19.9 m × 2.0 m. It spans 16.2 m. The fundamental vertical flexural vibration mode has natural frequency of 2.44 Hz, modal mass of 7700 kg and damping ratio around 0.45% [2]. Both bridges are slender simply supported structures with overhangs on both sides. The mode shape for the vertical flexural mode that is studied in this paper is a typical wave with movement between the supporting points being 180° out-of phase with the movement of overhangs [1, 2]. The slenderness (span length to truss height or cross section depth) ratio is 35 for FRPB and 46 for CSCB. The two bridges are shown in Fig. 14.1. Next two sections explain and compare performance of the two bridges under pedestrian and jogger excitation, respectively. This is followed by summary of main conclusions from the study.

14.2 Walking

Vibration response to dynamic excitation by a person walking was calculated using a stochastic model of the dynamic force developed by Racic and Brownjohn [3]. The excitation by one person at a time only was considered in this paper. Pacing frequencies starting from extremely low 1.2 Hz to extremely high 2.6 Hz were considered (in frequency steps of 0.1 Hz). For each frequency 100 forces with different waveforms were generated to represent intra-subject variations. The mean and standard deviation for peak vibration response at the mid-span for each pacing frequency was found. The weight of the

---

S. Živanović (✉)
College of Engineering, Mathematics and Physical Sciences, University of Exeter, Exeter, UK
e-mail: s.zivanovic@exeter.ac.uk

J. M. Russell
School of Engineering, University of Warwick, Coventry, UK

V. Racic
Department of Civil and Environmental Engineering, Politecnico di Milano, Milan, Italy

Fig. 14.1 Photographs of the two footbridges. Left: FRPB. Right: CSCB

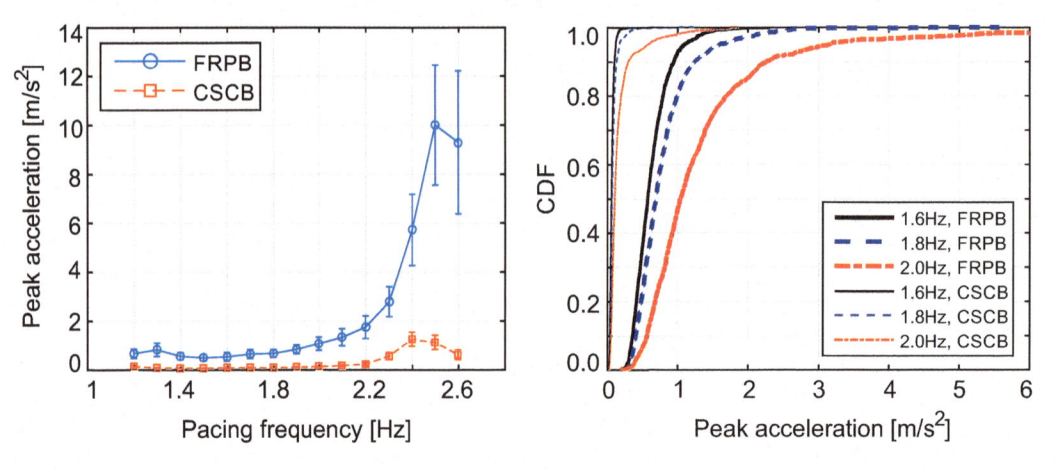

Fig. 14.2 Walking excitation: Left: Peak acceleration (mean ± STD) vs. pacing frequency. Right: CDF for peak acceleration responses

pedestrians is assumed to be 700 N; this parameter has not been treated as random in order to isolate the influence of the most significant locomotion parameter—pacing rate—on the vibration response. The influence of the pedestrian weight is simple to gauge later on, if necessary, due to the response being directly proportional to the pedestrian weight.

The left part of Fig. 14.2 shows the mean values of the peak response as a function of pacing frequency for the two bridges, as well as mean ± STD, where STD represents the standard deviation of the peak response. Both bridges are most vulnerable to pacing rates between 2.4 and 2.6 Hz since the forces in this range are either close to the natural frequency or matching it. The FRPB is exceptionally responsive, with the strongest mean peak acceleration of 10 m/s^2 overcoming the acceleration of gravity. The strongest mean response of CSCB is comparatively much lower at 1.2 m/s^2. The main cause of the difference in the response is that the modal mass of the FRPB is one order of magnitude lower than for CSCB. The differences in damping ratio and natural frequencies further combine to influence the response, but this influence is relatively small compared with the influence of the modal mass.

Whilst the walking at a frequency close to the natural frequency results in the strongest response of the two bridges, these frequencies are not frequently occurring in human population. To account for the probability of walking at any particular frequency, the vibration response was simulated for 1000 individuals. Pacing frequency for each person was drawn from a normal distribution N (1.6 Hz, 0.2 Hz), where 1.6 Hz and 0.2 Hz represent the mean value and the standard deviation, respectively. This particular distribution represents a relatively slow pedestrian population, which might be encountered on a bridge in, say, a leisure park. In addition, populations with two additional pacing distributions: N (1.8 Hz, 0.2 Hz) and N (2.0 Hz, 0.2 Hz), where the latter represent a population in hurry, are considered. A cumulative distribution function (CDF) for the peak responses on both bridges are shown in the right part of Fig. 14.2. As expected, when the mean pacing frequency is closer to the structural natural frequency, the response is larger. However, the probability of exciting the resonance on

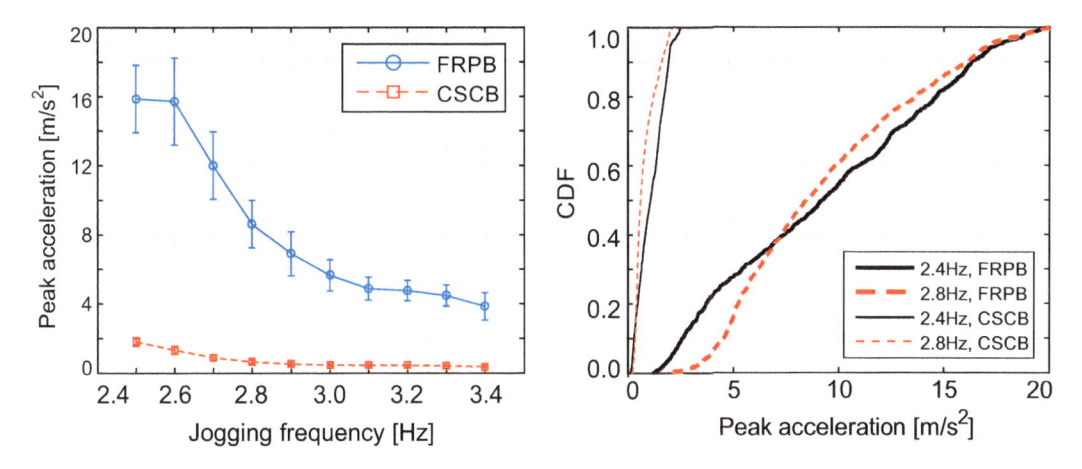

Fig. 14.3 Jogging excitation: Left: Peak acceleration (mean ± STD) vs. jogging frequency. Right: CDF for peak acceleration responses

FRPB even in the case of the fastest pedestrian population is still not very large, and despite maximum response exceeding 10 m/s^2 (these values are not presented in the figure), the probability of exceeding, say, 2.5 m/s^2 is less than 10%.

14.3 Jogging

Similar information to that in Fig. 14.2 is shown in Fig. 14.3 in relation to single jogger's excitation. The jogging force is generated using the model developed in [4]. The jog frequency is varied from 2.5 to 3.4 Hz, and 100 individuals are simulated for each case. The response of FRPB is again one order of magnitude larger than that of CSCB. The response even to non-resonance frequencies is high and at least 4 m/s^2 due to higher energy of the jogging action compared with walking. Considering population of 1000 joggers (traversing the bridge one at a time) with the jog frequency drawn from N (2.4 Hz, 0.25 Hz) and N (2.8 Hz, 0.25 Hz), the CDFs (on the right-hand side of the figure) illustrate that the large majority of joggers would cause vibrations larger than even the most relaxing acceleration limit of 2.5 m/s^2.

14.4 Conclusions

The paper compares responses of an FRP and a non-FRP bridge to pedestrians and joggers. The response in the first vertical flexural mode is shown only. It demonstrates that for all loading scenarios considered, FRPB is one order of magnitude more responsive than CSCB, primarily due to having modal mass that is almost 12 times lower. Having damping ratio that is 3.8 time larger is a positive feature, but it, in comparison, has less influence on the vibration response. The results show the importance of considering probability of occurrence for pacing rates. In addition, the FRP bridges are more vulnerable to jogging excitation even when the resonance does not occur. This type of analysis should, in general, be conducted for all vibration modes that are of interest. Responsiveness of the modes above 5 Hz should also be checked, since they are likely to be excited by weaker (than first) harmonics of the excitation force due to low modal masses of FRP structures. Note that both bridges studied here have been designed to be lively for research purposes, and the actual structures of similar span are almost certain to vibrate less. Still comparison of the relative behavior is revealing. Given the high vibration levels generated on the FRPB, it is almost certain that both pedestrians and joggers would interact with the structure in some way. The actual vibration of this bridge, therefore, might be different from the predications presented in this study. The next step in this research is to quantify if and how the experimental responses differ from those predicted, and develop a model to describe the interaction phenomenon.

Acknowledgements This research work was supported by the UK Engineering and Physical Sciences Research Council [grant number EP/M021505/1: Characterising dynamic performance of fibre reinforced polymer structures for resilience and sustainability].

References

1. Russell, J.M., Mottram, J.T., Živanović, S., Wei, X.: Design and performance of a bespoke lively all-FRP footbridge. Proceedings of IMAC-XXXVII, Orlando, Florida, 28–31 January (2019)
2. Dang, H.V., Živanović, S.: Influence of low-frequency vertical vibration on walking locomotion. J. Struct. Eng. **142**(12), 04016120 (2016)
3. Racic, V., Brownjohn, J.M.W.: Stochastic model of near-periodic vertical loads due to humans walking. Adv. Eng. Inform. **25**(2), 259–275 (2011)
4. Racic, V., Morin, J.B.: Data-driven modelling of vertical dynamic excitation of bridges induced by people running. Mech. Syst. Signal Process. **43**(1–2), 153–170 (2014)

Chapter 15
A Study of Suspension Bridge Vibrations Induced by Heavy Vehicles

Jonas Thor Snæbjörnsson, Thomas Ole Messelt Fadnes, Jasna Bogunović Jakobsen, and Ove Tobias Gudmestad

Abstract The Lysefjord suspension bridge (Norway) is instrumented with accelerometers, anemometers, weather station and a GPS monitoring system, making it a full-scale test bed for research on dynamic bridge behavior and wind characteristics in a complex terrain. In this study we investigate bridge vibrations induced by heavy vehicles crossing the bridge. As the bridge is located on the outskirts of the urban environment, the traffic across the bridge is generally intermittent, allowing for investigating loading effects of individual heavy vehicles during days with low wind velocity. The findings suggest that heavy vehicles cause a considerable dynamic load effects, especially when entering and exiting the bridge, especially at higher modes of vibration with natural frequencies above 1 Hz.

Spectral analysis of the traffic induced acceleration time series agree with results from previous studies showing that the acceleration bridge response to vehicles consists of a combination of both low and higher frequency modes. Truck induced excitation has been used to estimate the modal parameters of the bridge, such as critical damping ratios for the vertical modes of vibration assuming a viscous-equivalent damping and a logarithmic decay of the bridge motion as the trucks exit the bridge. The identified damping ratios as well as natural frequencies are found to be comparable to values obtained from general ambient vibration data. However, there are significant variations in the results depending on measurement location and the driving pattern of the truck along the bridge.

Keywords Suspension bridge · Traffic · Vibrations · Acceleration · Damping

15.1 Introduction

Norway's coastline consists of long and deep fjords amongst mountainous terrain. This leads to challenges within road and transportation systems along the coast. Ferry transportation has been the solution for crossing the fjords which are up to 5 km wide and 1300 m deep. In 2012 the Norwegian Parliament presented a directive towards a ferry free coastline road (E39) between Kristiansand and Trondheim by 2030 [1]. In support of this goal, the Norwegian Public Roads Administration (NPRA) has been funding research work related to full scale monitoring of existing suspension bridges. The Lysefjord suspension bridge is instrumented with accelerometers, anemometers, weather station and a GPS system, making it a full-scale test bed for research on bridge dynamics and wind characteristics in a complex terrain [2]. A "full scale laboratory" of this kind provides the opportunity to study the characteristics and effects from wind in complex terrain on a suspension bridge, as well as allowing for an investigation into loading effects and response from the automobile traffic crossing the bridge.

The focus of the present work is to study bridge vibrations induced by individual, heavy vehicles during days of low wind velocity. These vehicles generate a forced vibration of the bridge over a broad frequency range, which can be used to estimate the modal parameters of the bridge. Then as the heavy vehicles exit the bridge, the bridge is left to swing freely without any other significant external loading and by filtering out the identified modal components of free vibrations, the critical damping ratios can be estimated. Heavy vehicles entering and exiting the bridge at high speed can also induce impulse loading on the bridge as they cross over uneven pavement surface of varying quality or the flexible joints between the approach road and

---

J. T. Snæbjörnsson (✉)
Department of Mechanical and Structural Engineering and Materials Science, University of Stavanger, Stavanger, Norway

School of Science and Engineering, Reykjavik University, Reykjavík, Iceland
e-mail: jonasthor@ru.is

T. O. M. Fadnes · J. B. Jakobsen · O. T. Gudmestad
Department of Mechanical and Structural Engineering and Materials Science, University of Stavanger, Stavanger, Norway

bridge, which may also include changes in the road geometry. The nature of the bridge response to impact loads of this type is known to depend on several parameters, both related to the bridge and the vehicle. Studying the bridge response caused by different vehicles traveling at different velocities can lead to a better understanding of these loading effects.

15.2 Background

Suspension bridges are a common type of cable supported bridge. They vary in length, with the longest main span being 2000 m at present. As the concept is proven and fairly well understood, it is one of the most feasible alternatives for infrastructure development such as the coastal highway E39 project along the west coast of Norway, where several fjords are to be crossed.

The Lysefjord Bridge was initially instrumented in 2013. In numerous studies undertaken since then, using the recorded full-scale data, the focus has mainly been directed towards the study of turbulence characteristics and the associated wind-induced structural response [3, 4]. Traffic-induced vibrations were briefly studied in [5], but mainly to understand and explain discrepancies in the buffeting response estimation. Further work on studying the frequency content of traffic induced vibrations of Lysefjord Bridge are found in [6], however, therein the focus is mainly on developing methods and criteria to eliminate time series dominated by traffic induced response in order to improve buffeting analysis results. In other words, extensive analysis of wind effects on the Lysefjord Bridge have been performed, utilizing both computational models and full-scale data, but detailed analysis focusing on traffic induced vibrations are still relatively unexplored, although in [7] the focus is on modelling traffic induced vibrations in order to identify the speed and weight of the vehicles crossing the bridge. In fact, the authors are unaware of any other full-scale studies on the effects of individual heavy vehicles crossing a suspension bridge available in the literature.

Ambient vibration measurements (AVM) of wind- and traffic-induced vibrations of suspension bridges are commonly used to evaluate bridge dynamics [8, 9]. The common objective of the field investigations reported in the existing literature is generally to identify and/or validate the natural frequencies of the bridges. Kim et al. [10] studies the effect of vehicle mass from heavy traffic on natural frequencies off three types of bridges, including a long span suspension bridge, and concludes that the vehicle mass has little effect on the natural frequencies.

Regarding the estimation of the damping ratios, Brownjohn [11] discusses different damping mechanisms and methods of analyzing physical test data and related accuracy for use in dynamic analysis of suspension bridges of different designs and span lengths. According to Brownjohn [11], forced vibration test (FVT) is considered to produce the best results in terms of quality of experimental data. One method for FVT used for relatively short span suspension bridges is described by Selberg [12], where 15–20 men jumped continuously in rhythm with a given natural frequency, coming to an immediate stop as the amplitude had reached its maximum and then the free decay oscillations were measured to estimate damping. As described previously, the idea of the present paper is to use traffic-induced vibration as a substitute for FVT, to estimate natural frequency and structural damping. For the Lysefjord Bridge, the aerodynamic and/or total damping has been studied based on wind induced response in [4], assuming a fixed structural damping ratio of 0.005 for every mode, although the analysis suggests some deviations from this value.

Mathematical and statistical analytical models for traffic-induced response of suspension bridges are presented in [13, 14], demonstrating various effects on the bridge response resulting from vehicle related parameters such as travelling speed, damping and spring force in the suspension system as well as the surface roughness of bridge deck. The bridge parameters presented in [13] are considered comparable to the Lysefjord Bridge and are used as basis for comparison with full-scale observations of heavy vehicle impact loading effects by Fadnes [15].

15.3 Instrumentation

The Lysefjord Bridge is a two-lane suspension bridge with a main span of 446 m approximately 55 m above sea level, and towers ranging approximately 102 m tall. Figure 15.1 gives an overview of the instrumentation on the bridge at the time of this study (winter-spring, 2017), showing the location of the six sonic anemometers, the Vaisala weather station and the GPS monitoring system, installed on the bridge. In the study presented herein, only the acceleration data is used. Four pairs of triaxial accelerometers are installed inside the bridge deck at four different locations along the span. Each pair has a sensor positioned near the hangers/edge of the deck, one on the east and one on the west side of the bridge, to capture the torsional response of the bridge. The positioning of the accelerometer pairs along the length of the bridge enables the capturing of both symmetrical and non-symmetrical modes of vibration. The sampling rate is 50 Hz, which is then decimated to 25 Hz in

15 A Study of Suspension Bridge Vibrations Induced by Heavy Vehicles

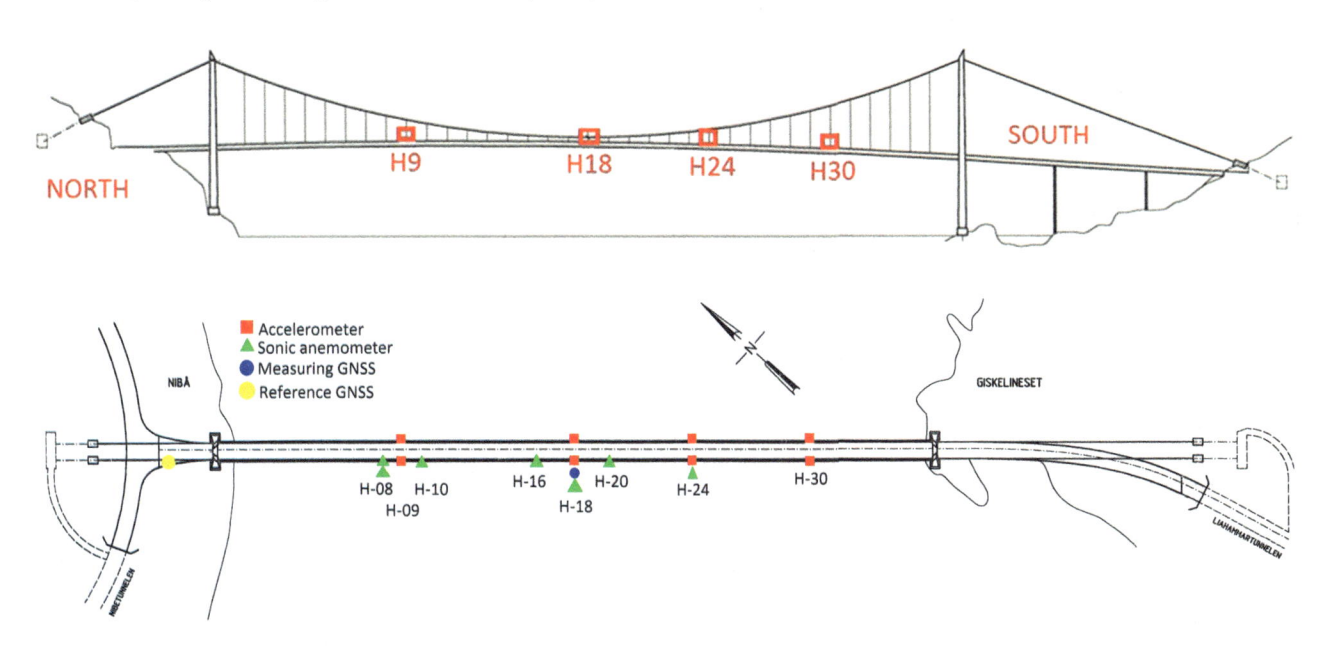

Fig. 15.1 The profile (top) and plan (bottom) of the Lysefjord Bridge, showing location of the installed instrumentation. The instrument locations are referred to a specific hanger number, i.e. H8–H30

Fig. 15.2 A screenshot from a video recording of the bridge during traffic monitoring in April 2017

the post-processing of the data. During the bridge observation periods studied herein, a video camera located about 1.75 km west of the bridge was used to constantly monitor the bridge span with vehicles passing. After each recording period, the videos are thoroughly reviewed, and individual recordings are manually synchronized in time with the recorded acceleration data, to allow an accurate correspondence between the response and the videos of the vehicles crossing the bridge. Figure 15.2 is a screenshot illustrating the video logging of the recorded traffic.

As the focus of this study is on traffic induced vibrations only, the acceleration data collection is limited to periods of low wind speeds. The registration of passing vehicles, requires the observer to be present at the location during the relevant period of data collection. Due to these practical limitations the data used herein is limited to 7 h of relevant data, recorded on three different occasions.

15.4 Response Analysis

The modal parameters are identified using the Frequency Domain Decomposition (FDD) [16]. The FDD relies on the single value decomposition of the matrix of auto and cross-power spectral density of the acceleration response to white noise excitation. If desired, damping and undamped natural frequencies can then be obtained by taking the equivalent single degree of freedom system representation back to the time domain by applying inverse Fourier transform. To identify the eigen-frequencies of the bridge deck, the peak picking method of the first singular values of the PSD matrix of the response is used.

The identified natural frequencies for key torsional and vertical modes of vibration are given in Table 15.1. The abbreviations VA, VS, TA, TS, HA and HS stands for: Vertical Asymmetric, Vertical Symmetric, Torsional Symmetric, Torsional Asymmetric, Horizontal Asymmetric and Horizontal Symmetric, respectively, and refer to the shape and direction of the bridge deck modes of vibration.

It is observed (see Fig. 15.3) that traffic excites several modes of vibration, across a broad frequency range at similar energy levels. In contrast, the wind mainly excites the low frequency modes of vibrations, where the main energy of the wind action is found, whereas at higher frequencies the wind possess less energy than traffic loading [3, 6]. It is also seen from this study that traffic excites vertical modes of vibration more effectively than torsional and horizontal modes of vibration.

To analyze the individual modal parameters, it is desirable to isolate the vibration at the natural frequencies of interest. In the present work, this is done by band-pass filtering the frequency band of interest using an eighth order Butterworth filter. The upper and lower cut-off frequencies are specified between 5 and 10% higher and lower than the target frequency, respectively.

It is also necessary to look at the modal acceleration response time histories for different vehicles arriving from different directions. The acceleration response from a set of chosen vehicle crossings are analyzed and illustrated to gain a better understanding of the vehicle induced bridge response. This is a preparation step for further analysis. Table 15.2 presents an overview of the vehicle observations from March 7, 2017. Data recorded on the bridge during these crossings are a part of the total data sample used for the response analysis.

The vehicle-induced bridge response is illustrated in Fig. 15.4 for every event listed in Table 15.2. From the events listed in Table 15.2, event no 1 (case 1) and 10 (case 2) are studied in more detail for illustration purposes.

Table 15.1 Identified natural frequencies for key torsional and vertical modes of vibration

Mode	Identified frequency [Hz]	Mode	Identified frequency [Hz]
VA1	0.22	TA1	1.24
VS1	0.29	TS1	2.19
VS2	0.41	TS2	3.24
VA2	0.59	TA2	3.96
VS3	0.85	TS3	4.29
VA3	1.16	TA3	4.78

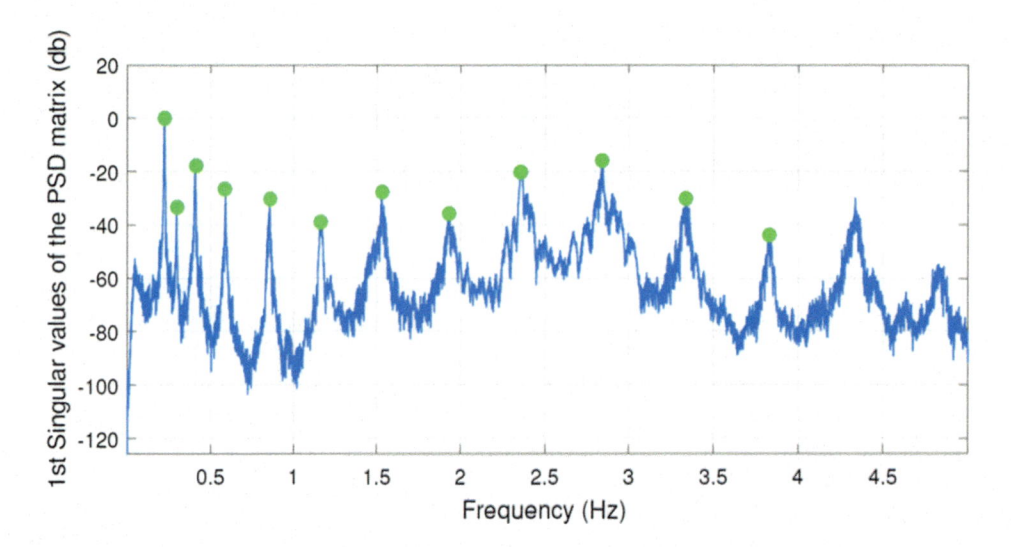

Fig. 15.3 PSD estimate of a traffic induced vertical acceleration response, showing the identified "peak-picking" frequency values

Table 15.2 Vehicle crossings on March 7th 2017

Event number	Time on bridge [HH:MM:SS]	Time off bridge [HH:MM:SS]	From direction enters bridge	Vehicle type (rough description)
1	12:40:18	12:40:46	North	Truck
2	12:43:49	12:44:56	North and South	Two trucks
3	12:56:12	12:56:41	South	Small Truck
4	13:01:45	13:02:09	South	Small Truck
5	13:12:08	13:12:28	South	Small buss
6	13:17:32	13:18:04	South	Buss
7	13:29:33	13:30:00	South	Truck
8	13:42:11	13:42:41	North	Truck (15 s after another truck)
9	13:46:40	13:47:13	North	Truck
10	14:05:55	14:06:16	South	Truck

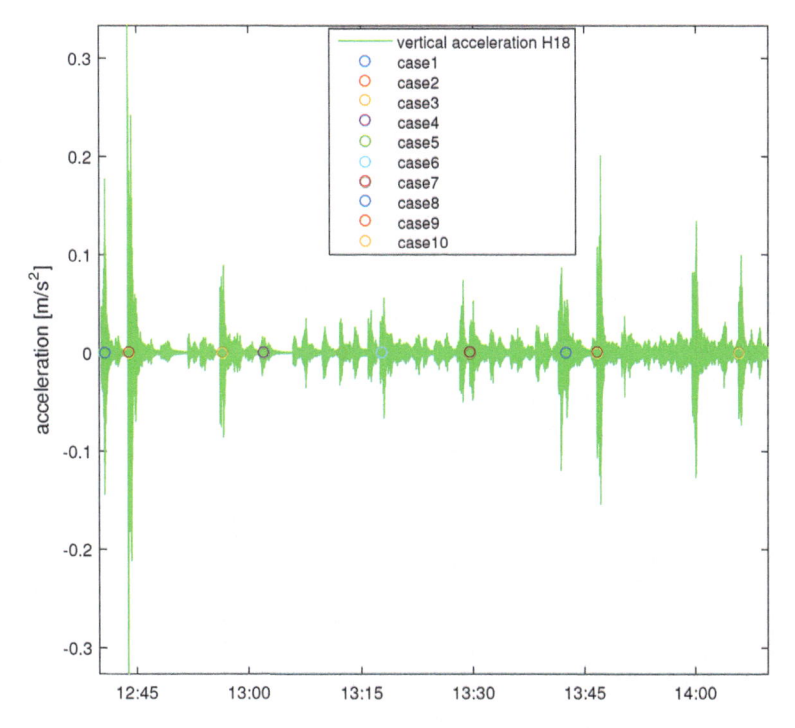

Fig. 15.4 Overview of unfiltered vertical acceleration response at mid span for the cases illustrated in Table 15.2. Case number in the legend refers to an event number in Table 15.2

In case 1, a garbage truck enters the bridge from the North side. Due to the intersection right before the bridge, the truck enters the bridge at a relatively low velocity (approximately 30 km/h), but accelerates to approximately 60 km/h during the first quarter span of the bridge. There are no vehicles on the bridge before, during or within 1 min and 22 s after the vehicle has exited the bridge. In case 2, a large postal truck enters the bridge from the South side of the bridge. The vehicle enters the bridge at a velocity approximated at 73 km/h and keeps this velocity until it reduces its velocity quite significantly during the last 50 m of the free span before it exits the bridge, as it has to turn either to right or left after leaving the bridge. There is one small vehicle on the bridge when the truck enters the bridge. Another small vehicle enters the bridge as the truck exits and then another about 20 s later.

The filtered acceleration response from mid-span is shown for mode VS3 for both Case 1 (event 1) and Case 2 (event 10) in Fig. 15.5. The dots representing "Case 1 on", "Case 2 on", "Case 1 off" and "Case 2 off" illustrate the time when the trucks in each case enter and exit the bridge, respectively.

For mode VS3, shown in Fig. 15.5, the maximum modal response in both cases is reached as the vehicles exit the bridge. The response amplitude in Case 2 is lower than for Case 1, which may be caused by the fact that the vehicle in Case 2 slows down as it exits the bridge. It is also noteworthy that the response, reaches its minimum as the vehicles are positioned approximately at the mid-span of the bridge, this could be an indication that the vehicle acts as a tuned-mass damper for this

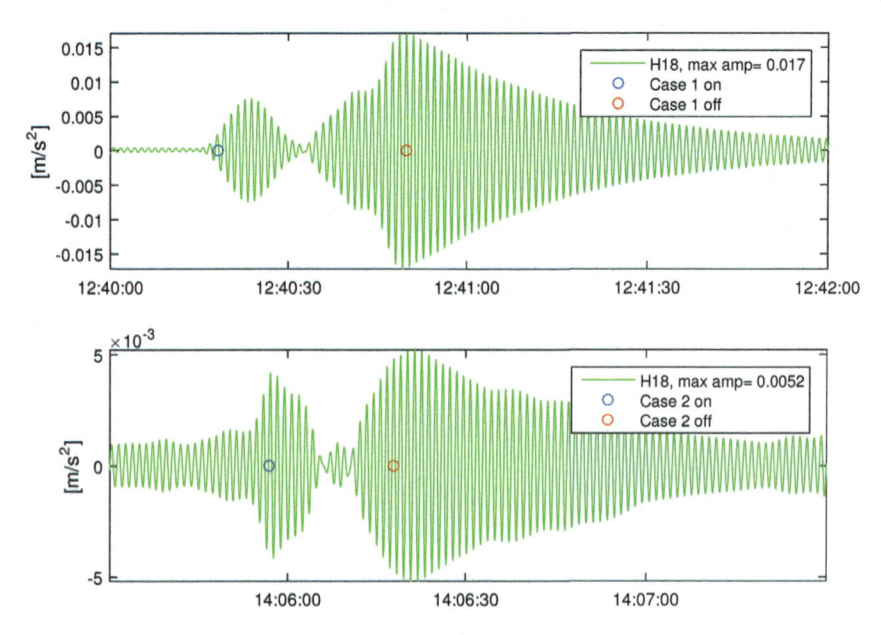

Fig. 15.5 Acceleration response at mid-span, band-pass filtered around 0.86 Hz (mode VS3). Top: Case 1—Truck approaching from North. Bottom: Case 2—Truck approaching from South

Table 15.3 Overview of maximum unfiltered acceleration amplitudes for the three trips made by the 50-ton truck over the bridge

Trip #	Vehicle velocity (km/h)	Entering from	H9 $[m/s^2]$	H18 $[m/s^2]$	H24 $[m/s^2]$	H30 $[m/s^2]$
1	55	South	0.29	0.21	0.23	0.25
2	30–50	North	0.38	0.30	0.43	0.30
3	23	South	0.17	0.20	0.14	0.17

mode of vibration (VS3) as it passes a certain part of the bridge. After the trucks have left the bridge a free vibration decay is observed. For Case 2, the decay is clearly distorted by the smaller vehicles entering the bridge as the truck exits.

On May 5, a 50-ton truck from a concrete factory in a nearby town, was driven over the bridge in a controlled fashion, at different speeds. The truck made three separate trips over the bridge. Table 15.3 summarizes the maximum acceleration values for unfiltered acceleration recorded during three separate trips of this 50-ton truck. The comparison between trip 1 and 2 indicates that the driving speed is a key parameter for the overall response amplitude. Unfortunately, trip 3 was disturbed by secondary traffic on the bridge, so the effect of low speed is partially inconclusive.

Figure 15.6, shows modal acceleration for modes VA1 and VA3 evaluated by band-pass filtering the response recorded at hanger 9 during Trip 1 with the 50-ton truck. The figure demonstrates the differences between individual contribution of the different modes, both with regard to frequency, amplitude and decay of free vibration. The modulation or clustering seen in the higher modes such as VA3 (Fig. 15.6) and VS3 (Fig. 15.5) is not seen in the lower modes i.e. VA1 and VA2. This may indicate that the vehicle-bridge interaction mainly occurs at frequencies around or above 1 Hz. It is also of interest that even though the truck enters from South, and Hanger 9 is close to the North end of bridge, the asymmetric vibrational modes shown in Fig. 15.6 are excited immediately as the truck enters the bridge. Similar behavior was observed for the other two Trips.

15.5 Damping Analysis

In the present work, the critical damping ratio, is estimated using viscous damping assumption based on free vibration decay. This is done by least-square fitting of an exponential decay function to the peaks of the free decay of filtered oscillations. In each case considered, the following steps are performed in order to estimate the modal damping ratios for each of the vehicles: Identify response frequencies; Find the best method to filter out the frequencies through trial and error; Find exact timing when free oscillation starts; Fit exponential decay function to peaks of free oscillation; Evaluate filtering and exponential decay fitting results.

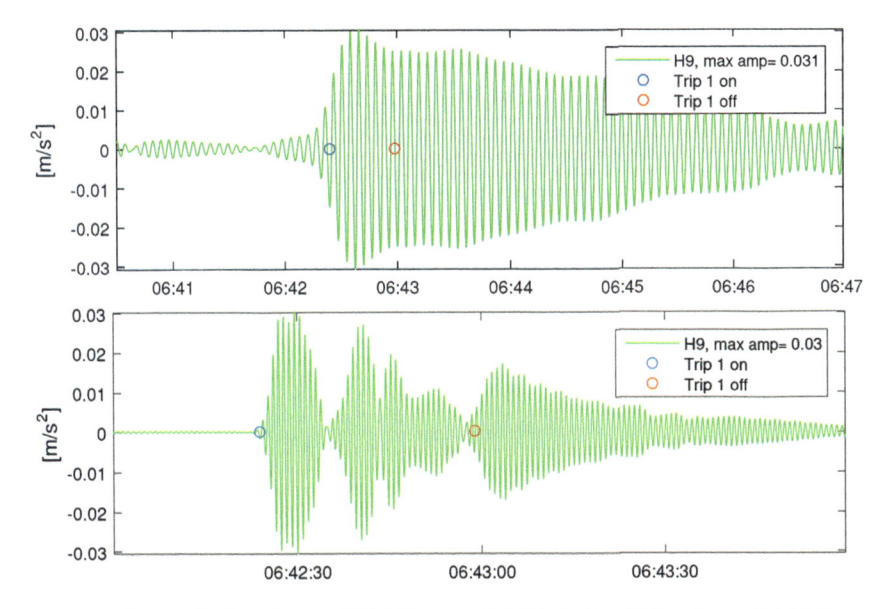

Fig. 15.6 Acceleration response at hanger 9, filtered around 0.22 Hz or mode VA1 (top) and filtered around 1.16 Hz or mode VA3 (bottom), for 50-ton truck entering the bridge from South at 50 km/h (trip 1 in Table 15.3)

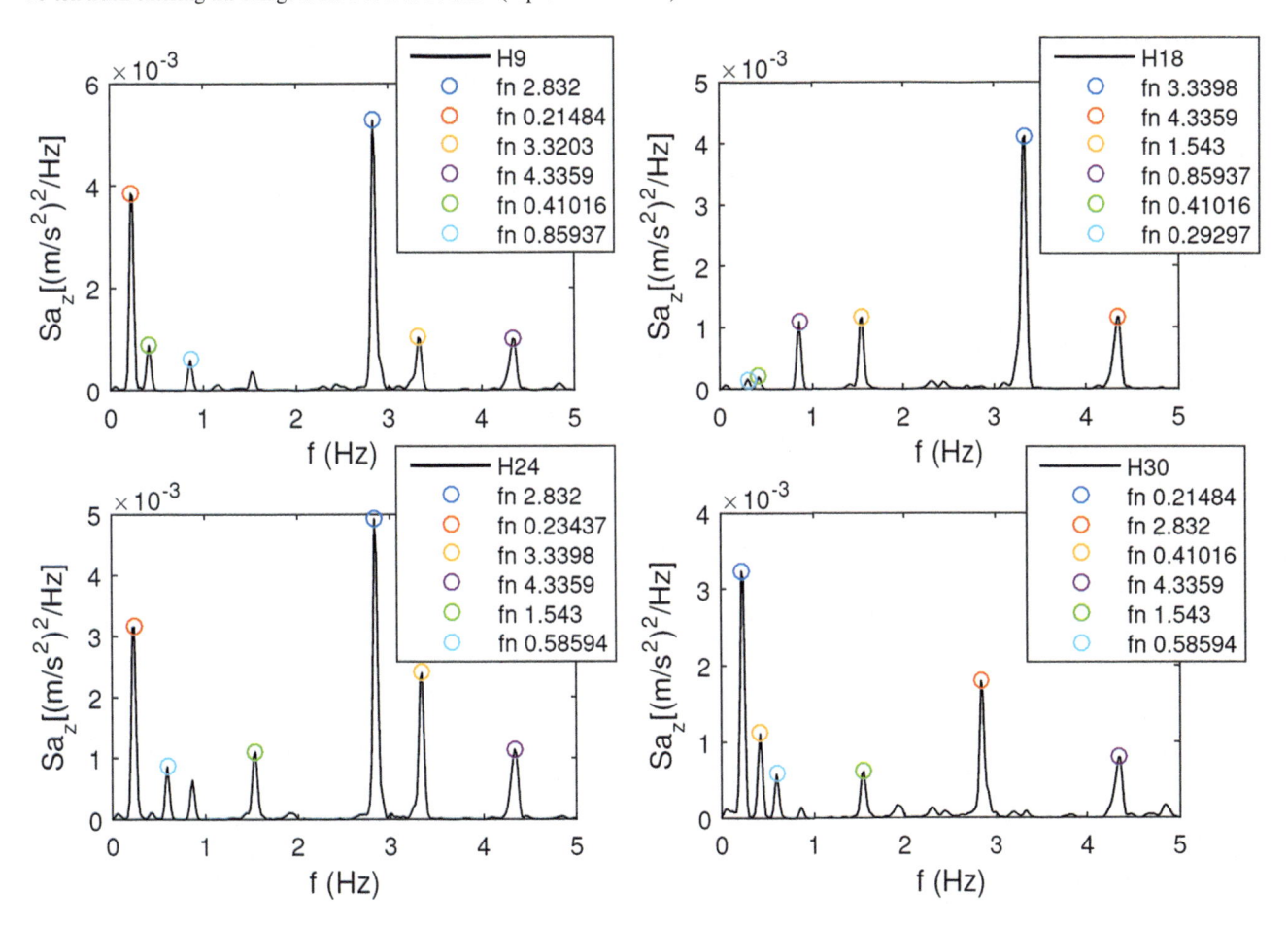

Fig. 15.7 Power spectral density function from individual sensor locations, for an example case, with an unknown truck on the bridge

The natural frequencies are identified using both the FDD method and through spectral analysis for each sensor individually (see Fig. 15.7). All the identified modes of vibration are of interest in terms of damping estimation. The information found through spectral analysis for the cases studied, are further used for isolating the frequencies in the next step.

Table 15.4 Estimated critical damping ratio for the second and third symmetric and antisymmetric modes of vibration

Mode of vibration	VS2 (0.41 Hz)	VA2 (0.59 Hz)	VS3 (0.85 Hz)	VA3 (1.16 Hz)
Mean value of damping	0.56	0.54	0.52	0.56
Standard deviation of damping estimate	0.32	0.19	0.18	0.24

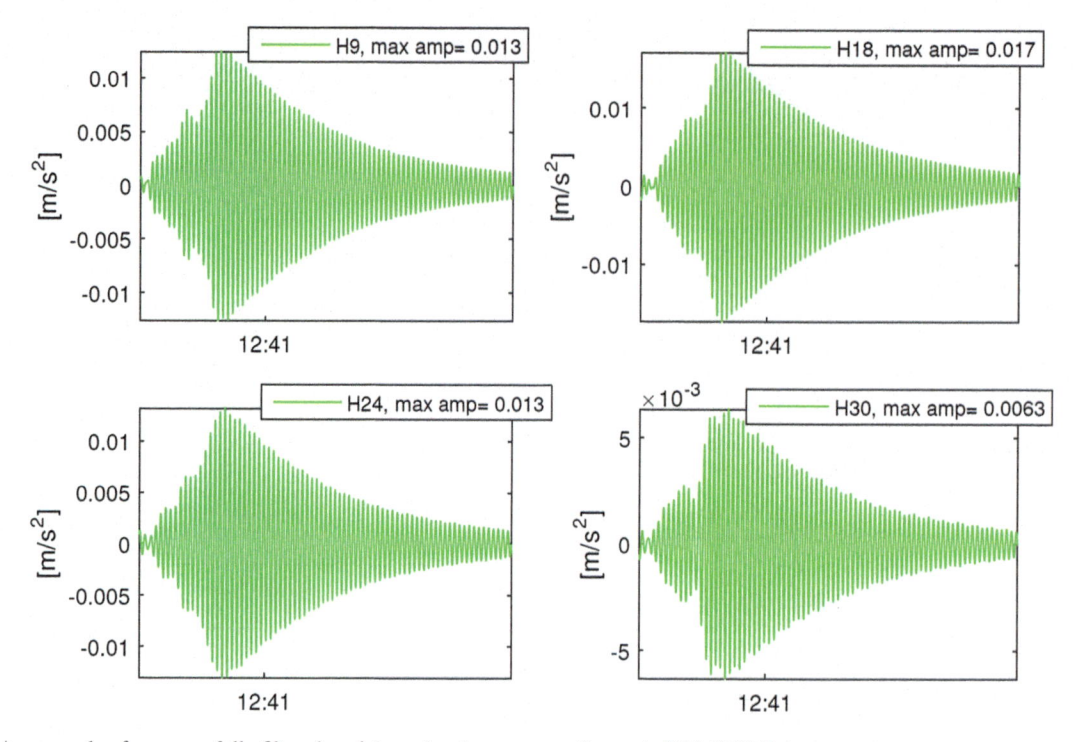

Fig. 15.8 An example of a successfully filtered modal acceleration response for mode VS3 (0.85 Hz)

Band-pass filtering is applied to the data to isolate the identified frequencies representing the different modes of vibration. Filtering is applied to records from individual sensors but also to a combined signal created by a Time Domain Decomposition (TDD) method. In short, TDD is similar to FDD, where it further converts the power spectral density functions from all sensors back to time domain, resulting in an acceleration time series that combines records from all sensors.

This approach was tested on 21 events, where a heavy vehicle was observed driving over the bridge. These 21 events were recorded on three different days in the spring of 2017. Table 15.4, summarizes the results for selected vertical modes of vibration.

The quality of the filtered modal acceleration response varies from case to case. One indication of "good" filtering is that the free oscillations after the vehicle has left the bridge looks relatively smooth, and is not distorted by other modes of vibration or secondary vehicles. Results vary from case to case, and in some instance none of the modes were filtered "well enough" to allow for reliable damping estimation based on free decay. Judging which modal filters are acceptable is a qualitative decision based on consideration of the free decay oscillations as well as a quantitative mean error function between the oscillation peaks and the fitted exponential decay function. The acceleration response illustrated in Fig. 15.8 is an example of a successful filtering results, as the free vibration decay is quite smooth. However, several cases did not produce satisfactory decay for the damping estimation for all the modes studied.

The relatively large standard deviation of damping seen in Table 15.4, is likely mainly due to varied accuracy of the exponential fitting to the filtered free oscillations, although parameters such as wind, temperature and amplitude of oscillation may also play a part in the overall variability [17, 18]. The wind velocity was varying from 0.5 to 5.7 m/s and the temperature was −1 °C for some the events studied but between 9 °C and 16 °C for others.

15.6 Conclusion

The present work is a full-scale study of traffic-induced vibrations of the Lysefjord Bridge based on observing vehicles crossing the bridge on days with low wind speed and correlating the observations with ambient vibration measurements from the bridge.

As a means of gaining an understanding of the bridge response to traffic load, spectral analysis of the acceleration time series are performed. The identified frequencies agree with those found in previous studies with little deviation.

The recorded acceleration response induced by heavy vehicles was studied. The trucks mainly excite vertical modes of vibration. The response appears to be relatively similar for trucks exiting and entering the bridge from each direction, but some dependence on velocity is observed. In case of a heavy vehicle entering the bridge at high speed, a larger high frequency "modulated" response is triggered during the vehicle passage over the bridge, compared to the case of vehicle entrance at a lower speed.

The damping ratios are estimated for selected vertical modes assuming viscous damping behavior by isolating the free decay oscillation of band-pass filtered signals and fitting an exponential decay function to the peaks. All in all the method is considered successful, and the estimated values seem reasonable in comparison to existing literature and the damping values previously estimated for the bridge using ambient vibration data. However, the variability of the results is significant, and is likely caused by distortion of the recorded signals caused by environmental effects such as wind and temperature as well as secondary traffic.

Acknowledgements This work has been supported by the Norwegian Public Road Administration, with Mathias Eidem as the main contact person. Thanks are due to Anette Ravndal, truck driver at Bjørn Hansen AS for her cooperation. Special thanks go to Etienne Cheynet at the University of Stavanger for his assistance in the data handling and manuscript preparation.

References

1. Norwegian Ministry of Transport and Communication. National Transport Plan 2014 – 2023 (2012)
2. Snæbjörnsson, J., Jakobsen, J.B., Cheynet, E., Wang, J.: Full-scale monitoring of wind and suspension bridge response. In: IOP Conference Series: 450 Materials Science and Engineering, vol. 276, no. 1, pp. 012007 (2017)
3. Cheynet, E.: Wind-Induced vibrations of a suspension bridge. PhD thesis, University of Stavanger (2016)
4. Cheynet, E., Jakobsen, J.B., Snæbjörnsson, J.: Buffeting response of a suspension bridge in complex terrain. Eng. Struct. **128**, 474–487 (2016)
5. Cheynet, E., Jakobsen, J.B., Snæbjörnsson, J.: Full scale monitoring of wind and traffic induced response of a suspension bridge. In: MATEC Web of Conferences, vol. 24. EDP Sciences, p. 04003 (2015)
6. Snæbjörnsson, J., Cheynet, E., Jakobsen, J.B.: Performance evaluation of a suspension bridge excited by wind and traffic induced action. In: Proceedings of 8th European Workshop on Structural Health Monitoring (EWSHM 2016), Bilbao, Spain (2016)
7. Cheynet, E., Snæbjörnsson, J., Jakobsen, J.B.: Identifying traffic induced vibrations of a suspension bridge: a modelling approach based on full-scale data. In: IMAC 37 (2019)
8. Apaydin, N.M., Kaya, Y., Safak, E., Alçik, H.: Vibration characteristics of a suspension bridge under traffic and no traffic conditions. Earthquake Eng Struct Dyn. **41**(12), 1717–1723 (2012)
9. Erdogan, H., Gülal, E.: Ambient vibration measurements of the Bosporus suspension bridge by total station and GPS. Exp Tech. **37**(3), 16–23 (2013)
10. Kim, C.-Y., Jung, D.-S., Kim, N.-S., Kwon, S.-D., Feng, M.Q.: Effect of vehicle weight on natural frequencies of bridges measured from traffic-induced vibration. Earthquake Eng and Eng Vibr. **2**(1), 109–115 (2003)
11. Brownjohn, J.M.W.: Estimation of damping in suspension bridges. In: Proceedings of the Institution of Civil Engineers - Structures and Buildings, vol. 104, no. 4, pp. 401–415 (1994)
12. Selberg, A.: Damping effects in suspension bridges. IABSE publications, Mémoires AIPC, IVBH Abhandlungen (1950)
13. Liu, Y., Kong, X., Cai, C., Wang, D.: Driving effects of vehicle-induced vibration on long-span suspension bridges. Struct. Control Health Monit. **24**(2), e1873 (2017)
14. Bryja, D., Sniady, P.: Stochastic non-linear vibrations of highway suspension bridge under inertial sprung moving load. J. Sound Vib. **216**(3), 507–519 (1998)
15. Fadnes, T.O.: A full-scale study on traffic induced vibrations of a suspension bridge. Master's thesis, University of Stavanger, Norway (2017)
16. Brincker, R., Zhang, L., Andersen, P.: Modal identification from ambient responses using frequency domain decomposition. In: Proceedings of the 18th International Modal Analysis Conference, IMAC XVIII (2000)
17. Cheynet, E., Snæbjörnsson, J., Jakobsen, J.B.: Temperature effects on the modal properties of a suspension bridge. Dynamics of civil structures, vol. 2. In: Proceedings of the 35th IMAC, A Conference and Exposition on Structural Dynamics 2017, Chapter 12, pp. 87—93. Springer, ISBN: 978-3-319-54777-0, https://doi.org/10.1007/978-3-319-54777-0_12 (2017)
18. Xia, Y., Chen, B., Weng, S., Ni, Y.-Q., Xu, Y.-L.: Temperature effect on vibration properties of civil structures: a literature review and case studies. J Civil Struct Health Monitor. **2**(1), 29–46 (2012)

Chapter 16
Design and Performance of a Bespoke Lively All-FRP Footbridge

J. M. Russell, J. T. Mottram, S. Zivanovic, and X. Wei

Abstract This paper presents an overview into the design approach and challenges in constructing a very lively 20 m experimental bridge of Fiber-Reinforced Polymer (FRP) components. FRP composite bridges are an attractive option for footbridge construction since their lightweight nature offers engineering advantages. However, when combined with their low stiffness, such structures can result in being lively to human-induced loading conditions. Our bespoke footbridge structure was designed to consider human-structure interaction problems, as well as to develop the understanding of the dynamic response of FRP structures. Comparisons are made between the predicted design behavior and measured responses. Measured vertical bending modes are 2.53 and 8.48 Hz, with a torsional mode at 3.36 Hz, agreeing with predicted design behavior. Walking tests on the bridge demonstrate that it responses strongly to typical human pacing frequencies and actions.

Keywords Vibration · FRP Bridge · Human-induced loading · Modal testing · Bridge design

16.1 Introduction

Lightweight footbridges are becoming increasing utilized as the reduction in material can lead to savings in fabrication and for installation. Glass Fiber-Reinforced Polymers, (FRPs) are an excellent option due to their low density and good strength to weight properties [1]. However, they typically have lower stiffness compared to traditional material, which when combined with their lower mass can lead to structures that are prone to vibration problems due to human induced loading [2, 3]. To investigate this problem further, an experimental footbridge was designed and constructed at the University of Warwick. It was intended to be as lightweight as possible, with a vertical natural frequency within the range of typical human walking. Pultruded FRP sections were used to create the structure. Modal testing was conducted on the bridge to determine the frequencies, damping, modal masses and mode shapes. Finally, single pedestrian tests at set walking frequencies were conducted and the acceleration response measured.

16.2 Bridge Design and Modal Properties

To achieve a flexible-lightweight footbridge with a low natural frequency, a shallow truss design is adopted. The truss has a depth of 475 mm between chord center lines and a span of 16.8 m. While this reduces flexural stiffness, it generates relatively high member forces, for which bolted connection strength controls the design. Overall the structure is 19.8 m long and 2.35 m wide, as shown in Fig. 16.1. Top and bottom chords are of back-to-back EXTREN® 152x41x6.4 FRP channel sections with vertical and diagonal elements of EXTREN® 50.8x6.4 FRP box section. Channel sections are used also for the transverse beams support the FRP decking of SafPlank (51 mm deep). Stainless steel bolts, with Nyloc nuts to reduce loosening from vibration actions, are used in all connections. Because sections of only 6.4 mm thickness are required, the total weight is

---

J. M. Russell (✉) · J. T. Mottram
School of Engineering, University of Warwick, Coventry, UK
e-mail: J.Russell.3@warwick.ac.uk

S. Zivanovic
College of Engineering, Mathematics and Physical Sciences, University of Exeter, Exeter, UK

X. Wei
School of Civil Engineering, Central South University, China

Fig. 16.1 Photograph of the FRP Bridge. Left. Full view of the structure. Right. Support conditions

Table 16.1 Measured modal properties

Mode description	Frequency (Hz)	Damping ratio (%)	Modal mass (kg)	Mode description	Frequency (Hz)	Damping ratio (%)	Modal mass (kg)
1st Vertical	2.53	1.69	650	1st Torsional	3.36	1.18	572
2nd Vertical	8.48	0.72	764.5	2nd Torsional	11.3	0.52	588
3rd Vertical	15.8	0.64	927.5	3rd Torsional	21.9	0.66	651

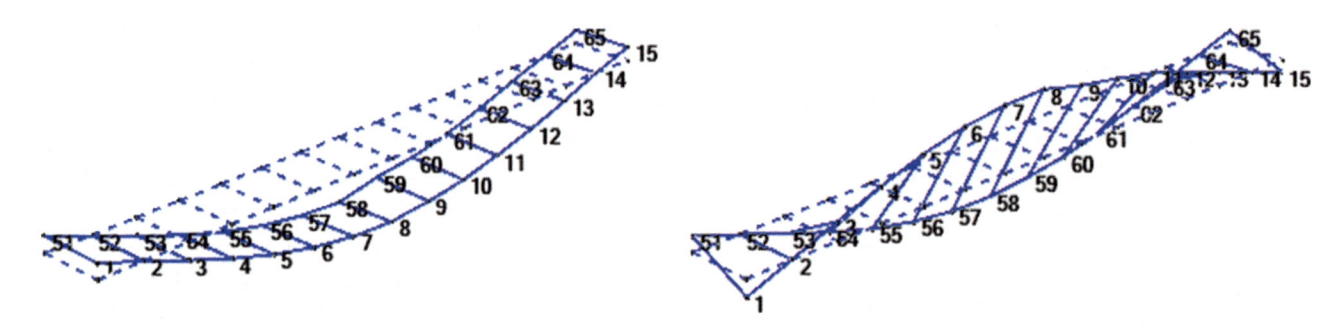

Fig. 16.2 Mode shapes Left. First vertical (2.53 Hz). Right. First torsional (3.36 Hz). Test points are shown. Supports are at TP 2/52, 14/64

very low at 1400 kg, or 71 kg/m. The structure sits on four steel bearing supports, including a roller at one end, to create a simply supported condition (see Fig. 16.1). Finite Element (FE) analysis was utilized to predict modal properties, deflections and member forces. Structural design is based on a maximum mid-span deflection of 100 mm, plus self-weight deflection, allowing for significant dynamic oscillations to occur. The key aspect is to ensure that the bolted connections has sufficient strength to prevent bolt bearing or net-tension failures in the FRP; to achieve this the channel's web in spliced joints was thickened with additional bonded FRP plating to increase local resistance. FE simulation predicted 2.5 Hz for first vertical bending mode, 3.6 Hz for first torsional and 8.7 Hz for second vertical. Because these resonant frequencies are in the range of interest for human-induced loading scenarios the design is considered fit for purpose.

In order to obtain the modal properties of the bridge, shaker testing was employed. An electromagnetic shaker was attached to the underside of the bridge by a thin strut. A PCB 208C03 load cell (nominal sensitivity 10 mV/lb) was used to measure the applied the force and Honeywell QA750 accelerometers (nominal sensitivity 1300 mV/g) used to measure the response. A chirp signal between 1 and 25 Hz was applied for 64 s, including a free decay period, and the vertical accelerations at 30 test points (TPs) were considered for calculating the frequency response functions. Table 16.1 gives the modal properties identified up to 22 Hz and Fig. 16.2 shows the first two mode shapes. The lowest vertical mode, at 2.53 Hz, is within the range of excitation due to human walking, and the close torsional mode also could respond to the second walking harmonic. Both these modes have high damping ratios compared to traditional structures, however their low modal masses means that they are easily excited. Finally, the good design and construction of the bridge and its supports mean that the mode shapes (Fig. 16.2) have a very smooth and predictable shape.

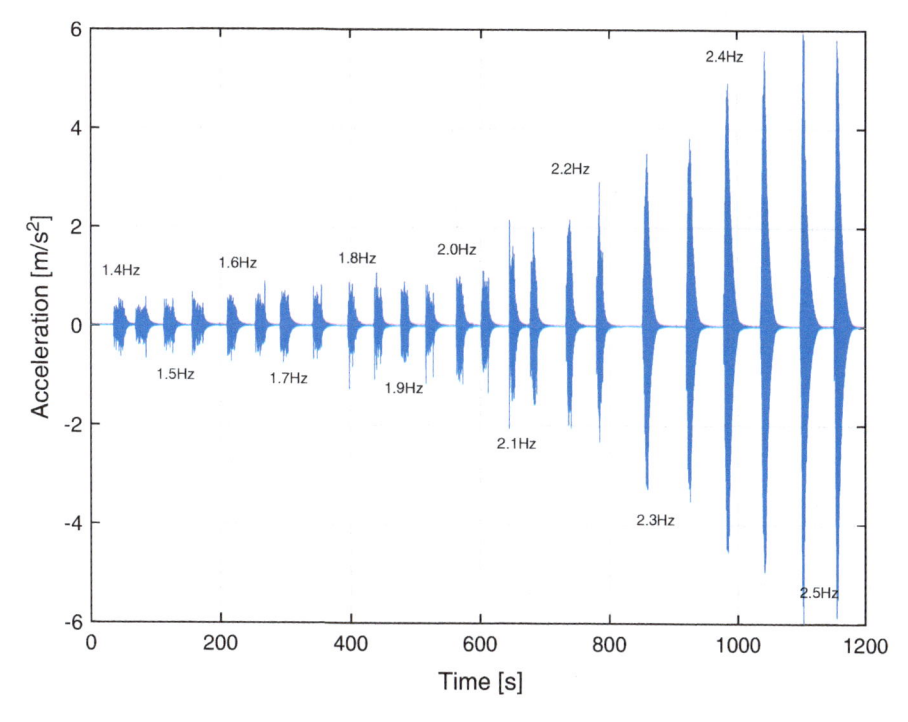

Fig. 16.3 Mid-span acceleration response from a single pedestrian walking at set pacing frequencies

16.3 Pedestrian Walking Testing

To consider the response of the bridge to human loading scenarios a single test subject walked across the centerline of the bridge at a set, metronome controlled, pacing frequency of 1.4 Hz. After the vibrations had died down, the test subject returned at the same frequency. Next the frequency was increased by 0.1 Hz and the test repeated up to a walking rate of 2.5 Hz. The vertical accelerations at the mid-span were recorded with a sampling rate of 512 Hz. The results were then low pass filtered at 20 Hz to focus on the modes of interest. Figure 16.3 show the results. From this it can be seen that with stepping rates of up to 2 Hz the peak acceleration was 1.11 m/s^2. However, beyond 2 Hz the vibration levels increase significantly and at 2.5 Hz (the first vertical bending mode) a peak acceleration of 6.00 m/s^2 was recorded. It is clear that this bridge responses to single walking events, and therefore is of great interest for human-structure interaction problems. The low mass of the bridge means that relatively small input forces result in significant accelerations. It is also expected that the influence of the pedestrian on the structure will have a noticeable effect on the modal properties, which will be investigate in future research.

16.4 Conclusion

A very lively, lightweight FRP footbridge was designed and constructed at the University of Warwick. The large displacements expected during human walking scenarios meant that the bolted connections had to be checked carefully to ensure sufficient strength. The modal properties of the bridge were predicted using finite element analysis and measured with shaker testing. The well-defined boundary conditions and good joint construction resulted in smooth mode shapes and three vertical and three torsional modes were identified up to 22 Hz. The lowest of these, a vertical mode at 2.53 Hz is within the range of human walking frequencies and, because of the low modal mass associated with the mode is very easily excited by pedestrians. This unique footbridge offers researchers an exceptional test structure for the characterization of human-structure interaction.

Acknowledgements This research work was supported by the UK Engineering and Physical Sciences Research Council [grant number EP/M021505/1: Characterising dynamic performance of fibre reinforced polymer structures for resilience and sustainability].

References

1. Wan, B.: Using fiber-reinforced polymer (FRP) composites in bridge construction and monitoring their performance: an overview. Adv. Compos. Bridg. Construct. Repair, 3–28 (2014)
2. Živanović, S., Feltrin, G., Mottram, J.T., Brownjohn, J.M.W.: Vibration performance of bridges made of fibre reinforced polymer. In: Catbas, F.N. (ed.) Dynamics of Civil Structures, vol. 4, pp. 155–162. Springer, Berlin (2014)
3. Živanović, S, Wei, X, Russell, J, Mottram, J.T.: Vibration Performance of Two FRP Footbridge Structures in the United Kingdom. Footbridge 2017, Berlin, Germany (2017)

Chapter 17
Convolutional Neural Networks for Real-Time and Wireless Damage Detection

Onur Avci, Osama Abdeljaber, Serkan Kiranyaz, and Daniel Inman

Abstract Structural damage detection methods available for structural health monitoring applications are based on data preprocessing, feature extraction, and feature classification. The feature classification task requires considerable computational power which makes the utilization of centralized techniques relatively infeasible for wireless sensor networks. In this paper, the authors present a novel Wireless Sensor Network (WSN) based on One Dimensional Convolutional Neural Networks (1D CNNs) for real-time and wireless structural health monitoring (SHM). In this method, each CNN is assigned to its local sensor data only and a corresponding 1D CNN is trained for each sensor unit without any synchronization or data transmission. This results in a decentralized system for structural damage detection under ambient environment. The performance of this method is tested and validated on a steel grid laboratory structure.

Keywords Convolutional neural networks · Real-time damage detection · Structural health monitoring · Structural damage detection · Wireless sensor networks

17.1 Introduction

The civil infrastructures are inevitably aging and engineers have always been interested in the level of aging by looking at the visible damage and/or trying to detect the invisible damage. Systematic monitoring of infrastructure has become a norm in time with the simultaneously emerging profession of Structural Health Monitoring (SHM) [1]. From visual inspection to sophisticated sensor usage, the field of Structural Damage Detection (SDD) techniques within the SHM context enabled engineers and facility owners make healthy decisions on infrastructure, for multiple disciplines. In parallel to improvements in sensor technology [2], the use of wireless sensors is adopted to create Wireless Sensor Networks (WSN) in SDD and SHM applications [3–8]. The mainstream "centralized" methods in WSN applications involve processing large amount of data which is infeasible. In this paper, the authors are introducing the use of One Dimensional Convolutional Neural Networks (1D CNNs) to create a "decentralized" technique for WSNs for efficient SDD applications in ambient dynamic environment [9]. In this technique, each sensor unit is trained locally in a decentralized way, running the raw ambient acceleration data on a steel laboratory structure at Qatar University.

17.2 Convolutional Neural Networks and the Laboratory Structure

The extensive vibration analysis [10–16], serviceability [17–25], suppression [26–32] and optimization [33–37] work conducted by the authors has motivated them to utilize their experience in SDD/SHM applications. The authors have been focused on vibration based SDD studies [38–43], and have lately introduced the use of CNNs in vibration based SDD field [44-47]. The structure of this study presented here is arguably the largest stadium structure built and instrumented in a

---

O. Avci (✉) · O. Abdeljaber
Department of Civil Engineering, Qatar University, Doha, Qatar
e-mail: oavci@vt.edu

S. Kiranyaz
Department of Electrical Engineering, Qatar University, Doha, Qatar

D. Inman
Department of Aerospace Engineering, University of Michigan, Ann Arbor, MI, USA

© Society for Experimental Mechanics, Inc. 2020
S. Pakzad (ed.), *Dynamics of Civil Structures, Volume 2*, Conference Proceedings
of the Society for Experimental Mechanics Series, https://doi.org/10.1007/978-3-030-12115-0_17

Fig. 17.1 The laboratory structure (left); the wireless units at the joints (right)

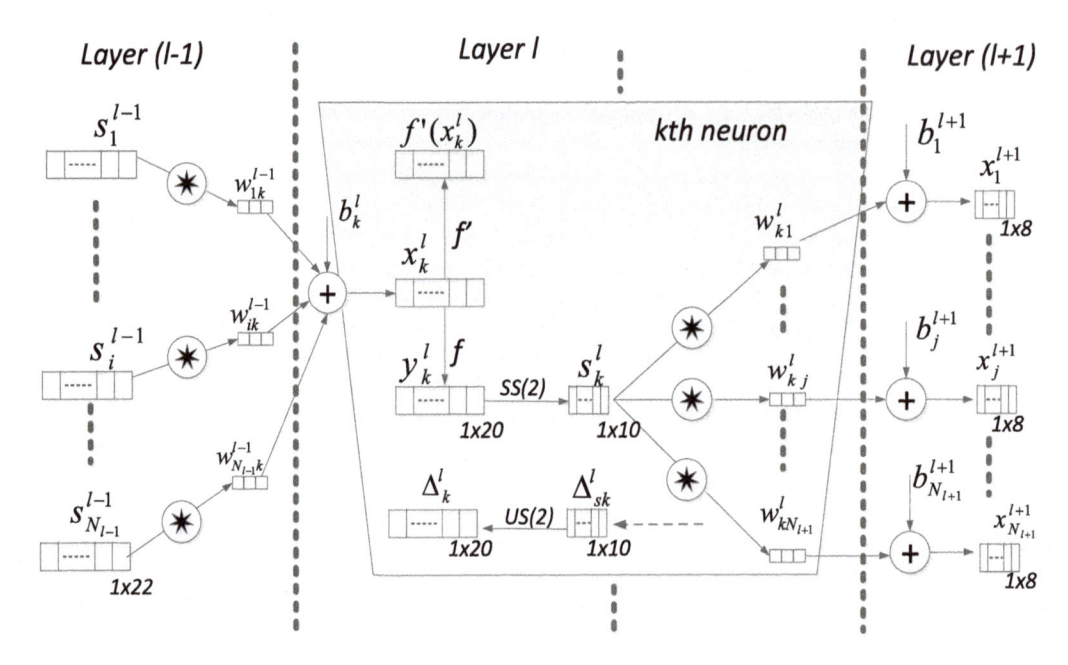

Fig. 17.2 Adaptive one dimensional convolutional neural network procedure

laboratory environment. The authors trained the CNNs based on the raw signals collected in ambient environment. The wireless sensors used in this study are state-of-the-art triaxial equipment sensitive enough to detect very low frequency signals. Since the sensing units are triaxial, the CNN algorithm runs for three orthogonal directions. The dimensions of the laboratory structure are 4.2 m × 4.2 m on the projected plan. The structure consists of 25 filler beams and 8 girders; the 10 wireless sensor units are placed at the joints, as shown in Fig. 17.1. The structure has been a test bed for various SDD and SHM studies by the authors with the interchangeable and/or removable filler beams [48]. So far, the authors mostly focused on loosening the bolts at specific connections for damage simulations. Since the bolt loosening is an extremely slight change in the overall stiffness of the structure; the authors noted that detecting this minor change in real time and wirelessly under ambient conditions would clearly verify the success of the 1D CNN SDD algorithm.

The adaptive 1D CNN mechanism has hidden neurons of the convolution layers that is processing the convolution and subsampling operations as depicted in Fig. 17.2. Combining the convolution and sub-sampling layers can be defined as the CNN layer [45, 49].

One dimensional forward propagation process in conjunction with the convolutional neural network layers can be summarized with the following (more details of this process can be found in [45–47]):

$$\mathbf{x}_k^l = \mathbf{b}_k^l + \sum_{i=1}^{N_{l-1}} \text{conv1D}\left(\mathbf{w}_{ik}^{l-1}, \mathbf{s}_i^{l-1}\right) \tag{17.1}$$

$$\mathbf{y}_k^l = f\left(\mathbf{x}_k^l\right) \quad \text{and} \quad \mathbf{s}_k^l = \mathbf{y}_k^l \downarrow ss \tag{17.2}$$

$$E_p = \text{MSE}\left(\mathbf{t}_i^p, \left[\mathbf{y}_1^L, \cdots, \mathbf{y}_{N_L}^L\right]\right) = \sum_{i=1}^{N_L} \left(\mathbf{y}_i^L - \mathbf{t}_i^p\right)^2 \tag{17.3}$$

$$\frac{\partial E}{\partial w_{ik}^{l-1}} = \Delta_k^l y_i^{l-1} \quad \text{and} \quad \frac{\partial E}{\partial b_k^l} = \Delta_k^l \tag{17.4}$$

$$\frac{\partial E}{\partial s_k^l} = \Delta s_k^l = \sum_{i=1}^{N_{l+1}} \frac{\partial E}{\partial x_i^{l+1}} \frac{\partial x_i^{l+1}}{\partial s_k^l} = \sum_{i=1}^{N_{l+1}} \Delta_i^{l+1} w_{ki}^l \tag{17.5}$$

$$\Delta_k^l = \frac{\partial E}{\partial y_k^l} \frac{\partial y_k^l}{\partial x_k^l} = \frac{\partial E}{\partial \text{us}_k^l} \frac{\partial \text{us}_k^l}{\partial y_k^l} f'\left(x_k^l\right) = \text{up}\left(\Delta s_k^l\right) \beta \, f'\left(x_k^l\right) \tag{17.6}$$

$$\Delta s_k^l = \sum_{i=1}^{N_{l+1}} \text{conv 1Dz}\left(\Delta_l^{l+1}, \text{rev}\left(w_{ki}^l\right)\right) \tag{17.7}$$

$$\frac{\partial E}{\partial w_{ik}^l} = \text{conv 1D}\left(s_k^l, \Delta_l^{l+1}\right) \quad \text{and} \quad \frac{\partial E}{\partial b_k^l} = \sum_n \Delta_k^l(n) \tag{17.8}$$

The back propagation algorithm procedure is provided in more detail in [45, 49].

$$w_{ik}^{l-1}(t+1) = w_{ik}^{l-1}(t) - \varepsilon \frac{\partial E}{\partial w_{ik}^{l-1}} \tag{17.9}$$

$$b_k^l(t+1) = b_k^l(t) - \varepsilon \frac{\partial E}{\partial b_k^l} \tag{17.10}$$

17.3 SDD, Localization and CNN Training

The primary purpose of the 1D CNN SDD work is to detect and locate the damage (bolt loosening) enforced on the structural joints during which each CNN is assigned to provide information on the status of the joint to the sensor dedicated to that joint. This is basically the decentralization of the SDD algorithm because each CNN is processing and getting trained independently at its home joint only [45–47].

The CNN training is all about data generation for undamaged and damaged acceleration data, as summarized in Fig. 17.3. For undamaged joint the signals:

$$\mathbf{U}_{E=1,J=1}^{p=1}, \ldots, \mathbf{U}_{E=1,J=n}^{p=1}, \mathbf{U}_{E=1,J=1}^{p=2}, \ldots, \mathbf{U}_{E=1,J=n}^{p=2}, \mathbf{U}_{E=1,J=1}^{p=3}, \ldots, \mathbf{U}_{E=1,J=n}^{p=3}.$$

To provide damaged/undamaged vectors necessary for CNN training, CNN_i^p:

$$\mathbf{Undamaged}_i^p = \left[\mathbf{U}_{E=1,J=i}^p \; \mathbf{U}_{E=2,J=i}^p \; \cdots \; \mathbf{U}_{E=i,J=i}^p \; \mathbf{U}_{E=i+2,J=i}^p \; \cdots \; \mathbf{U}_{E=n+1,J=i}^p\right] \tag{17.11}$$

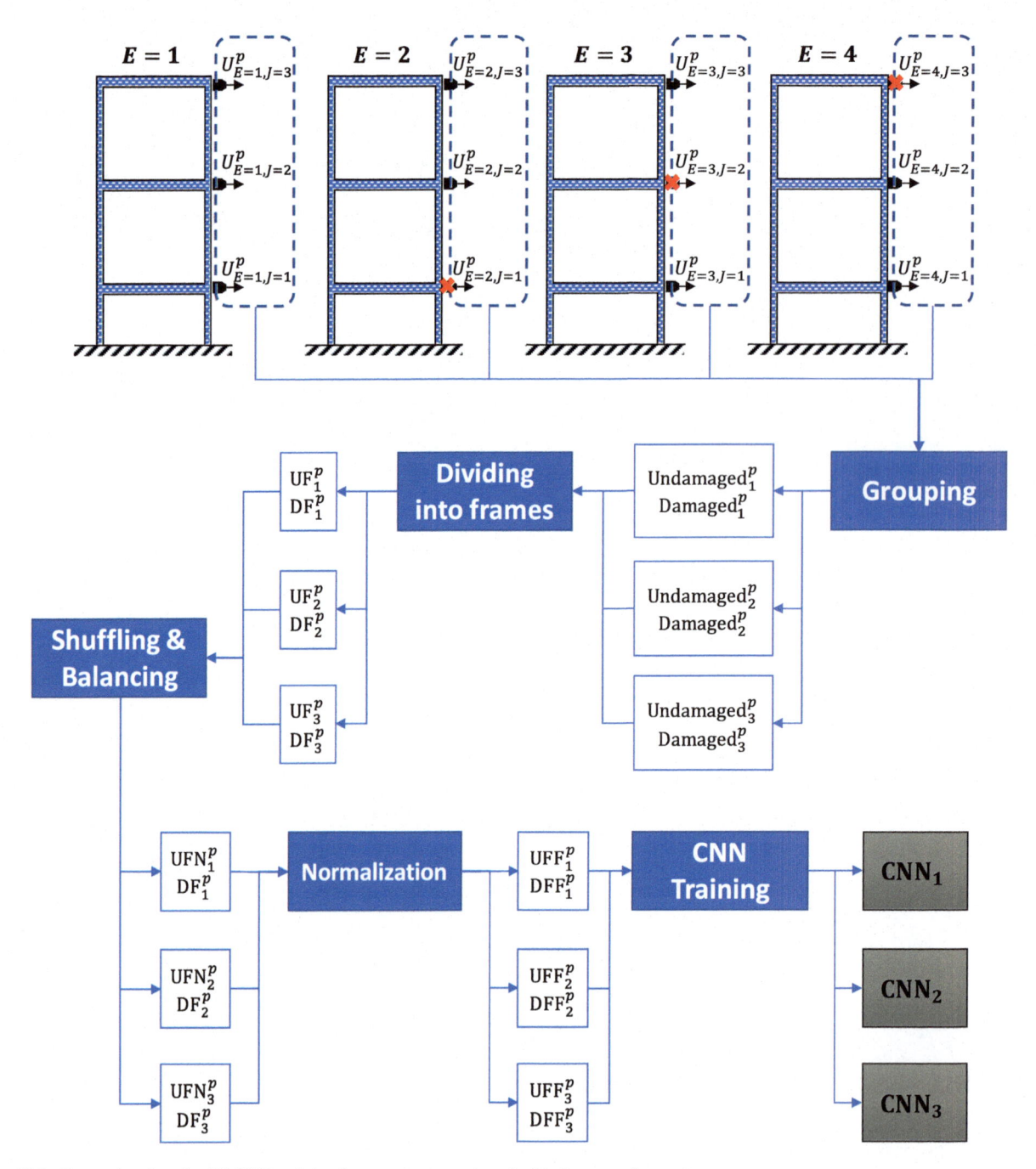

Fig. 17.3 Generating data for 1D CNN training for a structure equipped with three accelerometers

$$\mathbf{Damaged}_i^p = \left[\mathbf{D}_{E=i+1, J=i}^p \right] \tag{17.12}$$

According to [45–47], the procedure for joint i results in the following format:

$$\mathbf{UF}_i^p = \left[\mathbf{UF}_{i,1}^p \ \mathbf{UF}_{i,2}^p \ \cdots \ \mathbf{UF}_{i,n_{\mathrm{uf}}}^p \right] \tag{17.13}$$

$$\mathbf{DF}_i^p = \begin{bmatrix} \mathbf{DF}_{i,1}^p & \mathbf{DF}_{i,2}^p & \cdots & \mathbf{DF}_{i,n_{\mathrm{df}}}^p \end{bmatrix} \tag{17.14}$$

$$n_{\mathrm{uf}} = n \times \frac{n_T}{n_s} \tag{17.15}$$

$$n_{\mathrm{df}} = \frac{n_T}{n_s} \tag{17.16}$$

17.4 Attaining the Most Efficient CNNs

Calculating the classification error of each CNN, CNN_i^p can be formulated as [45–47]:

$$\mathrm{CE}_i^p = \frac{M_i^p}{T_i^p} \tag{17.17}$$

For SDD, computing the probability of damage (PoD$_i$) at the ith node can be formulated as:

$$\mathrm{PoD}_i = \frac{D_i}{T_i} \tag{17.18}$$

where D_i is the total number of frames sorted as "damaged", while T_i is the number of frames.

17.5 Experimental Testing on the Laboratory Structure

Training was done for thirty CNNs for ten joints shown in Fig. 17.1. The two stopping criteria for the back propagation training are set as; (i) 1% classification error; (ii) a maximum of 200 back propagation iterations. The training was terminated when either one was satisfied. Following this, after running the CNN algorithm, the eleven Probability of Damage results are presented in Fig. 17.4. Based on the plots, it is validated that the PoD distributions per 11 cases are successful at detecting and locating the structural damage. The fact that the real-time SDD operation is conducted efficiently with 1D CNNs for the ambient vibration environment, is the notable aspect of the decentralized WSNs described in this paper.

17.6 Conclusions and Future Work

In this paper, the one dimensional CNN algorithm is verified to be successful for detecting and locating the structural damage on a laboratory steel grid structure, wirelessly in real-time. Since the operation is done under ambient conditions with wireless sensors, the algorithm is promising to be used for massive civil infrastructure. For future work, there are items to be noted. The algorithm of this paper requires damaged data from a real structure; and it might be difficult to collect real damaged data from existing civil infrastructure. Therefore, more research needs to be conducted to simulate the structural damage in finite element models. If successful, this would help the training of CNNs without collecting real data from a real structure, which would enable the algorithm detect and locate structural damage much faster and in a more feasible way.

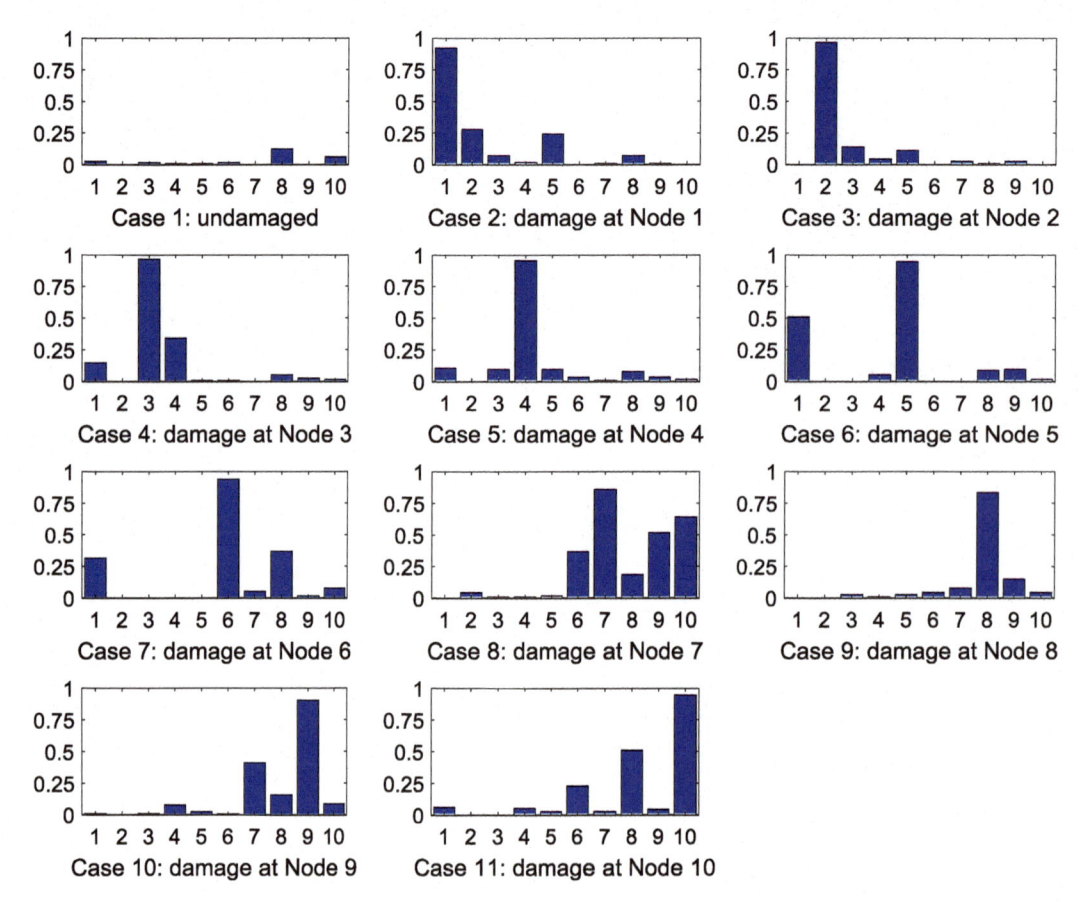

Fig. 17.4 Probability of Damage for 11 cases

References

1. Karbhari, V.M., Ansari, F. (eds.): Structural Health Monitoring of Civil Infrastructure Systems. Elsevier (2009)
2. Friswell, M.I., Adhikari, S.: Structural health monitoring using shaped sensors. Mech. Syst. Signal Process. **24**(3), 623–635 (2010). https://doi.org/10.1016/j.ymssp.2009.10.009
3. Wang, Y., Lynch, J.P., Law, K.H.: A wireless structural health monitoring system with multithreaded sensing devices: DESIGN and validation. Struct. Infrastruct. Eng. **3**, 103–120 (2007). https://doi.org/10.1080/15732470600590820
4. Straser, E.G., Kiremidjian, A.S., Meng, T.H., Redlefsen, L.: A modular, wireless network platform for monitoring structures. In: 16th International Modal Analysis Conference, pp. 450–456 (1998). https://doi.org/10.1016/S0920-5489(99)91996-7
5. Klis, R., Chatzi, E.N.: Vibration monitoring via spectro-temporal compressive sensing for wireless sensor networks. Struct. Infrastruct. Eng. **13**, 195–209 (2017). https://doi.org/10.1080/15732479.2016.1198395
6. Klis, R., Chatzi, E., Dertimanis, V.: Experimental validation of spectro-temporal compressive sensing for vibration monitoring using wireless sensor networks. In: Life-Cycle of Engineering Systems: Emphasis Sustainable Civil Infrastructructure—5th International Symposium on Life-Cycle Engineering IALCCE 2016 (2017). https://doi.org/10.1201/9781315375175-91
7. Kim, S., Pakzad, S., Culler, D., Demmel, J., Fenves, G., Glaser, S., Turon, M.: Wireless sensor networks for structural health monitoring. In: Proceedings of the 4th International Conference on Embedded Networked Sensor Systems, pp. 427–428. ACM (2006, October)
8. Kijewski-Correa, T.L., Haenggi, M., Antsaklis, P.: Multi-scale wireless sensor networks for structural health monitoring. In: Proceedings of the 2nd International Conference on Structural Health Monitoring and Intelligent Infrastructure (2005)
9. Mihaylov, M., Tuyls, K., Nowé, A.: Decentralized learning in wireless sensor networks. In: Belgian/Netherlands Artificial Intelligence Conference, pp. 345–346 (2009). https://doi.org/10.1007/978-3-642-11814-2_4
10. Inman, D.J., Singh, R.C.: Engineering Vibration, vol. 3. Prentice Hall, Englewood Cliffs (1994)
11. Bae, J.S., Kwak, M.K., Inman, D.J.: Vibration suppression of a cantilever beam using eddy current damper. J. Sound Vib. **284**(3–5), 805–824 (2005). https://doi.org/10.1016/j.jsv.2004.07.031
12. Pilkey, D., Inman, D.: A survey of damping matrix identification. In: Proceedings-SPIE . . . . (1998)
13. Sodano, H.A., Bae, J.S., Inman, D.J., Keith Belvin, W.: Concept and model of eddy current damper for vibration suppression of a beam. J. Sound Vib. **288**(4–5), 1177–1196 (2005). https://doi.org/10.1016/j.jsv.2005.01.016
14. Sodano, H.A., Bae, J.-S., Inman, D.J., Belvin, W.K.: Improved concept and model of eddy current damper. J. Vib. Acoust. **128**(3), 294–302 (2006). https://doi.org/10.1115/1.2172256
15. Inman, D.: Dynamics of asymmetric nonconservative systems. J. Appl. Mech. **50**(1), 199–203 (1984). https://doi.org/10.1115/1.3166991

16. Sodano, H.A., Inman, D.J., Belvin, W.K.: Development of a new passive-active magnetic damper for vibration suppression. J. Vib. Acoust. **128**(3), 318–327 (2006). https://doi.org/10.1115/1.2172258
17. Avci, O., Bhargava, A., Al-Smadi, Y., Isenberg, J.: Vibrations serviceability of a medical facility floor for sensitive equipment replacement: evaluation with sparse in situ data. Pract. Period. Struct. Des. Constr. **24**(1), 05018006 (2019)
18. Do, N.T., Gül, M., Abdeljaber, O., Avci, O.: Novel framework for vibration serviceability assessment of stadium grandstands considering durations of vibrations. J. Struct. Eng. **144**(2), 04017214 (2017)
19. Avci, O.: Nonlinear damping in floor vibrations serviceability: verification on a laboratory structure. In: Conf. Proc. Soc. Exp. Mech. Ser. (2017). https://doi.org/10.1007/978-3-319-54777-0_18
20. Younis, A., Avci, O., Hussein, M., Davis, B., Reynolds, P.: Dynamic forces induced by a single pedestrian: a literature review. Appl. Mech. Rev. **69**(2), 020802 (2017). https://doi.org/10.1115/1.4036327
21. Catbas, F.N., Celik, O., Avci, O., Abdeljaber, O., Gul, M., Do, N.T.: Sensing and monitoring for stadium structures: a review of recent advances and a forward look. Front. Built Environ. **3**, 38 (2017). https://doi.org/10.3389/fbuil.2017.00038
22. Avci, O.: Modal parameter variations due to joist bottom chord extension installations on laboratory footbridges. J. Perform. Constr. Facil. **29**(5), 04014140 (2014)
23. Avci, O.: Amplitude-dependent damping in vibration serviceability: case of a laboratory footbridge. J. Archit. Eng. **22**, 04016005 (2016). https://doi.org/10.1061/(ASCE)AE.1943-5568.0000211
24. Celik, O., Do, N.T., Abdeljaber, O., Gul, M., Avci, O., Catbas, F.N.: Recent issues on stadium monitoring and serviceability: a review. In: Conf. Proc. Soc. Exp. Mech. Ser. (2016). https://doi.org/10.1007/978-3-319-29763-7_41
25. Bhargava, A., Isenberg, J., Feenstra, P.H., Al-Smadi, Y., Avci, O.: Vibrations assessment of a hospital floor for a magnetic resonance imaging unit (MRI) replacement. In: Structural Congress 2013 Bridging Your Passion with Your Profession—Proceedings of 2013 Structural Congress (2013)
26. Avci, O., Setareh, M., Murray, T.M.: Vibration testing of joist supported footbridges. In: Structural Congress 2010 (2010). https://doi.org/10.1061/41130(369)80
27. Inman, D.J.: Vibration with Control. Wiley, New York (2006). https://doi.org/10.1002/0470010533
28. Avci, O., Davis, B.: A study on effective mass of one way joist supported systems. In: Structural Congress 2015—Proceedings of 2015 Structural Congress (2015). https://doi.org/10.1061/9780784479117.073
29. Davis, B., Avci, O.: Simplified vibration response prediction for slender monumental stairs. In: Structural Congress 2014—Proceedings of 2014 Structural Congress (2014). https://doi.org/10.1061/9780784413357.223
30. Davis, B., Avci, O.: Simplified vibration serviceability evaluation of slender monumental stairs. J. Struct. Eng. (United States). **141**, 04015017 (2015). https://doi.org/10.1061/(ASCE)ST.1943-541X.0001256
31. Al-Smadi, Y.M., Bhargava, A., Avci, O., Elmorsi, M.: Design of experiments study to obtain a robust 3D computational bridge model. In: Conf. Proc. Soc. Exp. Mech. Ser. (2012). https://doi.org/10.1007/978-1-4614-2413-0_29
32. Barrett, A.R., Avci, O., Setareh, M., Murray, T.M.: Observations from vibration testing of in-situ structures. In: Proc. Struct. Congr. Expo. (2006). https://doi.org/10.1061/40889(201)65
33. Avci, O., Abdeljaber, O., Kiranyaz, S., Inman, D.: Vibration suppression in metastructures using zigzag inserts optimized by genetic algorithms. In: Conf. Proc. Soc. Exp. Mech. Ser. (2017). https://doi.org/10.1007/978-3-319-54735-0_29
34. Abdeljaber, O., Avci, O., Inman, D.J.: Genetic algorithm use for internally resonating lattice optimization: case of a beam-like metastructure. In: Conf. Proc. Soc. Exp. Mech. Ser. (2016). https://doi.org/10.1007/978-3-319-29751-4_29
35. Abdeljaber, O., Avci, O., Inman, D.J.: Optimization of chiral lattice based metastructures for broadband vibration suppression using genetic algorithms. J. Sound Vib. (2015). https://doi.org/10.1016/j.jsv.2015.11.048
36. Abdeljaber, O., Avci, O., Inman, D.J.: Optimization of chiral lattice based metastructures for broadband vibration suppression using genetic algorithms. J. Sound Vib. **369**, 50–62 (2016)
37. Abdeljaber, O., Avci, O., Inman, D.J.: Active vibration control of flexible cantilever plates using piezoelectric materials and artificial neural networks. J. Sound Vib. **363**, 33–53 (2016). https://doi.org/10.1016/j.jsv.2015.10.029
38. Chaabane, M., Ben Hamida, A., Mansouri, M., Nounou, H.N., Avci, O.: Damage detection using enhanced multivariate statistical process control technique. In: 2016 17th Int. Conf. Sci. Tech. Autom. Control Comput. Eng. STA 2016—Proc. (2017). https://doi.org/10.1109/STA.2016.7952052
39. Abdeljaber, O., Avci, O.: Nonparametric structural damage detection algorithm for ambient vibration response: utilizing artificial neural networks and self-organizing maps. J. Archit. Eng. **22**(2), 04016004 (2016). https://doi.org/10.1061/(ASCE)AE.1943-5568.0000205
40. Abdeljaber, O., Avci, O., Do, N.T., Gul, M., Celik, O., Necati Catbas, F.: Quantification of structural damage with self-organizing maps. In: Conf. Proc. Soc. Exp. Mech. Ser. (2016). https://doi.org/10.1007/978-3-319-29956-3_5
41. Mansouri, M., Avci, O., Nounou, H., Nounou, M.: A comparative assessment of nonlinear state estimation methods for structural health monitoring. In: Conf. Proc. Soc. Exp. Mech. Ser. (2015). https://doi.org/10.1007/978-3-319-15224-0_5
42. Mansouri, M., Avci, O., Nounou, H., Nounou, M.: Iterated square root unscented Kalman filter for nonlinear states and parameters estimation: three DOF damped system. J. Civ. Struct. Heal. Monit. **5**, 493 (2015). https://doi.org/10.1007/s13349-015-0134-7
43. Avci, O., Abdeljaber, O.: Self-organizing maps for structural damage detection: a novel unsupervised vibration-based algorithm. J. Perform. Constr. Facil. **30**, 04015043 (2016). https://doi.org/10.1061/(ASCE)CF.1943-5509.0000801
44. Avci, O., Abdeljaber, O., Kiranyaz, S., Inman, D.: Structural damage detection in real time: implementation of 1D convolutional neural networks for SHM applications. In: Niezrecki, C. (ed.) Struct. Heal. Monit. Damage Detect, vol. 7. Proc. 35th IMAC, A Conf. Expo. Struct. Dyn. 2017, pp. 49–54. Springer, Cham (2017). https://doi.org/10.1007/978-3-319-54109-9_6
45. Abdeljaber, O., Avci, O., Kiranyaz, S., Gabbouj, M., Inman, D.J.: Real-time vibration-based structural damage detection using one-dimensional convolutional neural networks. J. Sound Vib. **388**, 154–170 (2017). https://doi.org/10.1016/j.jsv.2016.10.043
46. Abdeljaber, O., Avci, O., Kiranyaz, M.S., Boashash, B., Sodano, H., Inman, D.J.: 1-D CNNs for structural damage detection: Verification on a structural health monitoring benchmark data. Neurocomputing. **275**, 1308–1317 (2018). https://doi.org/10.1016/j.neucom.2017.09.069

47. Avci, O., Abdeljaber, O., Kiranyaz, S., Hussein, M., Inman, D.J.: Wireless and Real-Time Structural Damage Detection: A Novel Decentralized Method for Wireless Sensor Networks. J. Sound Vib. **424**, 158–172 (2018)
48. Abdeljaber, O., Younis, A., Avci, O., Catbas, N., Gul, M., Celik, O., Zhang, H.: Dynamic testing of a laboratory stadium structure. In: Geotech. Struct. Eng. Congr. (2016), pp. 1719–1728 (n.d.). https://doi.org/10.1061/9780784479742.147
49. Ince, T., Kiranyaz, S., Eren, L., Askar, M., Gabbouj, M.: Real-time motor fault detection by 1-D convolutional neural networks. IEEE Trans. Ind. Electron. **63**, 7067–7075 (2016). https://doi.org/10.1109/TIE.2016.2582729

Chapter 18
The Influence of Truck Characteristics on the Vibration Response of a Bridge

Navid Zolghadri and Kirk A. Grimmelsman

Abstract The measured vibration response of highway bridges is influenced by different factors including truck loads. The speed, length, number of axles and their spacing, weight and suspension type are all truck characteristics that are known to influence the vibration response of highway bridges. A bridge's structural characteristics and the pavement condition leading up to a bridge crossing also determine the vibration response due to truck loads. The specific influence of the various truck characteristics on the vibration response of highway bridges is generally difficult to quantify analytically or experimentally given that many of these truck characteristics are unmeasured. This paper presents the results of an investigation on the effects of measured truck characteristics on the measured vibration responses of a concrete multi-beam bridge. The bridge is being continuously monitored with triggered high-speed recordings of strains and vibrations during truck crossing events. A weigh-in-motion system located in the roadway in front of the bridge measures the weight, speed and other characteristics of every truck that crosses the bridge. The effects of the different truck characteristics on the measured vibration response of the structure were evaluated in both time and frequency domains. The results of these evaluations are presented and discussed, and the most influential truck characteristics are identified.

Keywords Bridges · Vibration · Prestressed concrete · Structural health monitoring · Weigh-in-motion

18.1 Introduction

The vibrations caused by heavy trucks have a significant impact on the serviceability and performance of the bridge structure. Design of a bridge can also be effected by considerations that need to be taken with regards to the impact of the vibrations due to the increase in the weight and speed of vehicles. One of the design considerations reflects in dynamic load allowance (DLA) which increases the effects of static load on the bridge by certain percentages for different limit states [1]. Different vehicles cause bridges to vibrate because of different reasons such as irregularities in the road surface. The vibration is a result of the complex dynamic interaction between the bridge structure and the vehicle. This interaction can be influenced by different parameters including bridge dynamic properties such as natural frequencies and damping, vehicle vibration characteristics, vehicle speed, vehicle weight, etc. The vibrations of bridges have been investigated extensively [2–6]. The vibration measurements in most of these research studies were ambient vibration measurements without knowing more details about traffic properties. This study presents results of vibration monitoring of a bridge that was equipped with a weigh-in-motion (WIM) system. The WIM system measured the gross vehicle weight (GVW), speed, individual axle weights and spacing of any truck traveling over the bridge. The vibration measurements were also collected from different accelerometers when a truck traversed the bridge. The data from these two systems were matched accordingly and therefore, a set of data was available to study the variation of the vibration measurements with regards to various truck weights and speeds.

N. Zolghadri (✉) · K. A. Grimmelsman
Pennoni Associates Inc., Philadelphia, PA, USA
e-mail: nzolghadri@pennoni.com; kgrimmelsman@pennoni.com

18.2 Bridge Description and Data Collection Process

The bridge studied in this paper was a pre-stressed concrete bridge located on the westbound of I-84 over Nolin Road in Oregon (Oregon Bridge). This bridge consists of three simple spans which are 40, 80, and 40 ft. long respectively. Figure 18.1 only shows the two spans of the bridge which are instrumented with accelerometers. The structure of this bridge includes AASHTO Type III prestressed girders that are spaced at 9.25 ft. in the end spans and 7.4 ft. in the center span. Each end span has reinforced concrete diaphragms at each bent and the midspan location. The center span has reinforced concrete diaphragm beams at each bent and at approximately one-third and two-third points along the span. This bridge was constructed in 1969 and the 8-in. thick reinforced deck was reconstructed in 2004.

This bridge is instrumented with a Bridge Monitoring System (BMS) and a Weigh-in-motion (WIM) system. The BMS on Oregon Bridge includes 48 strain gages, 6 accelerometers, 4 tilt meters, 4 displacement gages, a GPS receiver, a weather station, and a camera. In this study, the data collected from the 6 accelerometers were merely used. The location of each accelerometer can be found in Fig. 18.1. These accelerometers are from TE Connectivity with a range of ± 2 g and a frequency range of DC to 200 Hz.

The BMS consists of four Measurement Control Units (MCUs) that are making measurements continuously and when a trigger is detected these MCUs will store the data. The trigger occurs when measured strain values exceeds a threshold. The event-based recordings occur when measured strain values exceed a predefined threshold value. The threshold strain values were established by trial and error adjustments such that trucks crossing the bridge initiate an event-based measurement and passenger vehicle crossings do not. This trigger threshold is currently defined as strain values recorded from either of the two selected strain gages on the right and left lane exceeds five micro strain from a moving average of 60,000 previous measurements. The number of moving-average points was also adjusted based on the available memory on the MCU and the functionality of the trigger. Currently, the recording after detection of a trigger consists of 2 s of pre-trigger measurements and 8 s of post-trigger measurements. The trigger signal is also sent to the bridge overview camera which signals it to capture an image of the truck and other traffic on the bridge at the time the event was detected.

The WIM system is a two-lane pavement system in the concrete roadway approaching the bridge. Two lanes of weighing sensors are installed in each lane with inductive loops. An overhead IP camera is also included to capture an image of the trucks traveling over the WIM system on the right lane. The WIM system provides detailed information including

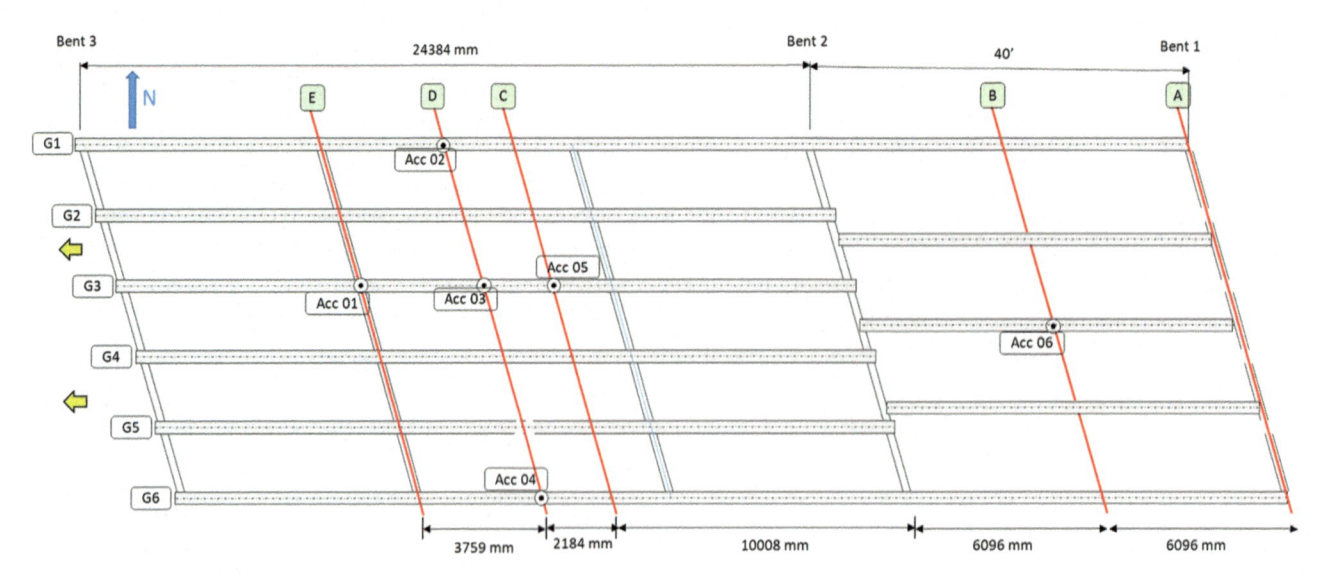

Fig. 18.1 Plan view of the bridge and sensors layout

Fig. 18.2 WIM and BMS images taken from trucks

GVW, speed, individual axle weights and spacing of every vehicle. The system also provides an FHWA classification for each vehicle. Two images are taken from each truck, one from BMS and one from WIM system (Fig. 18.2). These images will validate the matched event on the bridge monitoring with the trucks properties recorded on the WIM system.

Figure 18.3 shows samples of measurements from six different accelerometers when a trigger was detected. During this event the truck travelled over the right lane. It can be seen that Accelerometer 6 read the highest amplitude since it is in the end span with one less girder while Accelerometer 4 read the lowest amplitude as it was expected.

18.3 Analysis

The selected data for this analysis included 3577 events. These events were collected during only 48 h of data collection. Figure 18.4 shows the maximum amplitude of acceleration recorded by Acceleration 02, 03 and 04 located on three different girders at the same cross section. The average recorded maximum amplitude of acceleration for Accelerometer 2, 3 and 4 are 0.0724, 0.0952, and 0.0494. This shows how the girders on the sides vibrate with smaller amplitudes while girders in the middle are more prone to larger vibrations.

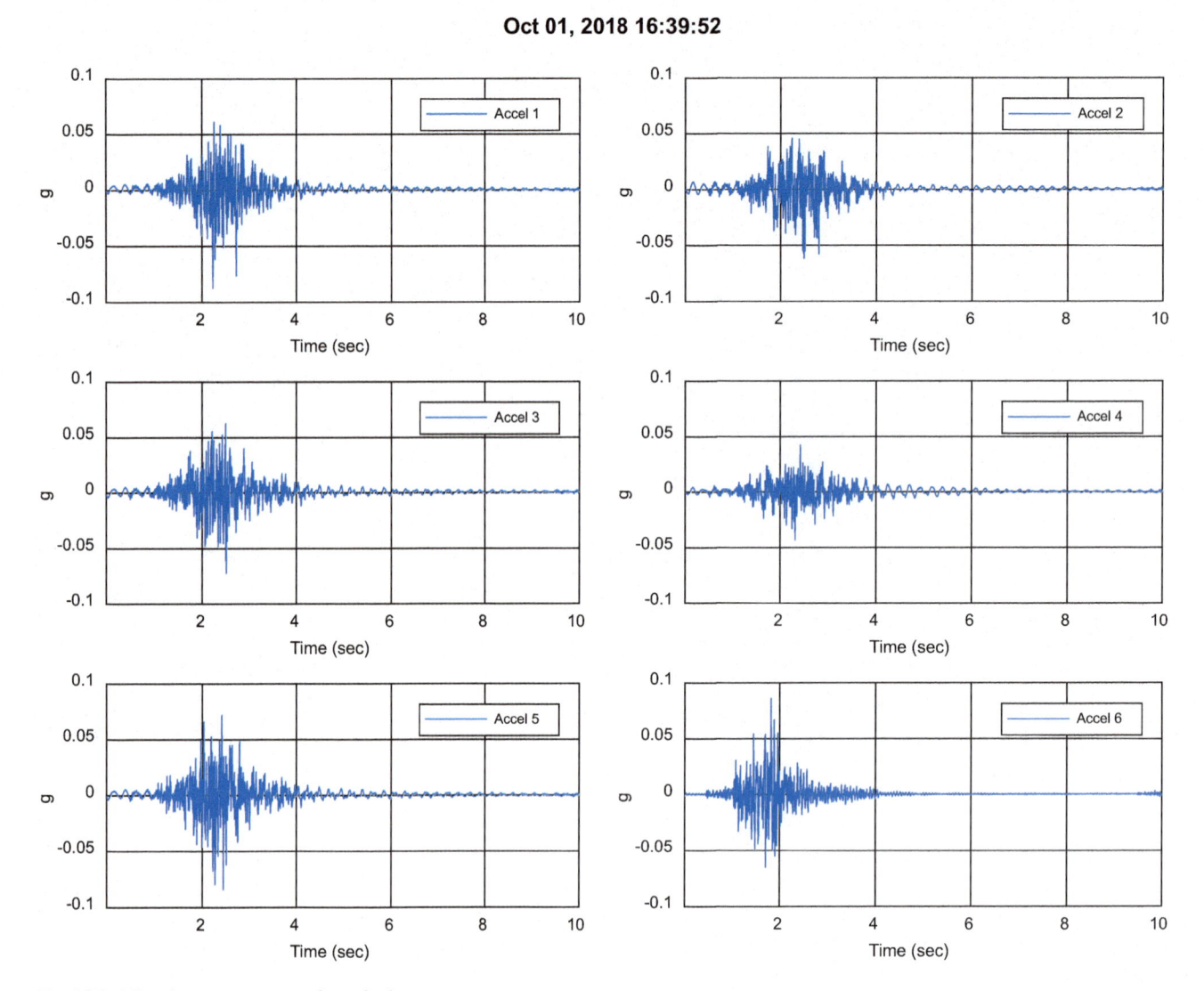

Fig. 18.3 Vibration measurements for a single event

The primary objective of this study was to evaluate the effects of speed on the vibration of bridges. Figure 18.5 shows the variation of root mean square (RMS) of acceleration from accelerometer 6 with regards to speed. The measured speed has a resolution of 0.62 mph. Data shows the effect of speed on the vibration at different locations are not significant. The other objective of this paper was to investigate the effects of GVW. Figure 18.6 shows the effect of GVW on the vibration of the bridge was also not considerable. One of the reasons is the stiffness of this bridge. This bridge has shown even very small strain values for heavy trucks. The other unknown parameter is the vibration characteristics of different trucks such as suspension properties. These characteristics may have a considerable impact that we cannot include in this study.

18.4 Conclusion

In this paper, the effect of gross vehicle weight (GVW) and speed of trucks on the vibration of a pre-stressed concrete bridge was presented. The bridge was instrumented with six accelerometers at different locations and whenever a truck travelled

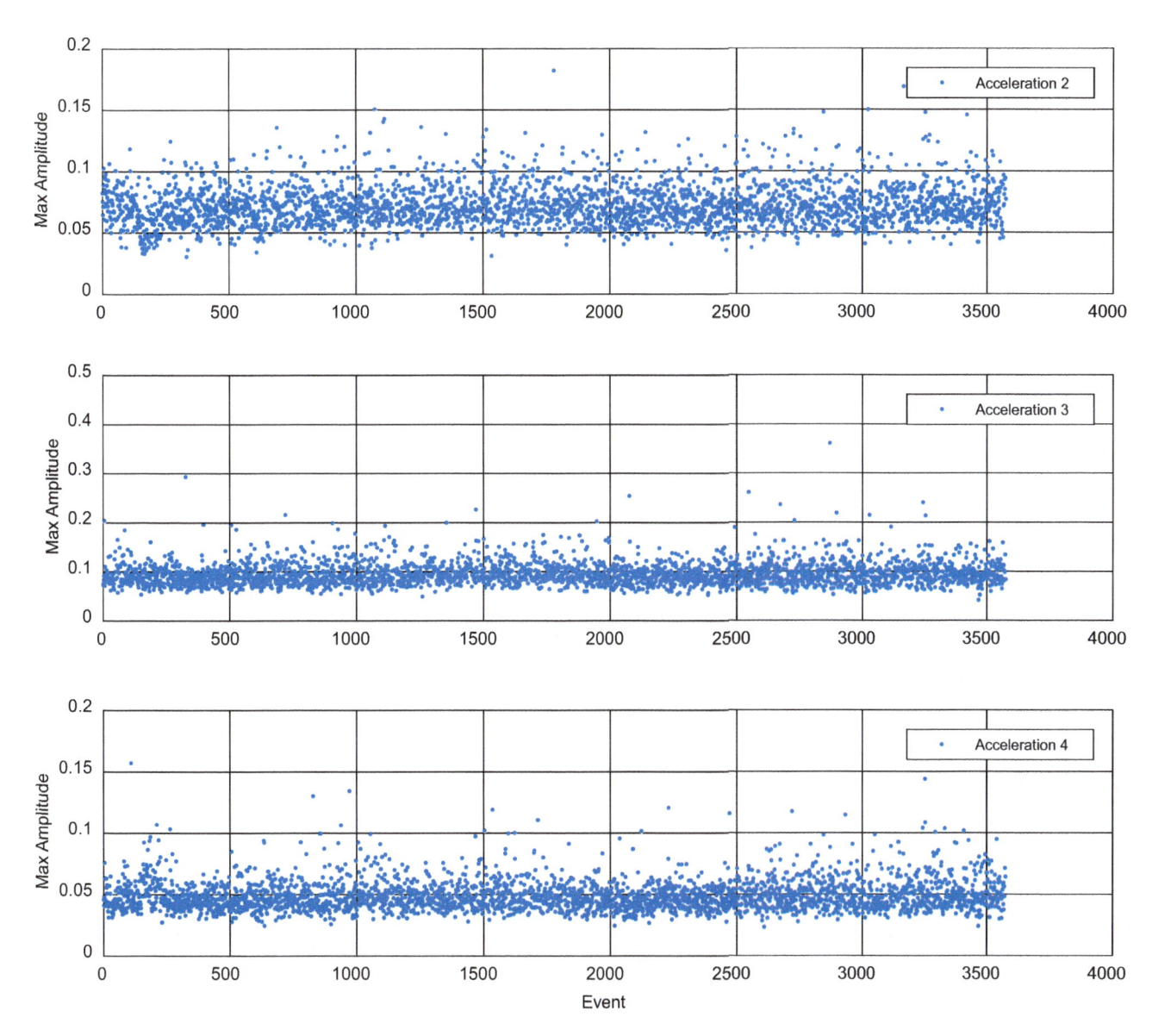

Fig. 18.4 Maximum amplitude of recorded acceleration during different events

over the bridge, the monitoring system recorded the data from these accelerometers. The data from 3577 trucks with known weights and speeds traveling over the bridge showed that the variation of the bridge vibration is not significantly impacted by GVW or speed of the heavy trucks. There are other parameters that have not been considered in this study such as vibration characteristics of trucks which may have a more significant influence on the vibration of the bridge. In the future work, more events may be investigated since the data collection on this bridge is continuously ongoing. The dynamic properties of the bridge can also be extracted from acceleration measurements and investigated in different condition. More sophisticated statistical analysis may also be necessary to make more meaningful conclusions.

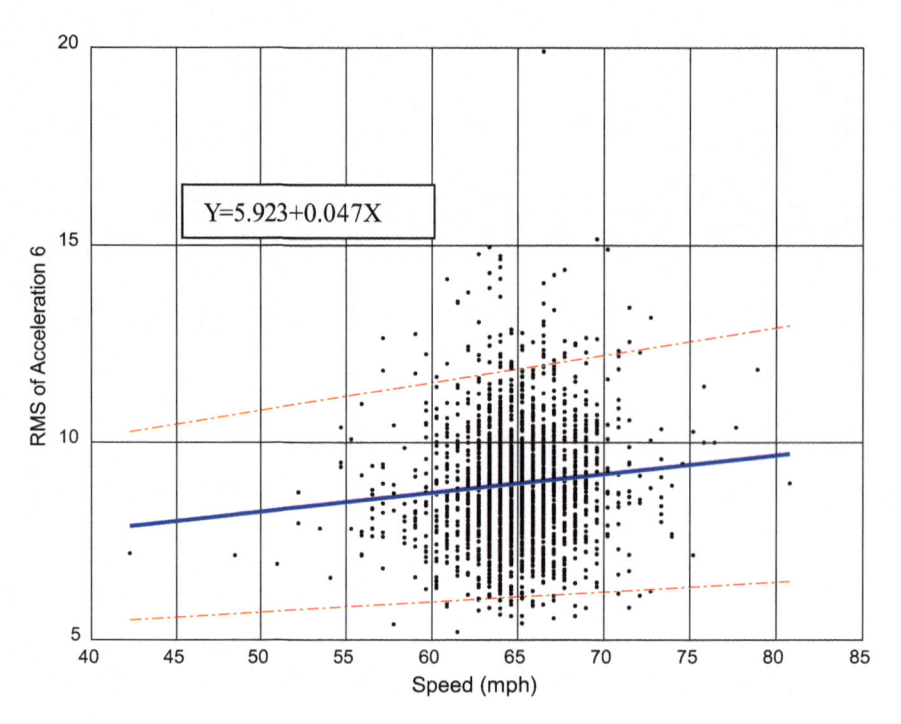

Fig. 18.5 Speed vs. RMS

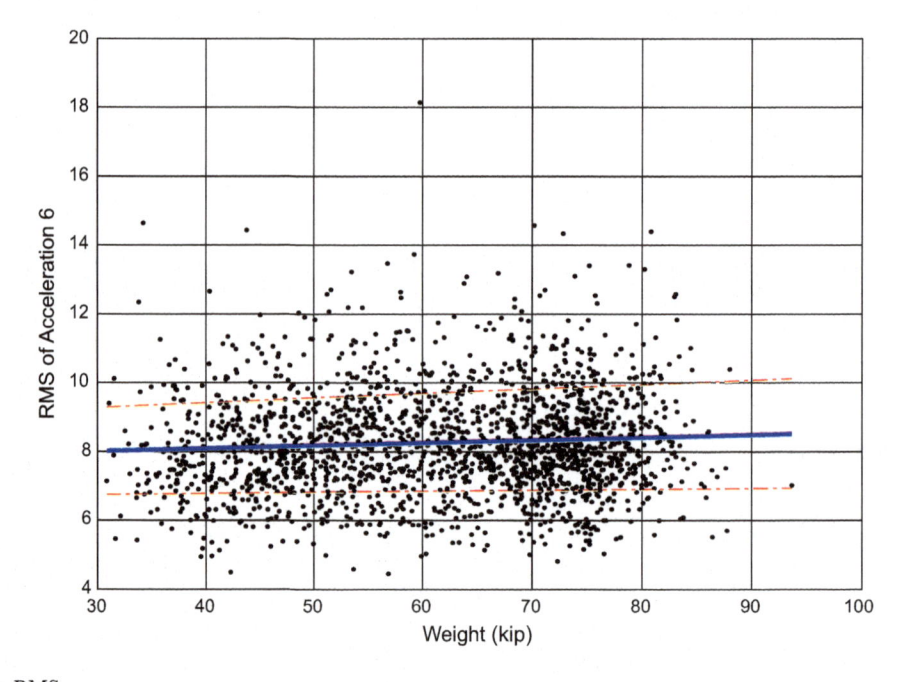

Fig. 18.6 Weight vs. RMS

References

1. AASHTO: LRFD Bridge Design Specifications, 7th edn. American Association of State Highway and Transportation Officials, Washington, DC (2014)
2. Au, F.T.K., Cheng, Y.S., Cheung, Y.K.: Vibration analysis of bridges under moving vehicles and trains: an overview. Prog. Struct. Eng. Mater. **3**(3), 299–304 (2001)
3. Adams, R.D., Cawley, P., Pye, C.J., Stone, B.J.: A vibration technique for non-destructively assessing the integrity of structures. J. Mech. Eng. Sci. **20**(2), 93–100 (1978)
4. Rytter, A.: Vibration Based Inspection of Civil Engineering Structures, Ph.D. Thesis. University of Aalborg, Denmark (1993)

5. Grimmelsman, K.A., Prader, J.B.: Evaluation of truck-induced vibrations for a multi-beam highway bridge. In: Barthorpe, R., Platz, R., Lopez, I., Moaveni, B., Papadimitriou, C. (eds.) Model Validation and Uncertainty Quantification, vol. 3. Conference Proceedings of the Society for Experimental Mechanics Series, Springer, Cham (2017)
6. Zolghadri, N., Halling, M., Barr, P.: Effects of temperature variations on structural vibration properties. In: Conference on Geotechnical and Structural Engineering Congress 2016. pp. 1032–1043 (2016)

Chapter 19
Experimental Evaluation of Low-Cost Accelerometers for Dynamic Characterization of Bridges

Kirk A. Grimmelsman and Navid Zolghadri

Abstract The design of structural health monitoring systems for bridges often requires that the measurement objectives be carefully balanced with associated cost and logistical considerations. Using accelerometers to measure and track the dynamic characteristics of a bridge is a frequent objective with many potential benefits; however, instrument-grade accelerometers are often the most expensive sensors employed in a structural health monitoring system for a bridge. This often leads to compromises in terms of the number of accelerometers deployed, or a reduced emphasis for vibration measurements in the hierarchy of measurement objectives for the project. Various low-cost MEMS accelerometers are commercially available and can potentially complement or serve as alternatives to instrument-grade accelerometers for structural health monitoring of bridges. Such MEMS accelerometers are used for wireless accelerometer and sensing systems, and the accelerometer chip is integrated on a circuit board with additional electronics for power, data acquisition and communications. In the more common design of structural health monitoring system employing wired sensors of various types, a low-cost MEMS accelerometer that provides analog outputs, and that also has very minimal packaging and electronics integration requirements is most desirable. A study was initiated to experimentally evaluate the performance and capabilities of a low-cost accelerometer that met these requirements for bridge vibration measurements. The low-cost accelerometer's performance was compared to a more conventional instrument-grade accelerometer. Measurements were recorded by both sensors from a simple physical model and from a full-scale highway bridge structure and evaluated for different vibration measurement scenarios. The capabilities, limitations and optimal application scenarios for the low-cost accelerometers in the context of bridge vibration measurements are presented and discussed.

Keywords MEMS accelerometer · Laboratory measurements · Field measurements · Bridge vibration

19.1 Introduction

There is a significant emphasis in Structural Health Monitoring (SHM) research applications on using bridge vibration measurements for structural identification, performance evaluation and condition assessment objectives. In bridge monitoring practice; however, the emphasis is more decidedly directed towards strain and displacement measurements for characterizing bridge structural performance and condition. Unless there is a specific vibration problem to be evaluated that requires the use of accelerometers for the characterization, the sensors are unlikely to be employed for most routine bridge SHM applications. This tendency to forgo the use of accelerometers for many bridge SHM applications can be partially explained by a general lack of appreciation in the bridge engineering practice for the many possible benefits of measuring and tracking the dynamic characteristics of bridges for assessing their long-term performance and condition. The actionable information generated by quantifying the dynamic characteristics of a bridge is often less direct than what can be obtained from monitoring structural stresses or displacements. Cost is another possible reason for not employing accelerometers in routine bridge SHM applications. The cost of an instrument-grade accelerometer and its associated signal conditioning and data acquisition hardware can be much larger on a channel per channel basis relative to the other types of instrumentation used in bridge monitoring applications. Indeed, in cases where a bridge owner may be convinced of the benefits of vibration measurements, the final deployment of sensors is often constrained to something less than robust instrumentation design by cost considerations and the need to deploy other types of instrumentation.

K. A. Grimmelsman (✉) · N. Zolghadri
Pennoni Associates Inc., Philadelphia, PA, USA
e-mail: Kgrimmelsman@pennoni.com

The advent of MEMS type accelerometers in the marketplace has mitigated the cost issue with these sensors to a significant degree. Many of the low-power wireless structural monitoring systems that have been developed and employed in bridge monitoring applications in recent years [1, 2] have only been possible and practical due to the availability of these types of sensors. More recently, there have been numerous attempts to employ smart phones and similar devices that include MEMS type accelerometers to characterize the dynamic response of bridges and other structures [3, 4]. A wide variety of digital and analog output MEMS accelerometers are used for such systems and they have widely varying operational and performance characteristics. These two approaches have very specific SHM application scenarios where they can be justified, but they are generally not the accepted practice for long-term bridge SHM application utilizing mixed forms of instrumentation. In addition, the MEMS accelerometers utilized for these devices are rarely systematically characterized in terms of the operational and performance characteristics relative to more conventional instrument-grade accelerometers. It follows that there is merit in identifying a low-cost MEMS accelerometer that may be practical and suitable for use in long-term bridge SHM applications with mixed sensor types, and systematically characterizing its operational and performance characteristics. Such information would be of great value to bridge owners and engineers and could prompt more widespread acceptance in practice for including dynamic characterization measurements in bridge SHM applications.

This paper presents the results of an investigation to characterize the performance characteristics of a low-cost MEMS accelerometer that was identified as potentially practical solution for such bridge SHM applications. The MEMS accelerometer evaluated in this study is inexpensive, provides an analog output that can be easily integrated with conventional wired SHM data acquisition hardware and systems used to monitor a variety of other types of sensors, and requires a minimal degree of electronic skills and expertise to prepare them for use in real bridge monitoring applications. The low-cost accelerometer was evaluated relative to instrument-grade accelerometers in the laboratory and in the field under normal operating conditions for a highway bridge.

19.2 Objectives and Scope

The objective of this study is to evaluate the performance of a low-cost MEMS type accelerometer relative to more conventional instrument-grade accelerometers in both the laboratory and under actual field conditions for a highway bridge structure. The frequency range of interest for the vibrations of most bridge structures can be safely assumed to be below 50 Hz and the necessary amplitude range is generally less than ± 2 g. The MEMS accelerometer selected for this evaluation has nominal performance characteristics that meet these criteria.

The low-cost MEMS accelerometer selected is a triple axis sensor with analog outputs that can easily be integrated with data acquisition hardware for measuring a variety of different sensors such as strain gages and displacement sensors. The MEMs accelerometer is designed around the ADXL335 MEMs sensor from Analog Devices. The accelerometer was purchased for a unit cost of approximately 15 dollars and comes already mounted to a small printed circuit board with additional electronics and header pinouts. The accelerometer can be powered by up to 5 V DC and has a nominal analog output range of ± 3 g. The analog outputs of the accelerometer are ratiometric with the internally regulated 3.3 V DC power. The typical sensitivity is 330 mV/g for each axis, and XYZ filter capacitors give the accelerometer a 50 Hz bandwidth for each direction. The broadband resolution for the accelerometers is 30 μg RMS. The noise density for the X and Y output directions is 150 $\mu g/(Hz)^{0.5}$ RMS and is 300 $\mu g/(Hz)^{0.5}$ for the Z output direction. A small screw terminal block was soldered on the header pinouts to permit lead wires to be easily connected to the accelerometer. A 5 V DC power supply was connected to the accelerometer to provide power for the laboratory testing. Only one output direction for the accelerometer was evaluated for the laboratory study. The Z output direction was evaluated since its noise density is the worst case for the accelerometer. A photograph of the accelerometer with the screw terminals attached is shown in Fig. 19.1.

Two different instrument-grade accelerometers were used in this study. The instrument-grade accelerometer used for the laboratory study was a PCB Piezotronics Model 393A03 ceramic shear ICP accelerometer. This a uniaxial accelerometer with a nominal sensitivity of 1000 mV/g, a measurement range of ± 5 g, and a frequency range of 0–200 Hz. The unit cost for this accelerometer is 575 dollars. The instrument-grade used in the bridge monitoring evaluation was a PCB Piezotronics Model 3741E122G MEMS DC response accelerometer. This is a uniaxial accelerometer with a differential output and has a nominal sensitivity of 2000 mV/g, a measurement range of ± 2 g, and a frequency range of 0–200 Hz. The broadband resolution for the accelerometer is 0.1 mg. The unit cost for this accelerometer is 530 dollars. A limited number of these accelerometers were employed in the bridge monitoring application and were primarily selected because of their performance specifications and because they could be easily integrated with the data acquisition hardware and equipment that was being used for other sensors on the bridge including strain gages and displacement transducers.

Fig. 19.1 Low-cost MEMS accelerometer with screw terminals attached

The details of the laboratory and field measurement evaluations of the low-cost MEMS accelerometer and the results are presented and discussed in the following sections of this paper.

19.3 Laboratory Evaluation

The low-cost MEMS accelerometer was installed next to the instrument-grade accelerometer on the armature of an APS 113 vibration exciter for testing. A total of 10 different harmonic excitation cases were employed with the vibration exciter to evaluate the accelerometer performance: (1) 5 Hz sinusoidal excitation, (2) 10 Hz sinusoidal excitation, (3) 25 Hz low amplitude sinusoidal excitation, (4) 25 Hz medium amplitude sinusoidal excitation, (5) 25 Hz high amplitude sinusoidal excitation, (6) 50 Hz low amplitude sinusoidal excitation, (7) 50 Hz medium amplitude sinusoidal excitation, (8) 50 Hz high amplitude sinusoidal excitation, (9) 65 Hz sinusoidal excitation, and (10) 100 Hz sinusoidal excitation. Although the frequency band for the low-cost accelerometer is 50 Hz, several excitation cases that exceeded this frequency limit were included to evaluate its performance in these ranges. The low-cost and instrument-grade accelerometers were both connected to National Instruments NI-9234 vibration input module for testing. The instrument-grade accelerometer was powered by the module and low-cost accelerometer was externally powered. The measurements from both accelerometers were sampled at 2000 Hz for all excitation cases. The measurement data were collected for approximately 3 min for each excitation case.

Typical time domain signals from the shaker testing are shown in Fig. 19.2. Several observations are apparent from the time domain plots. First, the shaker output was rather noisy, especially at the lower frequencies. The noise is due to a mechanical issue with this specific shaker, which is rather old and not in the best physical condition. The mechanical performance issues with the shaker leads to multiple harmonics for both accelerometers when the measurements are transformed into the frequency domain. Small differences in the amplitude and phase of the acceleration signals measured by the low-cost and instrument grade accelerometers is also observable from the time domain signals.

Power spectrums were computed for the accelerometer measurements recorded from each excitation case. The power spectrum results were computed using Welch's method for the data sampled at 2000 Hz using a block size of 16,384 points and a 50% overlap. The resulting frequency resolution was 0.1221 Hz. The cross spectrum between each accelerometer was also computed for each excitation case using the same block size and percent overlap parameters. The cross spectrum was used to compute the phase lag between the low-cost and instrument grade accelerometers at the excitation frequencies. Finally, the magnitude squared coherence was computed between the two accelerometers for each excitation case. The coherence between the two accelerometers was found to be equal to 1.0 at the forcing frequency for all excitation cases. The results of the frequency domain analysis of the laboratory measurements are summarized in Table 19.1.

Representative examples of the power spectrums for the 10 Hz and 50 Hz medium amplitude excitation cases are shown in Fig. 19.3. The power spectrums show the presence of multiple harmonics of the excitation frequency for both the low-cost and instrument-grade accelerometers, which is related to condition of the shaker as discussed earlier. There is also a harmonic located at 60 Hz for both accelerometers and this has a slightly larger amplitude for the low-cost accelerometer. This indicates some electrical power noise was present in the experimental setup.

A difference in the amplitude (dB power) between the instrument-grade and low-cost accelerometers was observed at the excitation frequency for each test case. The amplitude of the low-cost accelerometer is observed to always be smaller than that for the instrument-grade accelerometer. The ratio of the low-cost to instrument-grade power at the forcing frequency

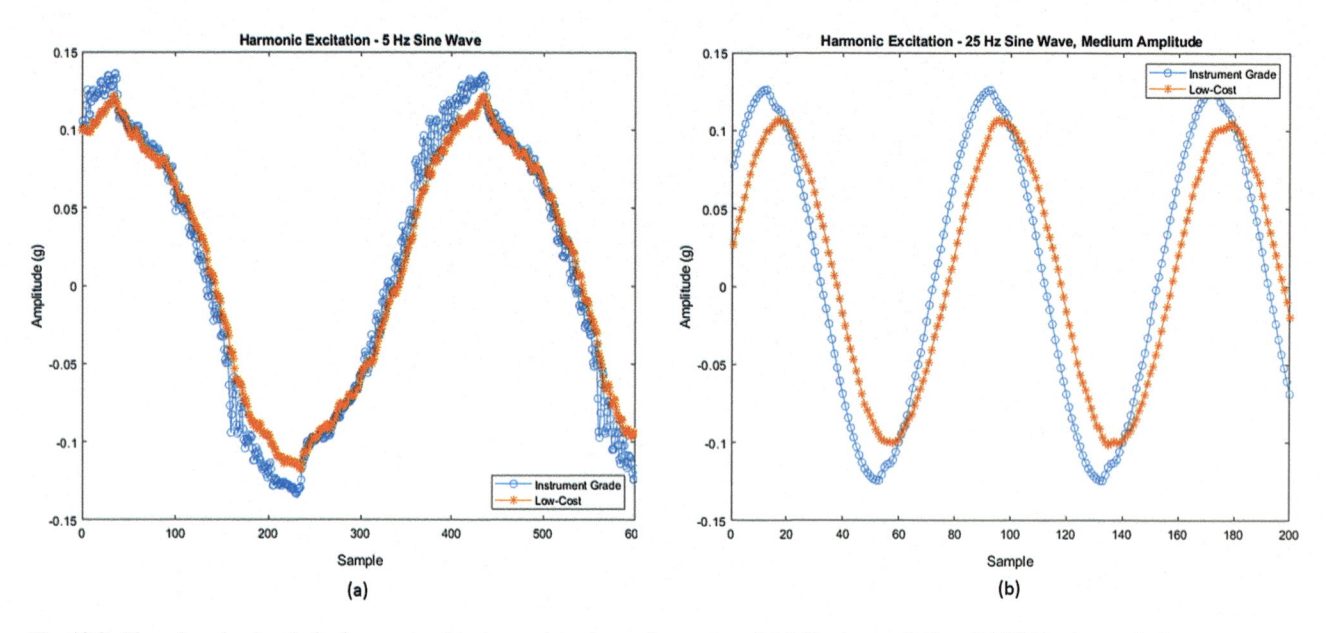

Fig. 19.2 Time domain signals for low-cost and instrument-grade accelerometers: (**a**) 5 Hz sine excitation, (**b**) 25 Hz sine excitation

Table 19.1 Summary results from laboratory evaluation of low-cost and instrument-grade accelerometers

Excitation case	Instrument-grade accel. freq. (Hz)	Low-cost accel. freq. (Hz)	Power ratio at natural freq. (low-cost to instrument-grade)	Phase lag (radians)
5 Hz Sine	5.005	5.005	1.042	0.1204
10 Hz Sine	10.01	10.01	1.048	0.1863
25 Hz Sine, Low Amplitude	25.02	25.02	1.055	0.4169
25 Hz Sine, Medium Amplitude	25.02	25.02	1.068	0.4177
25 Hz Sine, High Amplitude	25.02	25.02	1.097	0.4186
50 Hz Sine, Low Amplitude	50.05	50.05	1.109	0.7416
50 Hz Sine, Medium Amplitude	50.05	50.05	1.098	0.7422
50 Hz Sine, High Amplitude	50.05	50.05	1.162	0.7412
65 Hz Sine	64.94	64.94	1.150	0.8582
100 Hz Sine	99.98	99.98	1.274	1.161

was computed for each excitation case as shown in Table 19.1. The amplitude ratio becomes larger with increasing excitation frequency. The computed amplitude ratios also indicate there is some relationship to the amplitude of the shaker excitation, with a higher amplitude excitation producing a slightly larger amplitude ratio than for lower amplitude excitation.

The phase lag between the two sensors was also extracted at the excitation frequency from the cross spectrum of the two accelerometers in each excitation case and is summarized in Table 19.1. The results indicate that phase lag increases with increasing frequency. The phase lag between the two accelerometers tends to become much larger for frequencies above 25 Hz. The phase lag has a minimum value of 0.1204 radians at 5 Hz and a maximum value of 1.161 radians at 100 Hz. The nominal frequency band for the low-cost accelerometer is 50 Hz, so the phase lag should not be expected to be particularly good beyond this band.

19.4 Field Evaluation

The low-cost accelerometer was also installed on a typical highway bridge structure along with a limited number of instrument-grade accelerometers for characterizing the dynamic response of the structure. The bridge is being monitored by the authors to study the effects of truck loads on the long-term performance of the bridge. The instrument-grade accelerometers were primarily included in the instrumentation plan developed for the bridge to characterize the dynamic effects of truck loads crossing the bridge. The authors were able to add the low-cost accelerometers to the structure to

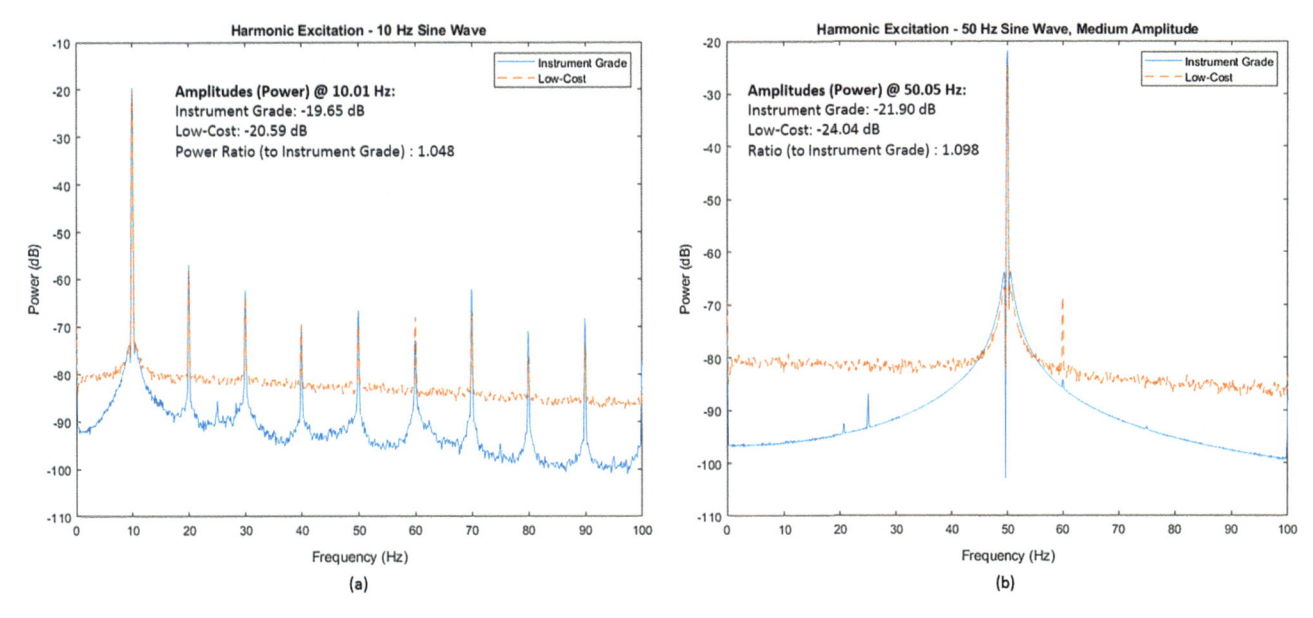

Fig. 19.3 Power spectrums for low-cost and instrument-grade accelerometers: (**a**) 10 Hz sine excitation, (**b**) 50 Hz sine excitation

assist in tracking natural frequencies, mode shapes and damping ratios since their cost was negligible in relation to the total cost of the monitoring system. The bridge is primarily instrumented with vibrating wire strain gages and vibrating wire displacement transducers which are used to measure the live load and thermal stresses of the bridge girders and the thermal displacements of the bearings. The monitoring system records the strain and accelerometer measurements at 100 Hz for 12 s whenever a truck crosses the bridge. These event-based recordings capture the forced- and free-vibration responses of the bridge superstructure due to truck loads. The bridge monitoring system also records continuous measurements from these sensors at 100 Hz for 5 min at 4 a.m. and 2 p.m. each day.

The bridge that is being monitored is a three-span continuous steel multibeam bridge with a cast-in-place reinforced concrete composite deck. The bridge carries two westbound lanes of state highway traffic over a small river. The end spans of the bridge are both 10,668 mm-long and the middle span is 15,240 mm in length. Figure 19.4 illustrates the locations of the low-cost and instrument-grade accelerometers on the steel girders in the middle span. Additional accelerometers will be installed in the end span of the structure in the future.

All the accelerometers installed on the bridge are wired for uniaxial measurements of the vertical vibration responses only. A low-cost and instrument-grade accelerometer are installed side-by-side of each other on Girder 3 at the midpoint of the middle span. Some of the low-cost accelerometers are oriented with the sensor's X-output oriented vertically, while others are oriented with the sensor's Z-output oriented vertically. The low-cost accelerometers were all installed in small plastic enclosures and fully potted with epoxy to ensure their long-term survivability in the field. All accelerometers were bonded to the steel girders using epoxy.

Two sets of the 5-min-long continuous measurements from the accelerometers on the bridge were evaluated for this study. The continuous measurements were both recorded at 2 p.m. local time in July and October 2018. Power spectrums were computed from the time domain measurements of the accelerometers using Welch's method with a block size of 2048 points and a 50% overlap. The resulting frequency resolution was 0.048 Hz. The power spectra from 0 to 25 Hz from the low-cost and instrument-grade accelerometers that are located side-by-side in the middle span of the bridge are shown in Figs. 19.5 and 19.6 for July and October, respectively.

The figures both show several natural frequencies of the bridge are identified by both the low-cost and the instrument-grade accelerometers. There is a slight shift in the frequencies for the common peaks observable in both figures which is likely due to thermal effects since the data were recorded from warm and cold weather months. The fundamental vertical mode of the bridge is located at about 7.5 Hz. A second mode can be observed at about 10.3 Hz from the low-cost accelerometers in both figures, while a peak at this frequency is only observable for the instrument-grade accelerometer from the October data. A third vertical mode can be observed at about 20 Hz for the low-cost and instrument-grade accelerometers in both figures. In general, the low-cost accelerometer appears to have a less noisy response than the instrument-grade accelerometer used in this case. The low-cost accelerometer also appears to perform quite adequately and consistently from a frequency identification perspective. This observation is very preliminary based on the amount of acceleration data that has been analyzed from

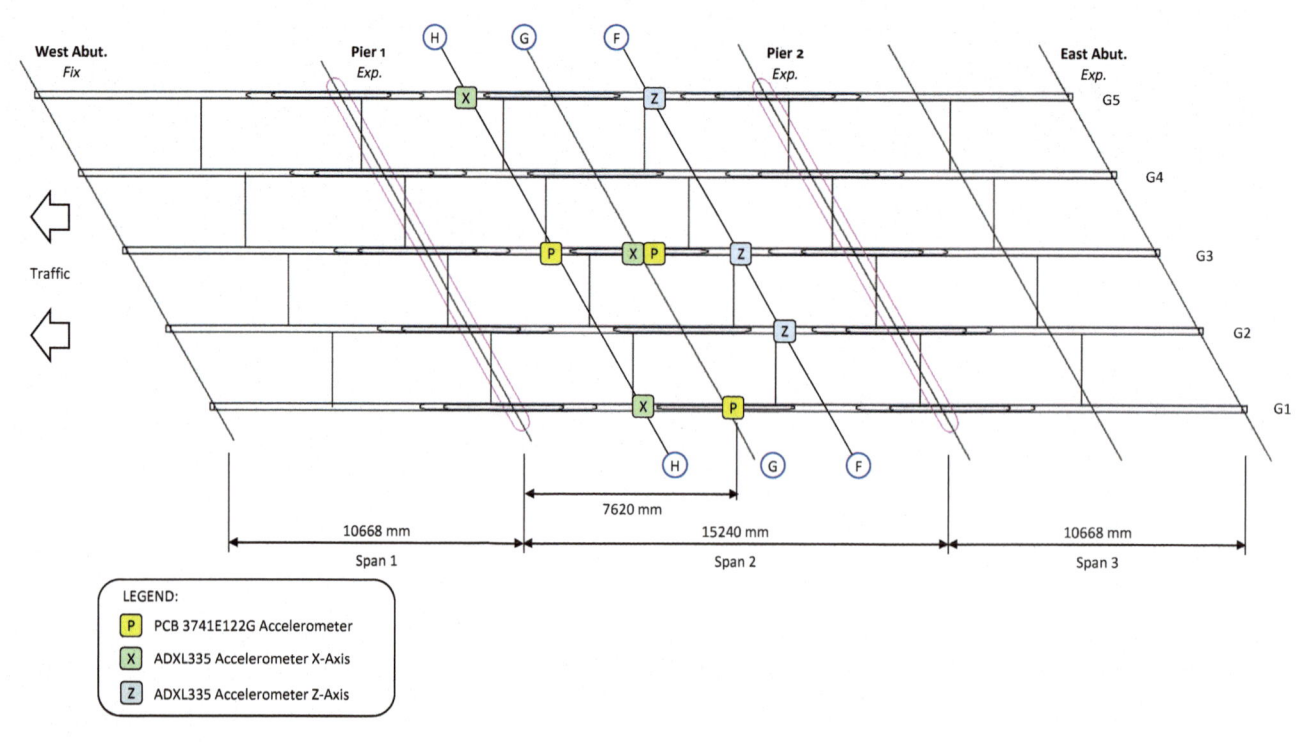

Fig. 19.4 Accelerometer locations on bridge superstructure

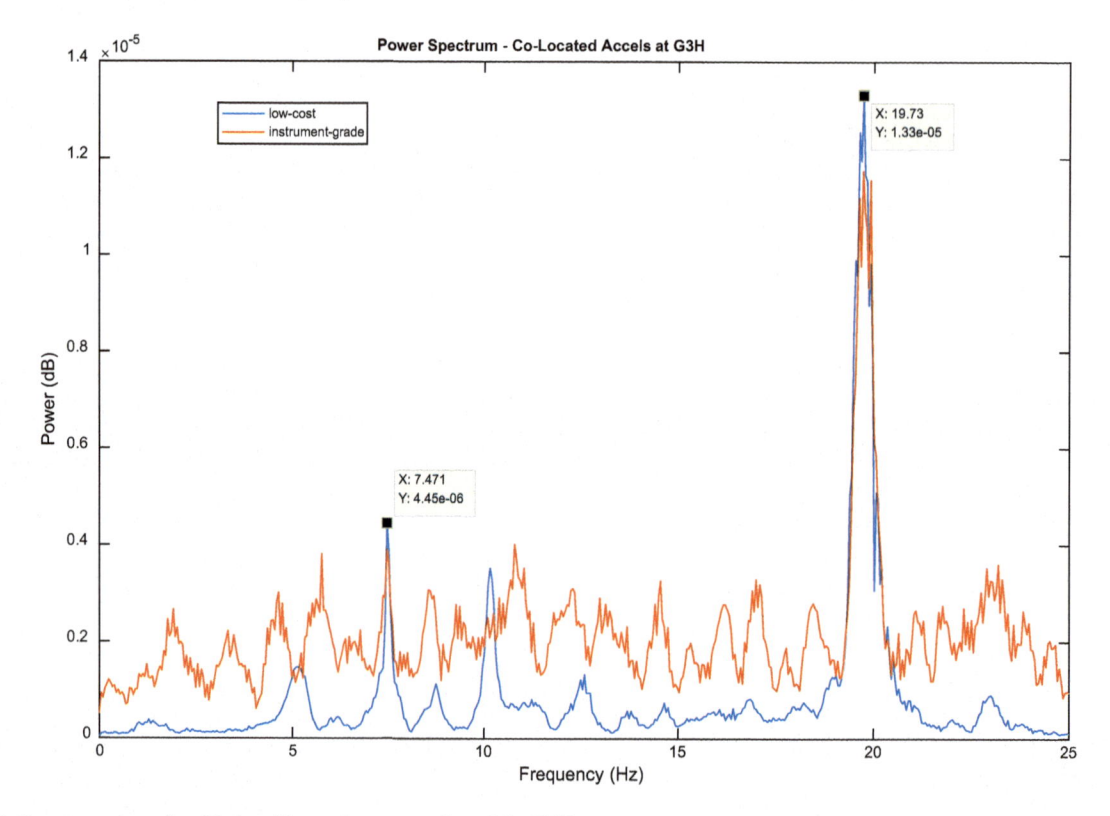

Fig. 19.5 Power spectrum for side-by-side accelerometers from July 2018 measurements

the bridge thus far. Additional ambient vibration data and the forced- and free-vibration measurements from truck crossing events needs to be evaluated to fully characterize the performance of the low-cost accelerometers, but the preliminary results look promising.

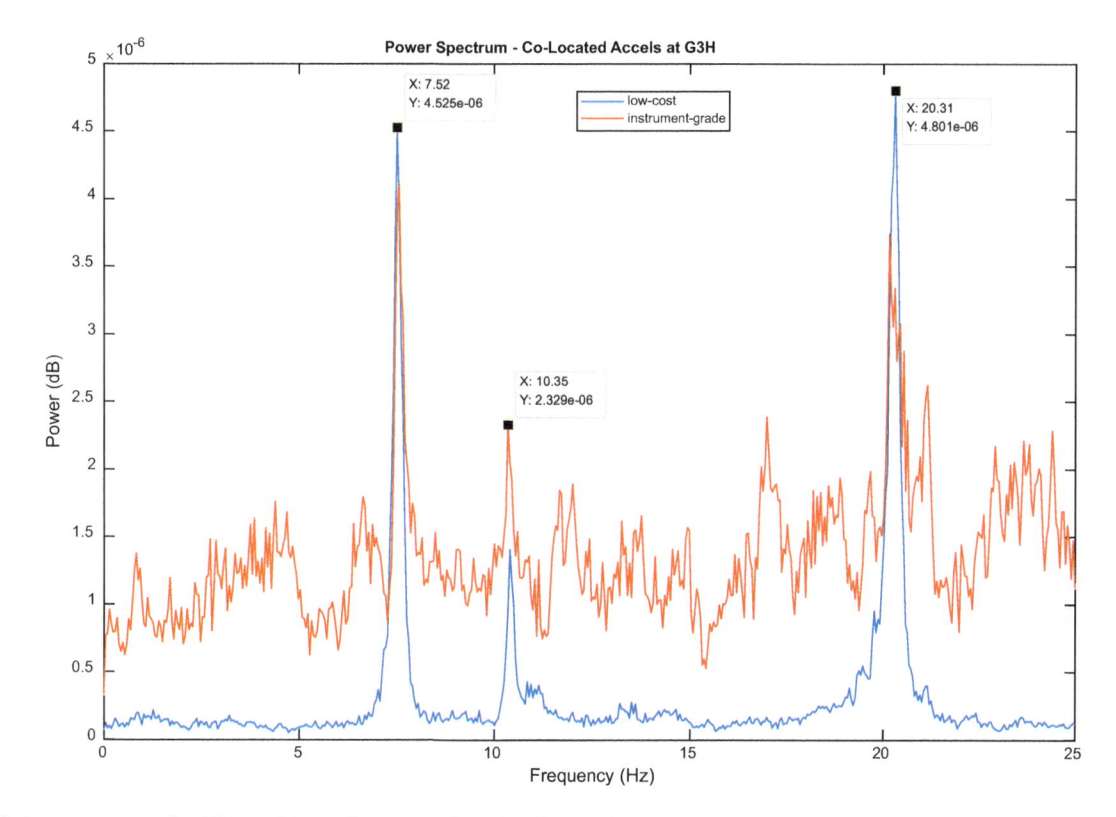

Fig. 19.6 Power spectrum for side-by-side accelerometers from October 2018 measurements

19.5 Conclusions

The performance of a low-cost MEMS accelerometer with analog outputs was evaluated for harmonic excitation cases in the laboratory and for ambient vibration measurements recorded from an in-service highway bridge. In both cases, the performance of the low-cost accelerometer was compared to that of a more expensive instrument-grade accelerometer. The laboratory results indicated that the low-cost accelerometer is quite capable of identifying natural frequencies to 100 Hz although its nominal frequency band is only 50 Hz. The low-cost accelerometer shows definite promise for use in measuring bridge vibrations and identifying their dynamic characteristics. Some additional evaluation work is necessary to completely characterize the performance capabilities of this sensor.

The vibration amplitudes identified by the low-cost accelerometer tended to be smaller than those measured by the instrument-grade accelerometer, and the difference becomes greater at higher frequencies. This could be an issue for estimating mode shapes if these low-cost accelerometers are mixed with other types of accelerometers to dynamically characterize a bridge. There should not be an issue with identifying mode shapes if only the low-cost accelerometers are used to perform the measurements.

A phase lag between the low-cost and instrument grade accelerometer was also observed from the laboratory measurements. The phase lag increases with increasing frequency, especially beyond the nominal frequency band of the low-cost accelerometer. This would also be problematic for estimating mode shapes if mixed types of accelerometers are used with these to perform the measurements and would lead to some distortion in the mode shapes. It is less problematic if only the low-cost accelerometers are used for the measurements.

The field evaluation compared the low-cost accelerometer to a different type of instrument-grade accelerometer than was used for the laboratory study. In this case, both accelerometers were MEMS type sensors. Based on some preliminary analysis of these measurement data, the low-cost accelerometer results appear to be fairly good. The level of noise observed for this sensor was much less than for the instrument-grade accelerometer that was installed side-by-side with it on the bridge. The low-cost and instrument-grade accelerometers should be analyzed using the forced- and free-vibration measurements that are recorded every time a truck crosses the bridge. These measurements should be used to evaluate and compare natural frequencies, mode shapes and damping rations. A significant amount of ambient vibration measurement data has been recorded from the bridge using the low-cost and instrument-grade accelerometers. This data will also be analyzed to identify the dynamic characteristics of the bridge, and the results compared with those identified from the forced- and free-vibration measurements. The bridge monitoring system includes a weather station, and this data can be evaluated in conjunction with the dynamic characteristics identified from the vibration measurements.

References

1. Chung, H.C., Enotomo, T., Loh, K., Shinozuka, M.: Real-time visualization of bridge structural response through wireless MEMS sensors. In: Proceedings of SPIE 5392, Testing, Reliability, and Application of Micro- and Nano-Material Systems II (2004)
2. Lynch, J.P., Loh, K.J.: A summary review of wireless sensors and sensor networks for structural health monitoring. Shock Vib. Digest. **38**(2), 91–128 (2006)
3. Yu, Y., Ou, J., Li, H.: Design, calibration and application of wireless sensors for global and local monitoring of civil infrastructures. Smart Struct. Syst. **6**(5–6), 641–659 (2010)
4. Yu, Y., Han, R., Zhao, X., Mao, X., Hu, W., Jiao, D., Ou, J.: Initial validation of mobile-structural health monitoring method using smartphones. Int. J. Distrib. Sens. Netw. 1–4 (2015). https://doi.org/10.1155/2015/274391

Chapter 20
Theoretical and Experimental Verifications of Bridge Frequency Using Indirect Method

Shota Urushadze, Jong-Dar Yau, Yeong-Bin Yang, and Jan Bayer

Abstract According to the idea of indirect bridge frequency measurement proposed by Yang and co-workers (J Sound Vib 272(3–5), 471–493, 2004; Eng Struct 27(13), 1865–1878, 2005; Smart Struct Syst 13(5), 797–819, 2014), a moving test vehicle can be regarded as a message receiver to detect vibration data of the bridge that it passed. In the present study, an experimental setup will be carried out for indirect frequency measurement of a simply supported beam using a passing test vehicle with the feature of adjustable frequencies. The test vehicle is design as a single-degree-of-freedom unit in vertical vibration and guided by a set of tensile strings in self-equilibrium state so that it can drive the vehicle to move along the beam axis with full contact. To remain the test vehicle running over the beam at constant speed, this study proposed a set of cantilever spiral spring devices to adjust the frequency by regulating the arch length of the spiral springs. From the present experimental results, the indirect bridge inspection method is applicable to frequency monitoring of a bridge. Moreover, the harder stiffness adjusted by the spiral spring device can give a more accurate prediction for measuring bridge frequencies than the softer one.

Keywords Bridge health monitoring · Indirect method · Moving load · Vehicle-bridge interaction

20.1 Introduction

For conventional bridge structural healthy monitoring (SHM), a lot of sensors are installed on a bridge directly, which is cost-expensive and work-intensive. To simplify the bridge monitoring procedure in practice, Yang et al. [1] proposed a vehicle-bridge interaction (VBI) model to extract beam frequencies from the response of a passing sprung mass unit. This approach is referred to indirect monitoring method [2, 3]. As a test vehicle is traveling on a bridge, the passing car can be regarded as an active actuator to excite the bridge and also as a response receiver to capture the vibration data of bridge on which it moves [4, 5]. With this concept, an experimental setup for measuring the bridge frequency will be carried out for verification of the indirect method. From the present experimental results, the indirect method is verified to be an efficient and mobility in assessment techniques appropriate for bridge structural health monitoring.

20.2 Response of a Vehicle Running on a Simple Beam

Let us consider the simplified model shown in Fig. 20.1 for a sprung mass moving on a simple beam with smooth pavement. [6]. For the case of a moving load across the span of a beam, which is transient in nature, the response of the beam can be well simulated by considering only the fundamental mode of vibration [7].

In Fig. 20.1, the following parameters are adopted for the beam: m = mass per unit length, c = damping, L = span length, EI = flexural rigidity, and the following for the sprung mass unit: v = moving speed, M_v, = lumped mass, and k_v = spring

---

S. Urushadze (✉) · J. Bayer
Institute of Theoretical and Applied Mechanics, CAS CR, v.v.i, Prague, Czech Republic
e-mail: urushadze@itam.cas.cz

J.-D. Yau
Department of Architecture, Tamkang University, New Taipei City, Taiwan

Y.-B. Yang
School of Civil Engineering, Chongqing University, Chongqing, China

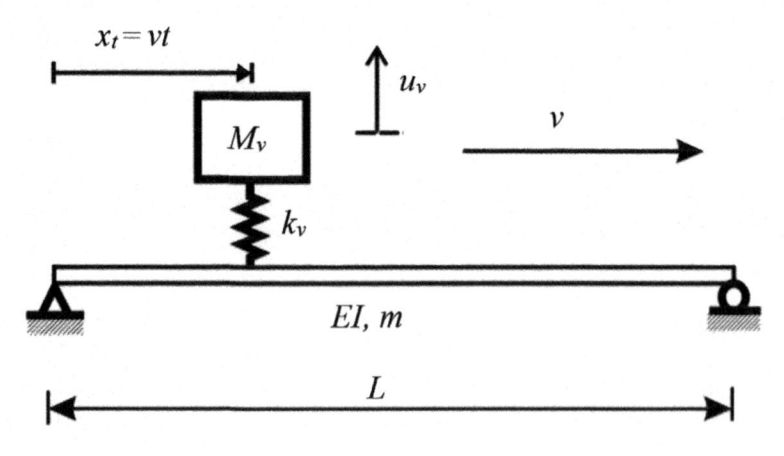

Fig. 20.1 Schematic diagram of a sprung mass moving on a beam

stiffness. We can write the equations of motion for the beam and the sprung mass moving over the beam as [1]:

$$m\ddot{u} + c\dot{u} + EIu'''' = -(p_0 - M_v \ddot{u}_v)\delta(x - vt) \quad 0 \leq t \leq L/v \tag{20.1}$$

$$M_v \ddot{u}_v + k_v u_v = k_v u(x_t, t) \tag{20.2}$$

where $(\bullet)' = \partial(\bullet)/\partial x$, $(\dot{\bullet}) = \partial(\bullet)/\partial t$, $u(x,t)$ = vertical deflection of the beam, u_v = vertical displacement of the sprung mass, $p_0 = M_v g$ = weight of the sprung mass, g = gravity acceleration, L = span length, $\delta(\bullet)$ = Dirac's delta function, and $x_t = vt$ = the position of the moving sprung mass on the beam. For a simply supported beam, the following boundary conditions are adopted:

$$u(0, t) = u(L, t) = 0, \quad EIu''(0, t) = EIu''(L, t) = 0 \tag{20.3}$$

From the viewpoint of practical bridges, the mass of a bridge mL is usually much larger than that of a running vehicle M_v, i.e. $M_v/mL \ll 1$ [8], As shown in reference [8], the numerical studies demonstrated that once the mass ratio of a coach to a simply supported bridge is smaller than 0.05, the dynamic effect of moving vehicles moving on the bridge could be neglected, which is the case studied in this paper. So the inertial force ($M_v \ddot{u}_v$) in Eq. (20.1) can be neglected and the deflection $u(x,t)$ of the beam subjected to a moving static force can be approximated as [1]

$$u(x, t) = \sum_{n=1}^{\infty} \Delta_s \left[\sin \frac{\pi v t}{L} - S_1 \sin \omega_b t \right] \times \sin \frac{\pi x}{L} \tag{20.4}$$

in which the speed parameter S_1 is defined as

$$\Delta_{s1} = \frac{-2 p_0 L^3}{EI \pi^4 (1 - S_1^2)} \tag{20.5}$$

$$S_1 = \frac{\pi v}{\omega_b L} \tag{20.6}$$

$$\omega_b = \left(\frac{\pi}{L}\right)^2 \sqrt{\frac{EI}{m}} \tag{20.7}$$

With the beam response shown in Eq. (20.4), one can obtain the displacement response of the sprung mass as

$$u_v(t) = \frac{\Delta_{s1}}{(1-S_v^2)} \left[\frac{1}{2} \left(1 - \cos \frac{2\pi vt}{L} \right) - S_v^2 \sin^2 \frac{\omega_v t}{2} \right]$$
$$+ \frac{\Delta_{s1} S_1}{2} \left[C_{b1} \cos \left(\omega_b t - \frac{\pi vt}{L} \right) + C_{b2} \cos \left(\omega_b t + \frac{\pi vt}{L} \right) + C_v \cos(\omega_v t) \right] \tag{20.8}$$

where

$$S_v = \frac{2\pi v}{\omega_v L} \tag{20.9a}$$

$$C_{b1} = \frac{-1}{1 - \left(\frac{\omega_b}{\omega_v} - \frac{S_v}{2}\right)^2} \tag{20.9b}$$

$$C_{b2} = \frac{1}{1 - \left(\frac{\omega_b}{\omega_v} + \frac{S_v}{2}\right)^2} \tag{20.9c}$$

$$C_v = -C_{b2} - C_{b1} \tag{20.9d}$$

As shown in Eq. (20.8), the vehicle response contains three important components: vehicle frequency (ω_v), bridge frequency (ω_b) and driving frequency ($2\pi v/L$). Let us consider the case that the moving speed of the test vehicle is restricted in very low speeds (<2m/s), then one can find the approximation of $\omega_b \pm \pi v/L \approx \omega_b$, from which Eq. (20.8) can be approximated as

$$u_v(t) \approx \frac{\Delta_{s1}}{(1-S_v^2)} \left[\frac{1}{2} \left(1 - \cos \frac{2\pi vt}{L} \right) - S_v^2 \sin^2 \frac{\omega_v t}{2} \right] - \frac{\Delta_{s1} S_v^2}{2} \left[1 - (\omega_b/\omega_v)^2 \right]^2 \cos(\omega_b t) \tag{20.10}$$

So the dominant frequency ω_b of the test beam would exist in Eq. (20.10) and one can detect the natural frequency of the test beam using the passing test vehicle. For this consideration, the flexible footbridges would not be considered using the present approach. Therefore, the indirect bridge monitoring method is useful to perform bridge health monitoring. In the following section, a VBI experimental setup will be introduced and used to verify the feasibility of the indirect method.

20.3 Experimental Setup of Indirect Method for Beam Frequencies

As shown in Fig. 20.2, a simple laboratory model for a test vehicle running on a simple beam was constructed for the indirect bridge frequency monitoring. The experimental simulation of a passing vehicle was carried out, in which the dimensions of the plexi-glass beam are as follows: length 2 m and mass of model without supports is 8.3 kg, see Fig 20.2 Vehicle model mass is about 400 g and logarithmic decrement of the beam model is $\vartheta = 0.294$. For this test, the vehicle to bridge mass ratio is 1/20. The response to the moving mass was measured during the passing over the beam and for the verification of dynamic parameters of the beam was also measured at several places. The experimental test set-up is shown in Fig. 20.3. The present experimental setup allows us to change the moving speed and stiffness parameters of the test vehicle.

20.4 Implementation of the Test Results

The purpose of an experiment was to try extract the fundamental frequency of the beam from the time history vertical vibration response of a moving vehicle during its passage over the beam. According to the present experiment test, the first natural frequency of the beam model is 6.348 Hz. Some interesting results were already measured from the vibration signal processing collected from the passing vehicle, in which the dominant frequency is shown in agreement with the first natural frequency of the beam. The experimental results are shown for different spring stiffness in Fig. 20.4 as well.

For the case hard spring where the vehicle passes through the beam at a speed of $v = 0.327$ m/s, the vertical displacements of the vehicle and the bridge midpoint obtained by the two approaches have been plotted in Fig. 20.5.

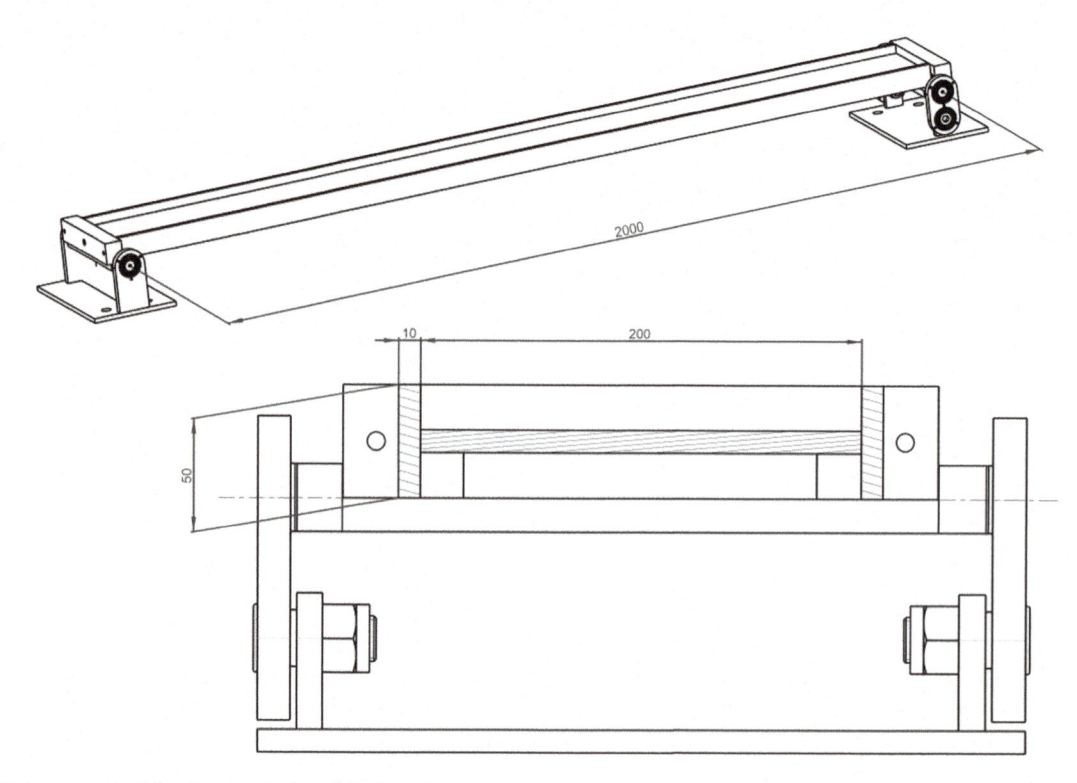

Fig. 20.2 Laboratory model and cross-section of the beam

Fig. 20.3 The test set up

As can be seen from Figs. 20.4 and 20.5, the solutions obtained by the experiments show first natural frequency accordance between the beam and vehicle response. But it was a problem to find agreement at a higher frequency, which was caused by the high damping coefficient of the beam model.

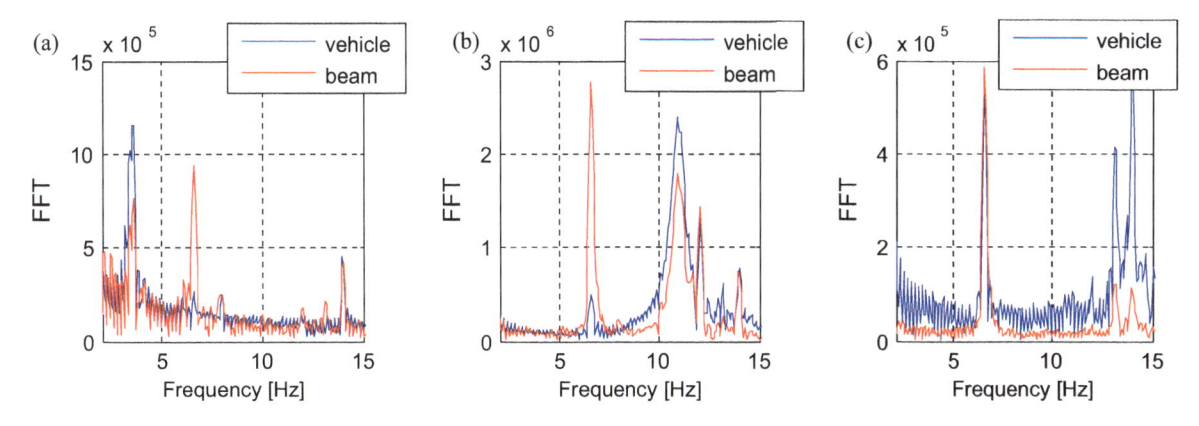

Fig. 20.4 Experimental results for beam frequency analysis with various springs; (**a**) soft spring; (**b**) medium spring; (**c**) hard spring

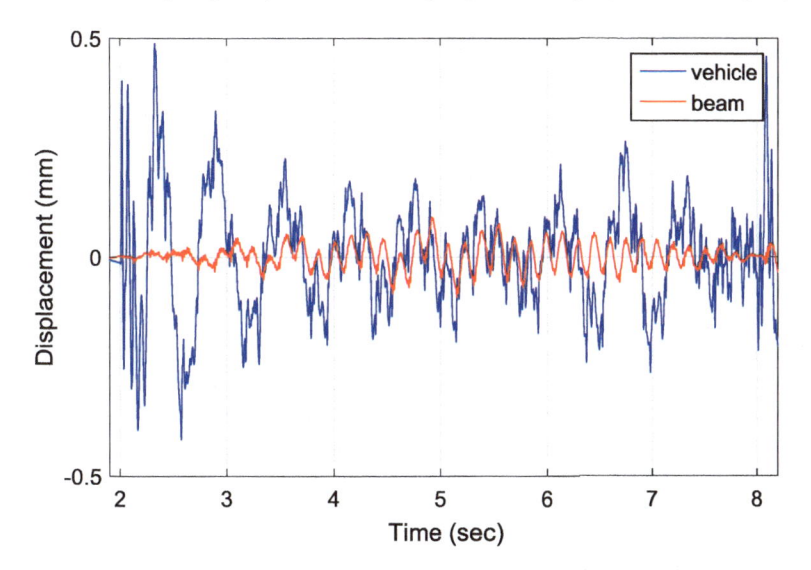

Fig. 20.5 Vertical displacement response of (*blue line*) vehicle and (*red line*) beam midpoint ($v = 0.327$ m/s)

20.5 Conclusion

This paper represents a preliminary experimental study on the feasibility of detecting the fundamental bridge frequency from the dynamic response of a vehicle passing over the bridge. From both the analytical and experimental studies, it is found out that the bridge frequency is included in and can be extracted from the vehicle acceleration spectrum. The results obtained from the tests were carried out in the laboratory ITAM AS CR shows the potential applications of using the proposed indirect method to identify the dynamic characteristics of the bridge, e.g. the natural frequency. Moreover, a harder spring equipped in the vehicle for the experiment can gives better predictions of the beam frequencies than the softer ones. Moreover, the present experimental results have verified the feasibility of using a passing test vehicle to detect dynamic information of a bridge, which can be regarded as a preliminary study using the indirect frequency measurement for bridges. Future research studies should be continued to address the experimental works, which are not covered in this preliminary study, including different roughness of pavement, damping and suspension mechanisms of the vehicle, measurements in situ, and so on.

Acknowledgements The supports of grants GA CR 17-26353J and joint project via grant MOST 106-2923-E-032-007-MY3 are gratefully acknowledged. Identification code of research project of the Institute of Theoretical and Applied mechanics is AVOZ 6838297.

References

1. Yang, Y.B., Lin, C.W., Yau, J.D.: Extracting bridge frequencies from the dynamic response of a passing vehicle. J. Sound Vib. **272**(3–5), 471–493 (2004)
2. Lin, C.W., Yang, Y.B.: Use of a passing vehicle to scan the bridge frequencies - an experimental verification. Eng. Struct. **27**(13), 1865–1878 (2005)
3. Yang, Y.B., Li, Y.C., Chang, K.C.: Constructing the mode shapes of a bridge from a passing vehicle: a theoretical study. Smart Struct Syst. **13**(5), 797–819 (2014)
4. Yang, Y.B., Yau, J.D.: Vertical and pitching resonance of train cars moving over a series of simple beams. J. Sound Vib. **337**, 135–149 (2015)
5. Yau, J.D., Yang, Y.B.: Vertical accelerations of simple beams due to successive loads traveling at resonant speeds. J. Sound Vib. **289**, 210–228 (2006)
6. Fryba, L.: Vibration of Solids and Structures under Moving Loads, 3rd edn. Thomas Telford, London (1999)
7. Biggs, J.M.: Introduction to Structural Dynamics, pp. 315–328. McGraw-Hill, New York (1964)
8. Yau, J.D., Fryba, L.: A quasi-vehicle/bridge interaction model for high speed railways. J Mech. **31**, 217–225 (2015)

Chapter 21
A Bayesian Inversion Approach for Site Characterization Using Surface Wave Measurements

Mehdi M. Akhlaghi, Babak Moaveni, and Laurie G. Baise

Abstract This paper presents a Bayesian inference method for the characterization of soil properties and stratigraphy for site response analysis using surface wave measurements. The method is evaluated using numerically simulated data for surface wave methods (e.g. SASW) and ambient noise methods (e.g. H/V) using horizontal, homogeneous, layered soil models. Wave propagation—both vertically propagating shear waves and horizontally propagating surface waves—through one-dimensional horizontally layered media and a linear soil constitutive model are assumed for the simulation of measured data. The inversion process is performed using error functions defined as the difference between selected features of model-predictions and measurements. The considered features include dispersion curves obtained from surface wave propagation and the estimated transfer functions from H/V spectral ratios of ambient noise assuming vertically propagating shear waves. The numerical study was performed to evaluate the proposed approach for horizontally layered sites where the number of layers, layer thicknesses, and layer properties are varied. Different levels of measurement noise, modeling errors, number of updating parameters, number of data sets, and error functions are considered. The numerically simulated sensor data are polluted with independent Gaussian white noise vectors at three different noise levels (0.5%, 1%, and 2% in terms of response root-mean-square) and their effects on bias and covariance of updating parameters are studied. Furthermore, the sensitivity of the results to the updating parameters (e.g., layer height, density, damping, and stiffness), available measurements (type, quantity, quality) and the used error functions (e.g., selected points on dispersion curves or H/V transfer function) are investigated. The value of information will be assessed for the inclusion of the ambient noise data in the Bayesian inversion in terms of reduction in site parameter uncertainty.

Keywords Inversion problem · Bayesian inference · Dispersion curve · H/V transfer function · Shear waves

21.1 Introduction

Spectral Analysis of Surface Waves (SASW) tests and H/V microtremor methods have gained more attention in site response prediction in recent decades. The main advantage of these methods compared to invasive methods like down-hole and cross-hole methods is their lower cost which is achieved by eliminating the dependency to the deep boreholes needed for the invasive methods. But one of the biggest challenges in using these approaches in response prediction of soil layers is how to interpret the results. The accuracy of these predictions is highly dependent on the characterization of soil properties and stratigraphy and also on the wave propagation formulation used in the process including both the soil constitutive model and the wave types considered for the analysis (most commonly assuming vertical propagation of Shear Waves through one-dimensional horizontally layered media—SH1D) [1]. Advances in the computational speed and efficiency of numerical algorithms in recent years have caused renewed efforts in improving the accuracy of these methods. Both dispersion curves from the SASW test and the transfer functions from the H/V microtremor data has been used for the inversion process [2–4]. In current research study, the data from both sources are combined to give us a more accurate estimate of the soil properties. Eventually, the effect of different noise levels in the measured data is investigated in the uncertainties of the estimated soil parameters.

M. M. Akhlaghi (✉) · B. Moaveni · L. G. Baise
Department of Civil and Environmental Engineering, School of Engineering, Tufts University, Medford, MA, USA
e-mail: mehdi.akhlaghi@tufts.edu

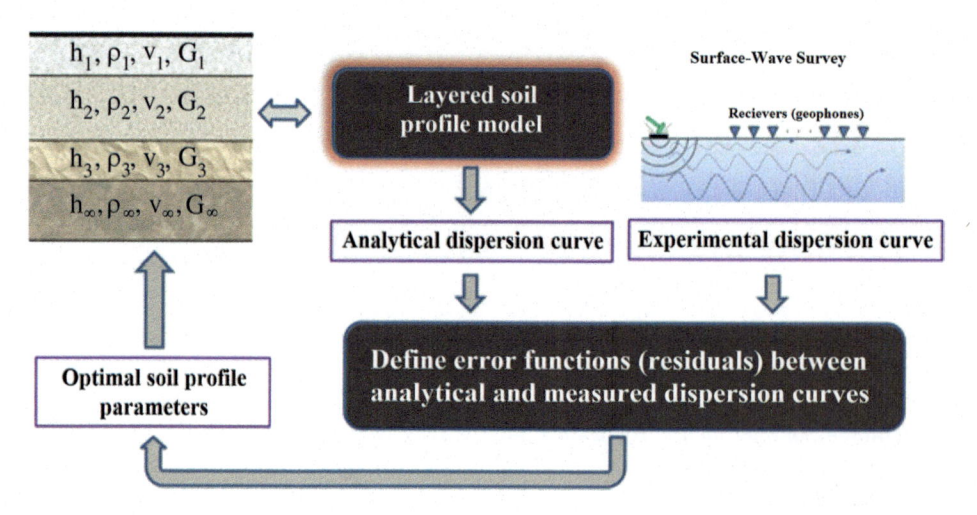

Fig. 21.1 Inversion process

21.2 Analysis

As a common approach in inversion problems for site characterization, error functions are defined as the difference between the values of selected parameters from in-situ measurements and the numerical model. The defined error functions in our study are based on the data from both the dispersion curves and the H/V spectral ratios:

$$e_f = \tilde{f} - f(\boldsymbol{\theta}) \tag{21.1}$$

$$\mathbf{e_y} = \tilde{\mathbf{y}} - \mathbf{y}(\boldsymbol{\theta}) \tag{21.2}$$

In these equations e_f is the fundamental natural frequency error function which is defined as the difference between the natural frequencies from the experimental H/V data and the model, $\tilde{f}$ and $f(\boldsymbol{\theta})$ respectively; and $\mathbf{e_y}$ is dispersion curve error function defined as the difference between the vector of selected points on the dispersion curve of the experimental data and the model, $\tilde{\mathbf{y}}$ and $\mathbf{y}(\boldsymbol{\theta})$ respectively. The goal of the inversion process, as shown in Fig. 21.1, is to minimize the value of these error functions through an updating process.

Bayesian inference is used in this study instead of the optimization stage common in the deterministic approaches. This results in a set of sample points that can represent the level of trust in our results for each parameter and so will help us to investigate the effects of uncertainties in the modeling and the data noise. This method uses the familiar Bayesian inference formula:

$$p\left(\boldsymbol{\theta}|\tilde{\mathbf{Y}}\right) \propto p\left(\tilde{\mathbf{Y}}|\boldsymbol{\theta}\right) p(\boldsymbol{\theta}) \tag{21.3}$$

where $\boldsymbol{\theta}$ is the vector of all soil profile paramters, $\tilde{\mathbf{Y}}$ is the vector of all natural frequencies and dispersion curve values from the experiments, $p\left(\tilde{\mathbf{Y}}|\boldsymbol{\theta}\right)$ is the likelihood function defined based on the error functions in (21.1) and (21.2) and $p(\boldsymbol{\theta})$ is the prior term, reflecting our initial understanding of the system. Markov chain Monte-Carlo sampling is used in the model updating process and effects of different choices of updating parameters and error functions on the efficency and accuracy of the method is investigated. Also a set of nemrically simulated errors are added to the data and the effect of this noise is studied for different scenarios.

Acknowledgements Partial support of this study by the United States Geological Survey Grants G18AP00034 is gratefully acknowledged. The opinions, findings, and conclusions expressed in this paper are those of the authors and do not necessarily represent the views of the sponsors and organizations involved in this project.

References

1. Thompson, E.M., Baise, L.G., Kayen, R.E., Guzina, B.B.: Impediments to predicting site response: seismic property estimation and theoretical simplifications. Bull. Seismol. Soc. Am. **99**(5), 2927–2949 (2009)
2. Xia, J., Miller, R.D., Park, C.B.: Estimation of near-surface shear-wave velocity by inversion of Rayleigh waves. Geophysics. **64**(3), 691–700 (1999)
3. Arai, H., Tokimatsu, K.: S-Wave velocity profiling by joint inversion of microtremor dispersion curve and horizontal-to-vertical (H/V) spectrum. Bull. Seismol. Soc. Am. **95**(5), 1766–1778 (2005)
4. Teaguea, D.P., Coxb, B.R., Rathje, E.M.: Measured vs. predicted site response at the Garner Valley Downhole Array considering shear wave velocity uncertainty from borehole and surface wave methods. Soil Dyn. Earthq. Eng. **113**, 339–355 (2018)

Chapter 22
Estimating Fatigue in the Main Bearings of Wind Turbines Using Experimental Data

Giovanni M. Fava, Sauro Liberatore, and Babak Moaveni

Abstract Axial bearing failures, in modern wind turbines, have been one of the major sources of unplanned maintenance expenses for the wind energy business. Although the causes of such failures are unclear, its statistics do increase in wind farms where the wind conditions range in considerably larger spans. It is common belief that torsional vibrations, combined with axial ones resulting from these extreme wind conditions are the major causes of high cycle, fatigue affecting the bearing. In this study, the strain measurements on blades are used to estimate the internal force time histories at the location of the main bearing. A rigid body equilibrium analysis was performed using moments and forces measured at the blades to estimate the force and moments acting on the low speed shaft. Experimental wind data was used to compare the difference between strong wind speed events and low wind speed events. The approach uses the blade bending-moments to isolate the axial excitations and is validated by measured data from an instrumented turbine with four interferometric strain sensors and four temperature sensors for each blade. The instrumented turbine is part of an onshore wind farm site, in Cohocton, NY, where wind conditions span in a large range and turbines are often placed in stop conditions. Preliminary data has shown some large excitations that could eventually lead to crack initiation.

Keywords Wind turbines · Stress estimation · Fatigue analysis · Main bearing monitoring

22.1 Introduction

The main bearing of a wind turbine is subjected to many varying loads and moments, the majority of which are cyclical in nature, as a result, continuous fatigue monitoring is imperative. Main bearing failure is often a reason for a loss in revenue and efficiency for a wind turbine [1]. Computing easily obtained, accurate fatigue estimation would allow for more time spent between bearing replacement and allow for detection of dangerous conditions for the bearing. Typical monitoring systems can be difficult to install and can be expensive [2].

Currently, there are multiple techniques for condition monitoring for bearings. The first method is to perform vibration analysis, this method is commonly used to find and trace the growth faults in the bearings [3]. Vibro-acoustics have also been shown to work, but it can be difficult to collect accurate sound readings from a nacelle [4]. Thermography can often be too expensive to be used frequently [5]. Shaft torque and torsional vibration measurements are generally expensive and can be difficult to install in newer turbines that have less free space in the nacelle [6]. Strain sensors placed on a wind turbine blade are often used to measure the blade moments, calculate fatigue and monitor wind turbine blades [7]. Bearing data in the frequency domain can be used to find faults in the bearing [8]. Fourier amplitude spectra are commonly used in structural health monitoring to find faults in structures, and this applies to bearings as well [9]. Main bearings are often spherical roller bearings in large wind turbines, but other bearings are also used [10].

This paper presents an analysis of the forces and moments applied to the main bearing of a clipper C96 turbine using measurements obtained from four strain sensors placed on each blade. A finite element model of the system was created using SAP2000 software [11] that can predict three dimensional moments and forces at the base of each blade as function of forces and moments applied to the main bearing for the low speed shaft. The aim of the paper is to provide a method of monitoring the fatigue applied to the main bearing of a wind turbine, using sensors that are easily accessible, and often used

---

G. M. Fava (✉) · S. Liberatore
Department of Mechanical Engineering, Tufts University, Newton, MA, USA
e-mail: giovanni.fava@tufts.edu

B. Moaveni
Department of Civil and Environmental Engineering, School of Engineering, Tufts University, Medford, MA, USA

to monitor other conditions, and as a result, may already be installed in a wind turbine. A comparison between strong wind conditions and weak wind conditions will also be performed, to show the results of the sensors.

22.2 Instrumentation of the Considered Wind Turbine

This study focuses on a Clipper C96 wind turbine that is part of a Cohocton wind farm in upstate New York. The Clipper C96 has a 96-m rotor diameter, with three blades each that measure 46.7 m long [12]. The blades of the turbine are made from fiberglass, and we assumed that the Young's modulus for the fiberglass was 65 Gpa [13]. Data was collected from four interferometric strain sensors placed on each blade. These sensors were paired with temperature sensors in the same location that would be later used to properly temperature compensate the raw data [14]. These sensors collected data at a frequency of 100 Hz and the turbine was monitored for nearly 6 consecutive months between 2014 and 2015. Other sensors were also placed on various points of the nacelle, measuring windspeed, as well as other variables at a rate of 0.5 Hz. These sensors were placed 1.5 m from the base of each blade in the pattern shown in Fig. 22.1. The sensors were placed with a tolerance of 10 cm from their intended location. A few assumptions were also made with regards to the placement of the sensors; it was assumed that the instrumentation cross-section measured by the sensors was 2.8 m away from the center point of the rotor hub. It was also assumed that the section was perpendicular to the blade, and not askew. In making the calculations, it was assumed that the structure was rigid. This assumption was made because the deflection involved was small compared to the size of the low speed shaft, rotor hub, and bottom 1.5 m of the blade. The turbine was rated to perform ideally at speeds of up to 11 meters per second [16]. At that speed the turbine could ideally produce 2.5 Megawatts. The data presented in this paper consist of a high wind speed time-history, and a low wind speed time-history. Due to the proprietary nature of some structural details in wind turbine manufacturing, several assumptions were made about the geometry of the turbine which may not be exact for the considered structure. No information about the main bearing was provided by Clipper, as a result we were unable to estimate the fatigue of the bearing, but those calculations could be made by using the resultant forces. Typically, main bearings do not experience certain moments, but these moments were still calculated to provide a more comprehensive understanding of what was occurring inside the low speed shaft.

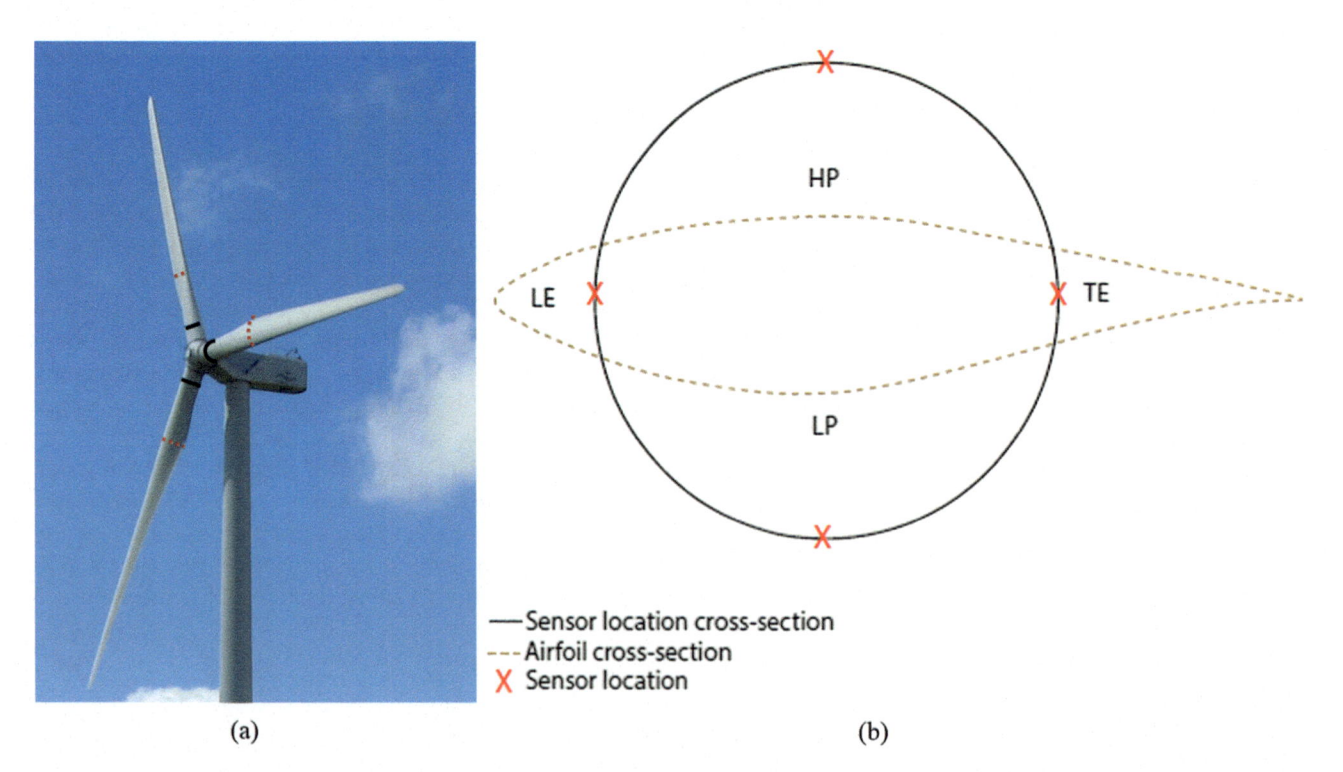

Fig. 22.1 Wind turbine cross-sectional view (*HP* high-pressure, *LP* low-pressure, *TE* trailing-edge, *LE* leading-edge point strain gauge). (**a**) Location of cross-sections [15] (**b**) Location of the sensors in black cross-section

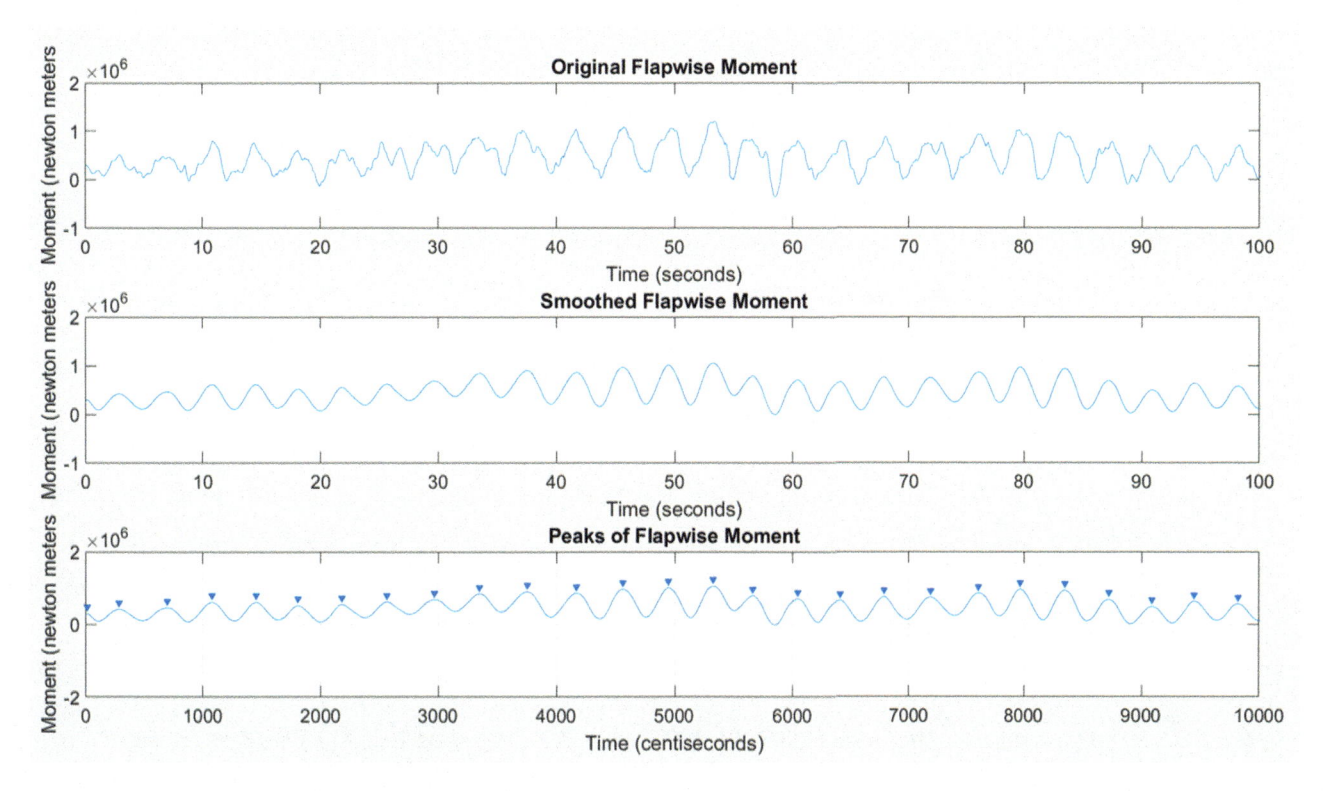

Fig. 22.2 Use of Flapwise moment to find rotation times

22.3 Calculation of Moments

Flapwise Moments were calculated by taking the temperature compensated strain values from the high-pressure strain sensor of a blade, subtracting the low pressure and then multiplying the result by a constant k provided by Clipper. This constant k varied from bladed to blade, and from different points.

$$M_{Flapwise} = (HP - LP) * k_1 \tag{22.1}$$

$$M_{Edgewise} = (TE - LE) * k_2 \tag{22.2}$$

In these equations, HP is the high-pressure point strain gauge, LP is the low-pressure point strain gauge, TE is the trailing-edge point strain gauge, and LE is the leading-edge point strain gauge.

The peaks of the Flapwise moments calculated are very cyclical. Each cycle occurs every full rotation of the blade. Smoothed Flapwise data was found created using MATLAB function "sgolayfilt", and the peaks were found using the function "findpeaks." These cycles can be explained by the gravity acting on the turbine blades when the blade is in different positions. The peaks were then correlated to an angle of the rotor hub, and that allowed for calibration of the peaks to a specific angle. The rotation of the blades was then assumed to not vary significantly during each cycle, and the angle of each blade was calculated this (Fig. 22.2).

22.4 Estimation of Internal Forces at Bearings

The statics equations used in the MATLAB calculations of this paper are listed below [17]. Several assumptions were made about the geometry of the turbine. First, that the radius of the rotor hub was assumed to be 1.3 m, based on subtracting the blade length from half of the rotor diameter to get 1.3 m. The distance from the center of the rotor hub to the main bearing was assumed to be 2 m, based on several images of the nacelle.

$$F_{x(bearing)} = -F_{v1} - F_{v2} - F_{v3} \tag{22.3}$$

$$\begin{aligned}F_{y(bearing)} &= -F_{n1}{}^*\cos(B) - F_{n2}{}^*\cos(B-120) - F_{n3}{}^*\cos(B-240) + F_{z1}{}^*\sin(B) \\ &\quad + F_{z2}{}^*\sin(B-120) + F_{z3}{}^*\sin(B-240)\end{aligned} \tag{22.4}$$

$$\begin{aligned}F_{z(bearing)} &= -F_{n1}{}^*\sin(B) - F_{n2}{}^*\sin(B-120) - F_{n3}{}^*\sin(B-240) + F_{z1}{}^*\cos(B) \\ &\quad + F_{z2}{}^*\cos(B-120) + F_{z3}{}^*\cos(B-240)\end{aligned} \tag{22.5}$$

$$M_{x(bearing)} = -M_{x1} - M_{x2} - M_{x3} - F_{z1}*(R+1.5) - F_{z2}*(R+1.5) - F_{z3}*(R+1.5) \tag{22.6}$$

$$\begin{aligned}M_{y(bearing)} &= -M_{y1}*\cos(B) - M_{y2}*\cos(B-120) - M_{y3}*\cos(B-240) + M_{z1}*\sin(B) + M_{z2}*\sin(B-120) \\ &\quad + M_{z3}*\sin(B-240) - F_{z1}*D*\cos(B) - F_{z2}*D*\cos(B-120) - F_{z3}*D*\cos(B-240) \\ &\quad - F_{n1}*D*\sin(B) - F_{n2}*D*\sin(B-120) - F_{n3}*D*\sin(B-240) - F_{v1}*(R+1.5)*\sin(B) \\ &\quad - F_{v2}*(R+1.5)*\sin(B-120) - F_{v3}*(R+1.5)*\sin(B-240)\end{aligned} \tag{22.7}$$

$$\begin{aligned}M_{z(bearing)} &= -M_{y1}*\sin(B) - M_{y2}*\sin(B-120) - M_{y3}*\sin(B-240) + M_{z1}*\cos(B) + M_{z2}*\cos(B-120) \\ &\quad + M_{z3}*\cos(B-240) - F_{v1}*(R+1.5)*\cos(B) - F_{v2}*(R+1.5)*\cos(B-120) \\ &\quad - F_{v3}*(R+1.5)*\cos(B-240) + F_{n1}*D*\cos(B) + F_{n2}*D*\cos(B-120) \\ &\quad + F_{n3}*D*\cos(B-240) - F_{z1}*D*\sin(B) - F_{z2}*D*\sin(B-120) - F_{z3}*D*\sin(B-240)\end{aligned} \tag{22.8}$$

where D = distance from the center of the rotor hub(origin), to the main bearing, R = radius of the rotor hub, and B = the rotor angle of the first blade of the turbine, where B = 0 means blade 1 is pointing vertically upward along the y-axis, and B = 90 means that the blade is pointing in the direction of the z-axis (Fig. 22.3).

22.5 Validation with Finite Element Model

Equations were validated with the results of an FE model in SAP when inputting local forces and moments to the ends of beams that were 120° apart. The beams in the model were given a Young's Modulus of 10^{20} to ensure that they would be very close to rigid. These forces and moments represent the recorded forces and moments acting on the turbine blade. The length of each beam was constant and represented the cumulative distance between the center of the rotor hub and the sensors. The beam in the Z axis of the SAP model was meant to represent the distance between the center of the rotor hub and the main bearing [11].

After various, differing forces and moments were applied to the ends of the beams, SAP analysis was preformed to find the resultant forces and moments acting at the origin (main bearing). The same moments and forces were applied in MATLAB to validate the equations written above. The results are shown in Table 22.1. The results of the SAP analysis confirmed the MATLAB equations. Note that the SAP diagram is oriented differently than the MATLAB equations, so the results of the SAP test were converted to the MATLAB orientation.

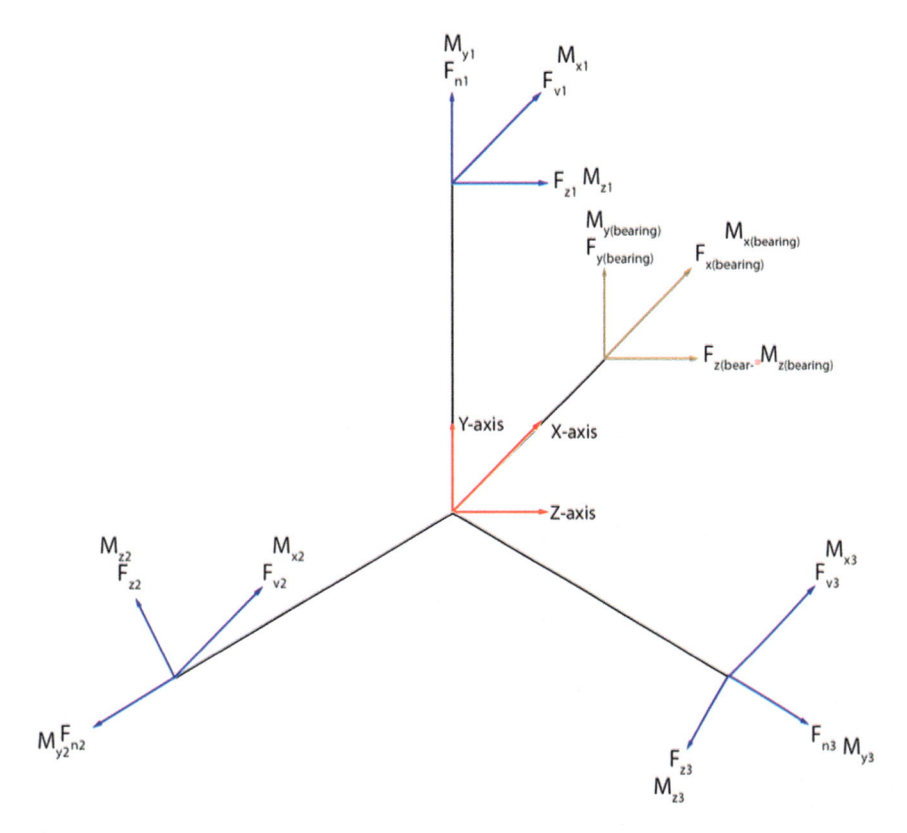

Fig. 22.3 Statics modal free-body diagram of forces and moments applied to turbine

Table 22.1 Comparison of MALTAB resultants at the bearing vs. SAP resultants

	Direction	SAP	MATLAB
Forces	x-axis	−28.000	−28.000
	y-axis	−37.990	−37.990
	z-axis	−9.356	−9.356
Moments	x-axis	58.400	58.400
	y-axis	−58.585	−58.585
	z-axis	37.875	37.875

22.6 Results

The plots below show the wind speed compared to the moments recorded at the base of the first blade. It can be easily seen that the Flapwise moment is correlated to the speed of the wind. As the wind speed drops, the Flapwise moment does the same. The Edgewise moment on the other hand does not have a clear correlation to the wind velocity. This phenomenon is likely caused by the changing pitch angle of the blade based on the wind velocity (Fig. 22.4).

The strong wind event has a different pattern than that of the weak wind event. The Edgewise moment appears to maintain its consistent cyclic pattern. The Flapwise moment in the strong wind condition appears to loosely follow the trends of the wind, but it has trouble surpassing the 50×10^5 N. This range shows that as the velocity of the wind surpasses 13 or 14 meter per second, the wind turbine is not able to get more energy from the wind. This result confirms the Clipper C96 wind turbine rating at 11 meters per second (Fig. 22.5).

The resultant forces in the F_x direction seems to follow a similar shape to that of the Flapwise moment. The reason why the values on the surface appear to differ in pattern is because the forces shown are the reaction forces, are the opposite direction to the forces acting on the shaft. The absolute value of the internal F_x, does match the wind speed. The F_y and F_z forces acting on the bearing are very cyclical in nature, and do not vary significantly with the wind (Fig. 22.6).

During the strong wind event, the internal F_x force appears to match the Flapwise moment calculated in blade 1. The F_y and F_z results from the strong and weak wind events are very similar in shape and magnitude. Is appears as if the internal F_x is the most affected by the wind speed (Fig. 22.7).

The weak event Fast Fourier transform has a distinct peak around 0.3 Hz and a smaller one around 0.5 Hz. The peaks are not very uniform (Fig. 22.8).

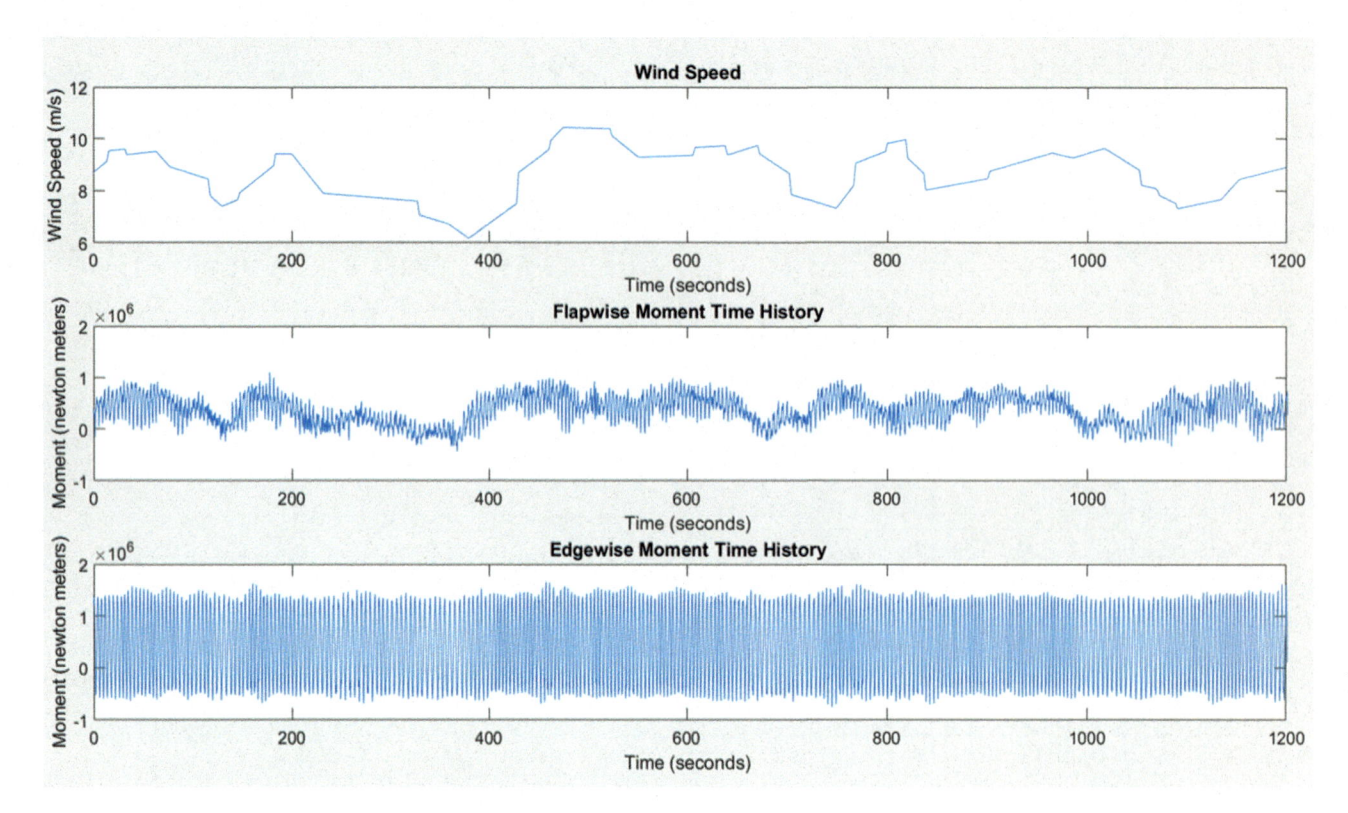

Fig. 22.4 Moment time-histories measured at the blade during a weak wind event

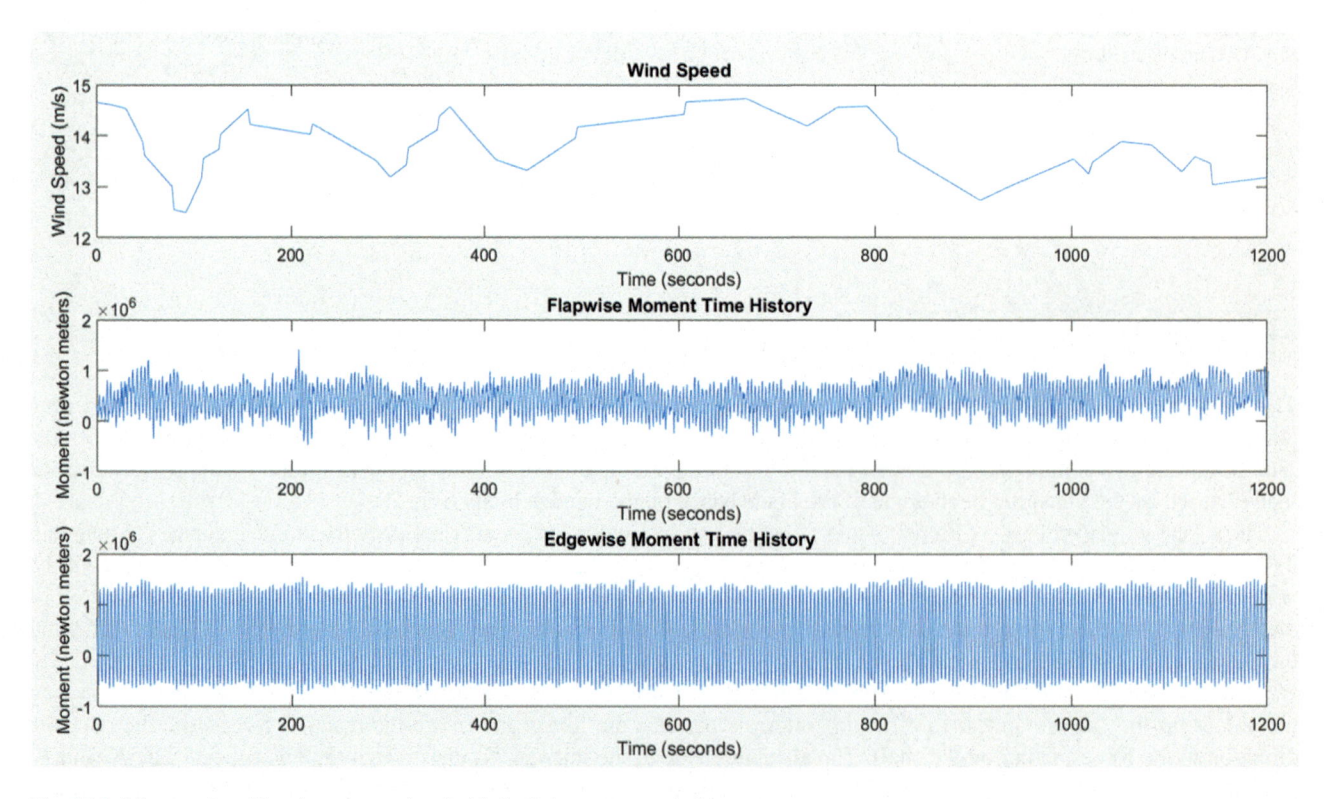

Fig. 22.5 Moment time-histories measured at the blade during a strong wind event

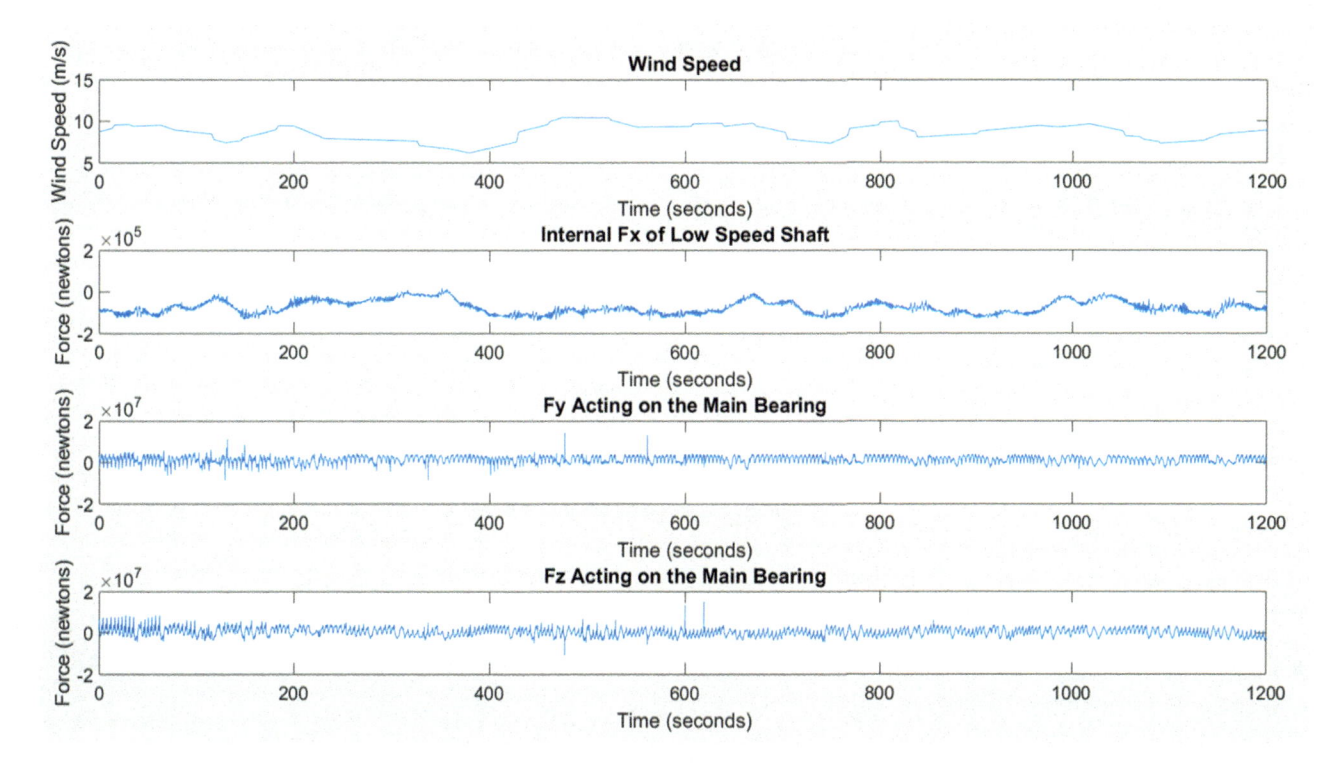

Fig. 22.6 Time-histories of forces calculated at the point of the main bearing during a weak wind event

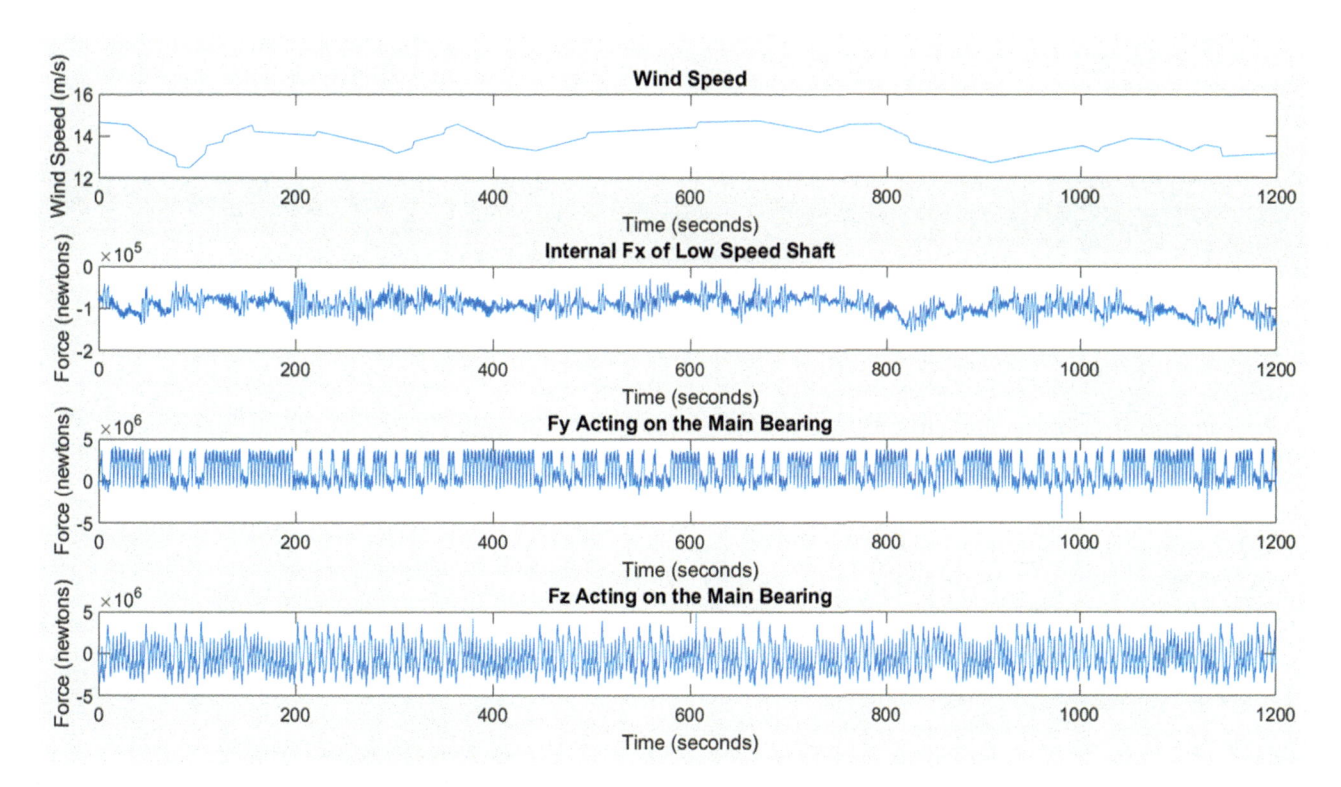

Fig. 22.7 Time-histories of forces calculated at the point of the main bearing during a strong wind event

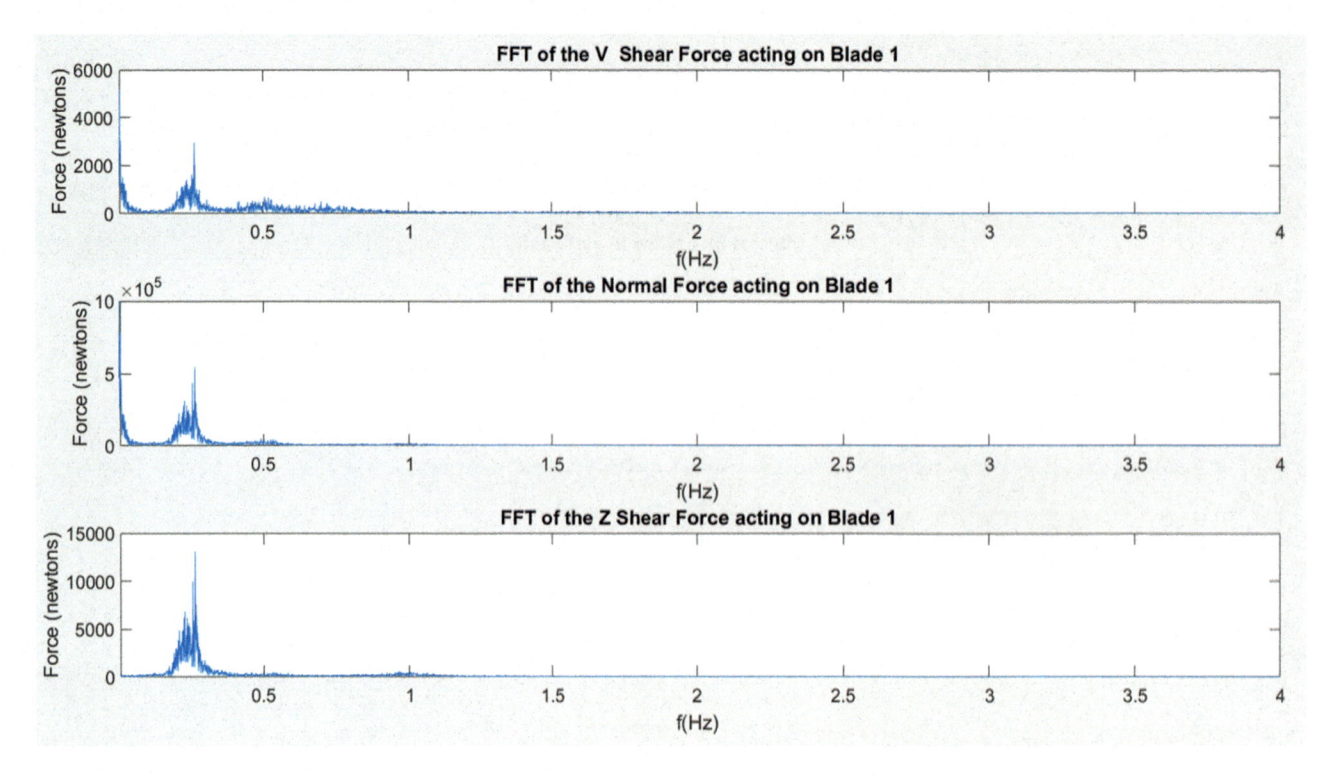

Fig. 22.8 Fast Fourier Transform of forces acting on the turbine blades at the sensor location during a weak wind event

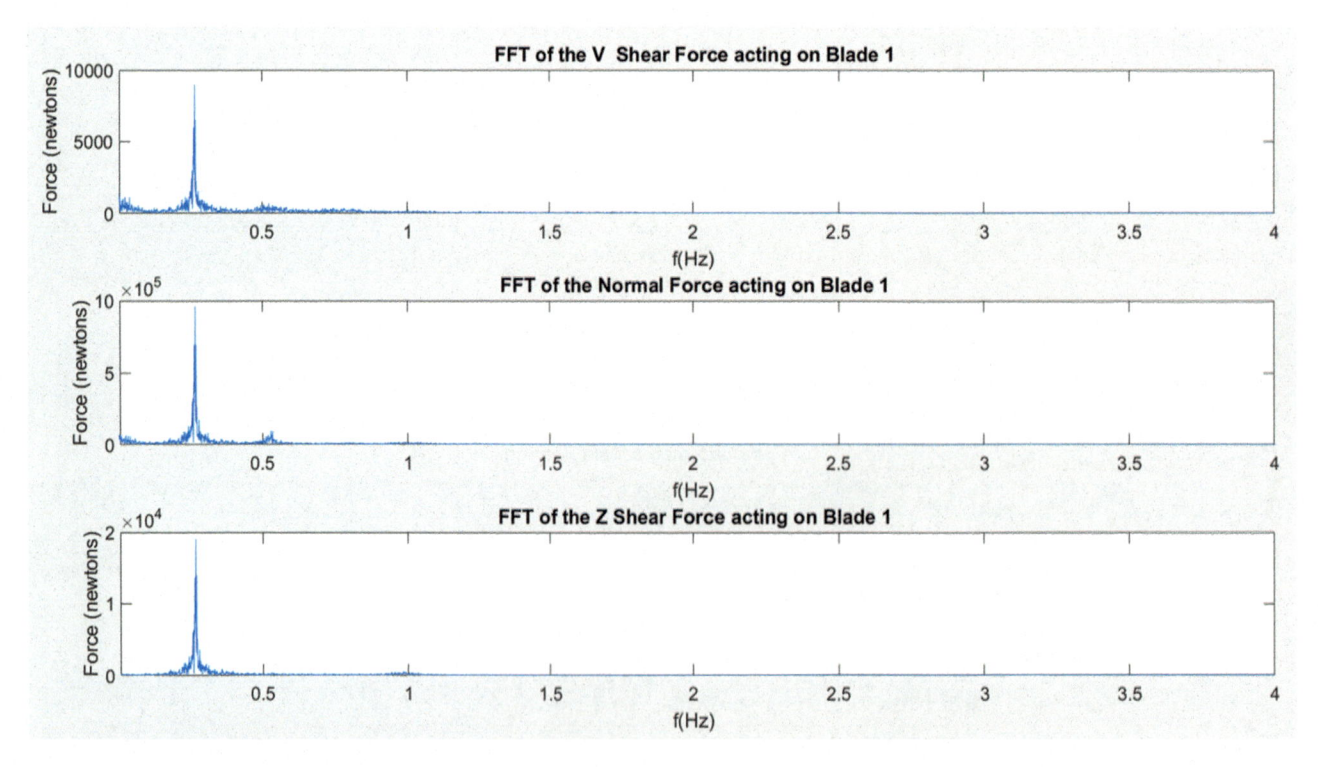

Fig. 22.9 Fast Fourier Transform of forces acting on the turbine blades at the sensor location during a strong wind event

The Fast Fourier Transform of the strong wind event reveals much more distinct and symmetrical peaks around 0.3. There is also a smaller, peak around 0.5 Hz. This peak is very similar to the peaks present in the weak wind event. The Fast Fourier Transform in the Z direction does not have a visible peak at 0.5 Hz in both graphs (Fig. 22.9).

22.7 Conclusion

This paper presents fatigue estimation capabilities for the main bearing of a wind turbine through strain sensors placed at the base of the turbine blades. The paper shows the ability of sensors placed on wind turbines to estimate the forces and moments applied to the low-speed shaft. The results of the strong wind event and the weak wind event show how wind turbines reach a maximum in power generation, and how Flapwise moment are the most proportional to wind speeds. Future research would need to be conducted to estimate fatigue on a particular bearing. The general concepts of this paper are widely applicable to many wind turbines, but with more information, a clear understanding of bearing fatigue can be understood. The methods of the paper can be applied to wind turbines that already have strain gauges mounted on the blades of a turbine, allowing for fatigue monitoring, as well as monitoring for dangerous weather conditions by showing the forces and moments imparted onto the main bearing of a wind turbine. The confidence of such estimates can be increased by a greater knowledge of the materials and geometries present in the turbine.

Acknowledgements The opinions, findings, and conclusions expressed in this paper are those of the authors and do not necessarily represent the views of the organizations involved in this project.

References

1. Wasilczuk, M., Gawarkiewicz, R., Bastian, B.: Analysis of failures of high speed shaft bearing system in a wind turbine. In: 9th International Conference on Tribology, pp. 3–4 (2018)
2. May, A., McMillan, D., Thons, S.: Economic analysis of condition monitoring systems for offshore wind turbine sub-systems, semantic scholar (2014)
3. Shakya, P., Darpe, A., Kulkarni, M.: Vibration based fault diagnosis in rolling element bearings: ranking of various time, frequency and time-frequency domain data based damage identification. Int. J. Cond. Monit. **3**, 1–10 (2013)
4. Mollasalehi, E., Wood, D., Sun, Q.: Indicative fault diagnosis of wind turbine generator bearings using tower sound and vibration: In: 24th International Congress on Sound and Vibration (2017)
5. Schultz, R., Verstockt, S., Vermeiren, J., Loccufier, M., Stockman, K., Van Hoecke, S.: Thermal imaging for monitoring rolling element bearings. In: Proceedings of the 12th International Conference on Quantitative InfraRed Thermography (2014)
6. Yang, W., Tavner, P., Crabtree, C., Feng, Y., Qiu, Y.: Wind turbine condition monitoring: technical and commercial challenges. Wind Energy **17**(5) (2012)
7. Iftimie, N., et al.: Wireless sensors for wind turbine blades monitoring. IOP Conference, vol. 209 (2017)
8. Sonawane, P., Kharate, N.K.: Fault diagnosis of windmill by FFT analyzer. Int. J. Innov. Eng. Technol. **4**(4), 47–54 (2014)
9. Greco, A., et al.: Bearing reliability—White Etching Cracks (WEC), Argonne National Laboratory, Energy Systems Division, NREL Gearbox Reliability Collaborative (2013)
10. Yagi, S.: Bearings for wind turbine. NTN Technical Review, no. 71 (2004)
11. SAP 2000 18, Computers and Structures, Inc., Berkeley, CA (2018)
12. Showers, D.: System Identification for the Clipper Liberty C96 Wind Turbine, Ph.D. Dissertation. University of Minnesota, Minneapolis (2014)
13. Clipper Windpower Plc, Liberty 2.5 MW Wind Turbine: Facts and Specifications (2009)
14. Hartman, D., Greenwood, M., Miller, D.: High Strength Glass Fibers, AGY, p. 3 (1996)
15. Neptull, Hesse, Germany (2012)
16. Kovacic, B., Kamnik, R., Štrukelj, A., Vatin, N.: Processing of signals produced by strain gauges in testing measurements of the bridges. Procedia Eng. **117**, 795–801 (2015)
17. MATLAB R2017a, MathWorks Inc., Natick, MA (2018)

Chapter 23
Cointegration for Detecting Structural Blade Damage in an Operating Wind Turbine: An Experimental Study

B. A. Qadri, M. D. Ulriksen, L. Damkilde, and D. Tcherniak

Abstract Environmental and operational variabilities (EOVs) are known to pose an issue in structural health monitoring (SHM) systems, as these variabilities can mask the effect of structural damage. Numerous approaches to remove, or, at least, mitigate, the effect of EOVs in SHM applications have been proposed and tested through numerical simulations and in experimental studies. One of the approaches that has exhibited promising potential is cointegration, which, in this particular SHM context, is a technique for singling out and removing common signal trends stemming from the EOVs. In the present paper, the cointegration technique is employed to mitigate the effect of certain EOVs in an experimental, vibration-based damage detection analysis of a wind turbine blade under operating conditions. In the experimental campaign, the installed SHM system was recording blade accelerations and different environmental and operational conditions over a 3.5-month period. In the period, one of the blades was treated in its reference state and in damaged states with a trailing edge opening of increasing size. Based on the available data from these different structural states, it is demonstrated how cointegration can be used to successfully detect the introduced damages under conditions not allowing for direct discrimination between damage and EOVs.

Keywords Damage detection · Cointegration · Environmental and operational variabilities · Wind turbine application · Structural health monitoring

23.1 Introduction

Vibration-based damage detection techniques have demonstrated promising results in numerous application studies, see, for example, [1–4]. However, challenges arise in real industrial applications, where environmental and operational variabilities (EOVs) can mask the effect of structural deterioration [5]. As an example, it is described in [6] how EOVs can account for at least 5% shifts in modal parameters; a percentage shift that, as a consequence of damage, typically will require a deterioration extent above what can be accepted in practice. This general issue with the EOVs is inevitable and call for a technique that is capable of differentiating between response changes due to EOVs and damage. One such approach, namely, Johanson cointegration, has been proposed for purging EOVs in some measured structural response in the context of structural health monitoring (SHM), see, for instance, [7–9]. The principle of the Johanson cointegration is to create a stationary relationship between a set of non-stationary measured signals, for example, acceleration measurements.

The scope of the present paper is to explore the use of cointegration in detection of structural damage in the blade of an operating Vestas V27 wind turbine (WT). The experimental campaign encompasses a measurement period of 3.5 months, in which the blade in question was instrumented with 11 piezoelectric accelerometers (distributed along the leading and trailing edge) and an electromechanical actuator (to excite a broad band of frequencies). During the experimental campaign, the WT was analyzed in a healthy/reference state and three damaged scenarios where a trailing edge opening was introduced gradually to the instrumented blade. In addition to the accelerations in the different structural states, the shifting environmental and operational conditions were recorded, hence allowing for exploring the applicability of the cointegration procedure in this particular damage detection application study.

B. A. Qadri (✉) · M. D. Ulriksen · L. Damkilde
Department of Civil Engineering, Aalborg University, Esbjerg, Denmark
e-mail: baq@civil.aau.dk; mdu@civil.aau.dk

D. Tcherniak
Brüel & Kjær Sound & Vibration, Nærum, Denmark
e-mail: dtcherniak@bksv.com

The paper is built as follows; after the present introduction, the employed damage detection method is outlined in Sect. 23.2. Subsequently, the experimental campaign is described in Sect. 23.3, followed by a presentation of the obtained damage detection results in Sect. 23.4. A brief conclusion in Sect. 23.5 closes the paper.

23.2 Methodology

In the present section, the method used for the damage detection is described. The theory outline is divided into four subsections, which each treats one of the governing components of the method. However, before these are treated, a general introduction to some of the fundamental principles is presented.

For the sake of convenience, let us assume that we have three monitored variables, which in this particular case could be accelerations. It is assumed that if one of the sensors is masked with EOVs, then it follows that all the utilized sensors are masked by the respective EOVs. The Johanson cointegration will form a linear relationship between the monitored variables and generate a series of stationary residuals, purged for the EOVs. The first step is to state the order of integration and the differencing of each monitored variable. It is said that the order of integration is directly coupled to the number of differencing; if the order of integration is p, then the monitored variables are differenced p times. The concept can be demonstrated by generating three non-stationary time series, with drift, and cointegrate the series of order one to generate a residual series; the procedure is sketched in Fig. 23.1.

It is seen in Fig. 23.1 that the stationary residual is purged for common trends and behaves independently of the influences that cause the signal to behave non-stationary. Let $\{y_i\}$ be a vector auto-regressive (VAR) model of a non-stationary time series and define

$$z_i = \{\beta\}^T \{y_i\}, \qquad (23.1)$$

where z_i is stationary and $\{\beta\}^T$ is the cointegrating vector that must be computed. Assuming temporarily that $\{y_i\}$ includes n monitored variables, then there is a maximum of n−1 linearly independent cointegrating vectors [7]. As a necessity for the Johanson cointegration method to work, the times series must, at least, share common trends to be cointegrated. One restriction of the method is that all the time series must be integrated of the same order. The first step in the Johanson cointegration is to state the order of integration. In this study, the Augmented Dickey-Fuller (ADF) test is used to ascertain the order of integration [7, 8]. The ADF test is based on a unit-root test for a time series that declares whether the signal behaves non-stationary or stationary [7]. The concept of the ADF can be demonstrated by a first-order auto-regressive (AR) model, with one lag (AR(1)),

$$y_i = a_1 y_{i-1} + \epsilon_i, \qquad (23.2)$$

where ϵ_i is assumed to be white Gaussian noise and the value of a_1 defines the root of the characteristic equation, which determines the stability and, therefore, its stationarity. If $|a_1| \geq 1$, the time series is non-stationary and if $|a_1| < 1$, the time series is considered stationary. If $a_1 = 1$, then Eq. (23.2) can be rewritten as

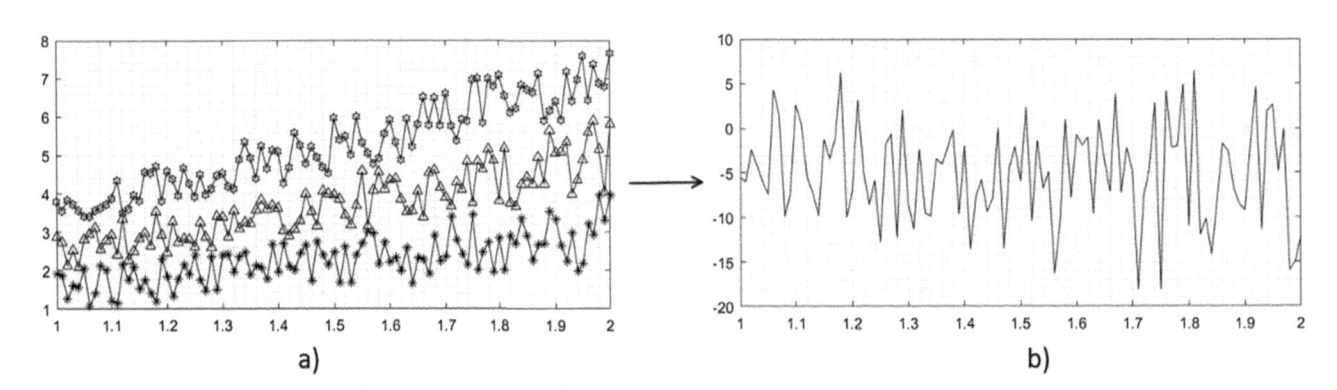

Fig. 23.1 (a) Three monitored non-stationary variables. (b) Cointegrated residuals being stationary

$$y_i = y_{i-1} + \epsilon_i \Rightarrow \Delta y_i = \epsilon_i \qquad (23.3)$$

in which the first difference, Δy_i, will be stationary. This would be called an integration of order one, denoted y_i $I(1)$.

23.2.1 Checking for Non-Stationarity by the Augmented Dickey-Fuller Test

The ADF follows by fitting data to a time series model described as

$$\Delta y_i = \rho y_{i-1} + \sum_{j=1}^{p-1} b_j \Delta y_{i-j} + \epsilon_i, \qquad (23.4)$$

where a number of p lags should be included to secure that ϵ_i is a white noise process. If $\rho = 1$, then the time series is a unit-root and, as such, non-stationary; this is normally checked by testing a null hypothesis of $\rho = 1$ and alternatively for $\rho < 1$.

23.2.2 Mitigating EOVs by Johanson Cointegration

The Johanson procedure can now be used to generate the residual series. The methodological premise is based on using a maximum-likelihood approach to estimate the parameters of a vector error-correction model (VECM) [10]. The VECM writes

$$\{\Delta y_i\} = [\psi]\{y_{i-1}\} + \sum_{j=1}^{p-1} [B_j]\{\Delta y_{i-j}\} + [\vartheta]\{D(t)\} + \{\epsilon_i\}, \qquad (23.5)$$

where $\{D(t)\}$ is the deterministic trend and $\{y_i\}$ is the vector with all n variables. $[\psi]$ is a matrix describing the long-run equilibrium between the variables and matrix $[B_j]$ provides short-run adjustments that return the process to equilibrium, if any drift occurs [7]. The Johansen procedure generates a number of linearly independent cointegrating vectors, and in this particular case, the main interest is to figure out the vector that represents the most stationary data. To find the desired cointegrating vector, the matrix $[\psi]$ can, if Δy_i and Δy_{i-j} are stationary, be decomposed into two $n \times r$ matrices,

$$[\psi] = [\alpha][\beta]^T, \qquad (23.6)$$

where $[\beta]^T$ contains the desired cointegrating vector, which, as described in [7], can be calculated from a generalized eigenvalue problem formulated for the product moments of residuals from the VECM regression. For details concerning this step, the reader is referred to [7, 8].

23.2.3 Dimensionality Reduction by Principal Component Analysis

Principal component analysis (PCA) is used to reduce the dimensionality of the features, in this study the cointegrating vectors, found in Sect. 23.2.3 by Johanson cointegration. The principle of PCA is, in this particular context, to project the data in $[\beta]$ into a set of principal scores, $[T]$. Consider the linear transformation

$$[T] = [\beta][P], \qquad (23.7)$$

where $[P]$ contains the principal components and $[\beta]$ is the matrix containing cointegrated accelerations. By maximizing the variance, and to achieve the lowest redundancy, the transformation matrix P can be constructed such that the covariance matrix of $[T]$, that is,

$$[C_T] = \frac{1}{n-1}[T]^T[T], \qquad (23.8)$$

is diagonal. By rearranging and substituting Eq. (23.7) into Eq. (23.8), $[C_T]$ can be expressed as

$$[C_T] = \frac{1}{n-1}\left[P^T\right]\left[\beta^T\right][\beta][P] = \left[P^T\right]\left[C_\beta\right][P], \qquad (23.9)$$

with $[C_\beta]$ being the covariance matrix of $[\beta]$. The linear transformation can be expressed, as seen in [11], by

$$\left[C_\beta\right][P] = [P][\Lambda], \qquad (23.10)$$

where $[\Lambda] = diag\,(\lambda_1, \lambda_2, \ldots, \lambda_m)$ contains the eigenvalues.

23.2.4 Detection Metric by the Q-Statistics

The Q-statistics is employed as the damage metric to quantify the discordance between the statistical baseline representing the healthy structural state and realizations from the current, potentially damaged state. The principal components, outlined in Sect. 23.2.4, are employed as features, thus the Q-statistics writes

$$Q_i = \{\beta_i\}\left([I] - [P][P]^T\right)\left\{\beta_i^T\right\}. \qquad (23.11)$$

To distinguish between a healthy and a damaged sample, a threshold is simply chosen as the maximum Q_i-value obtained in the training of the baseline model. In this way, if a sample from the current testing state exceeds the threshold, the system is declared damaged.

Summary of the Steps for Applying the Damage Detection Technique

1. Organize a measurement data matrix with each column populated by vibration recordings from a particular sensor.
2. Run an Augmented Dickey-Fuller test for each sensor signal in the data matrix to test for non-stationarity and subsequently determine the order of integration.
3. Use the Johanson cointegration to calculate a feature vector for the data matrix presented in step 1. A unit-root test can be performed for the generated feature vector to check for stationarity by going one step back.
4. Reduce the dimensionality of the feature vector with PCA.
5. Perform all four previous steps for training a statistical baseline representing the healthy state and for testing with measurements from the current structural state.
6. Use the Q-statistics to quantify discordance and declare the current structural state damaged if Q_i exceeds the threshold.

23.3 Experimental Work

As stated in Chap. 1, the damage detection study concerns a Vestas V27 WT. One of the blades was instrumented with accelerometers and a mechanical actuator, as depicted in Fig. 23.2. The WT is located at the Technical University of Denmark (DTU), Department of Wind Energy, and has been used in many SHM-related studies [12, 13]. The WT is a relatively old model, with a 27 m rotor diameter and with an efficiency of 225 kW rated power. The blades of the WT are relatively stiff and only operates in two regimes, namely, 32 and 43 rotations per minute (RPM).

The actuator, which is installed near the blade root and is covered by a waterproof lid, is a simple device that works by releasing a plunger towards the structure, hence exposing it to mechanical energy. The induced vibrations, of which an example is presented in Fig. 23.3, are captured through the installed piezoelectric accelerometers, which are distributed along the leading and trailing edge. The position of the mechanical actuator, damage and sensors can be seen in Fig. 23.4.

The measurement campaign was performed over a 3.5-months period, in which WT initially operated in its reference/healthy configuration. Based on measurements in this state, the statistical baseline model is computed. After a period of time, the WT was introduced to a 15 cm long opening of the trailing edge and measurements were subsequently captured for this state. The same procedure followed for a 30 cm long and 45 cm long opening, respectively; Fig. 23.5 shows the 45 cm

Fig. 23.2 Vestas V27 WT and the mechanical actuator marked with a yellow circle

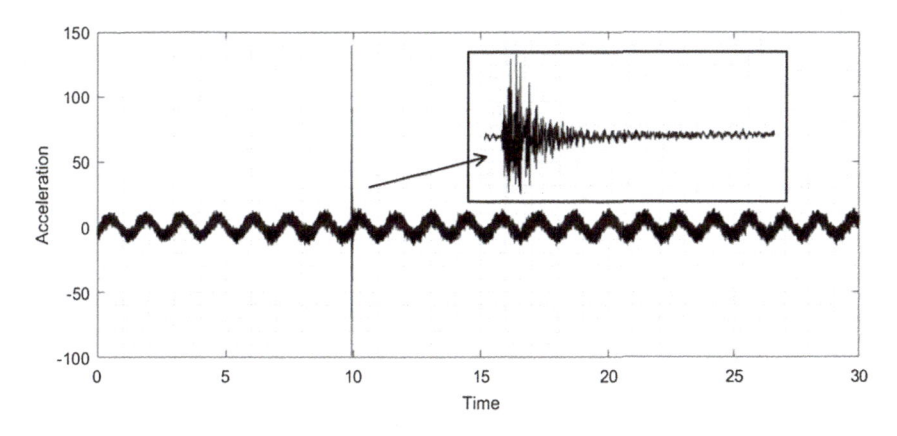

Fig. 23.3 Response captured in one sensor and a zoomed illustration of the vibration caused by the mechanical actuator

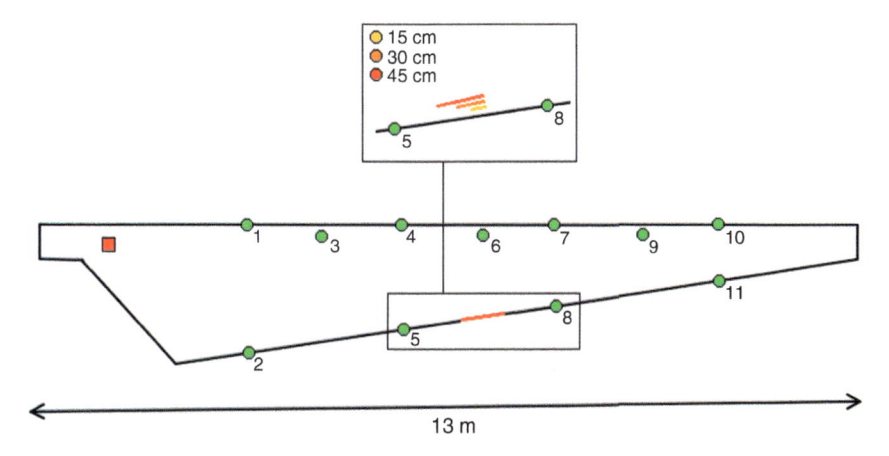

Fig. 23.4 Sketch of the WT blade, along with the sensor distribution (green dots), the actuator (red square) and an insert of the three openings

Fig. 23.5 WT blade with a 45 cm long opening of the trailing edge

long opening. During the entire campaign, the weather data was simultaneously recorded from a meteorological mast. These data describes the wind speed and direction at different altitudes, temperature, atmospheric pressure and precipitation. A full description of the measurement campaign can be found in [12, 13].

23.4 Damage Detection

In the damage detection analysis, focus is on discriminating between the healthy state and the damaged one with the 15 cm long opening. In this regard, a total of 204 measurement sequences are extracted for an operating speed of 43 RPM, where 100 measurements are from the reference state and 104 are measurements from the state with the 15 cm long opening. In the present study, only three out of the eleven accelerometers are, in accordance with the conclusions drawn in [12, 13], employed to further challenge the damage detection methodology. In particular, the three sensors used are 1, 3 and 4.

In Fig. 23.6, the results obtained by discarding the cointegration procedure in the damage detection analysis are presented. In this way, the features in the Q-statistics are taken as accelerations subjected to PCA, and, as evidenced, this conventional outlier analysis approach fails to consistently discriminate between damage and EOVs.

The results obtained by following the approach outlined in Sect. 23.2 (including Johanson cointegration) are shown in Fig. 23.7. It is clearly seen how the approach succeeds in detecting the introduced trailing edge opening, hence demonstrating the merit of mitigating the effects of the EOVs from the employed damage detection features.

23.5 Conclusion

The present paper documents an experimental damage detection study, in which a trailing edge opening of a WT blade is detected during turbine operation. The applied damage detection method is based on outlier analysis; with processed acceleration measurements as features and the Q-statistics as discordance measure/damage metric. More specifically, the features are composed of acceleration signals filtered for EOVs using cointegration and reduced, in terms of dimensionality, through PCA. The detection results evidence the merit of mitigating the EOVs, as implementation of the cointegration step facilitates detection of the trailing edge opening under conditions not allowing for direct discrimination between damage and EOVs.

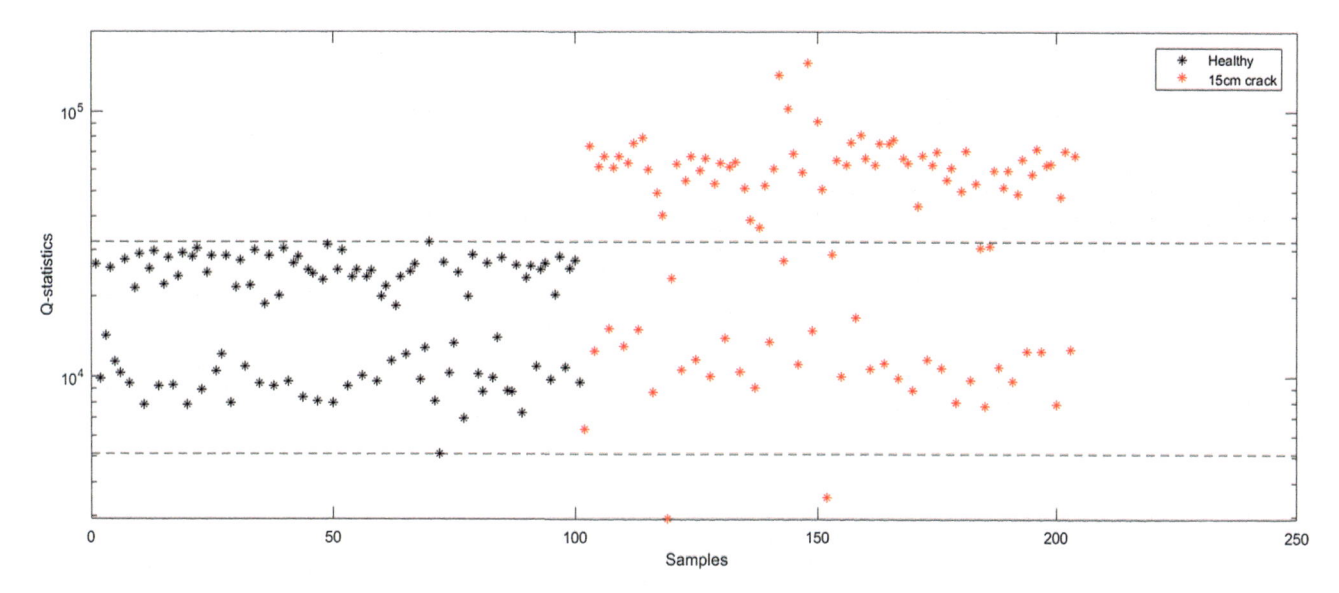

Fig. 23.6 Damage detection of a 15 cm long opening without the Johanson cointegration

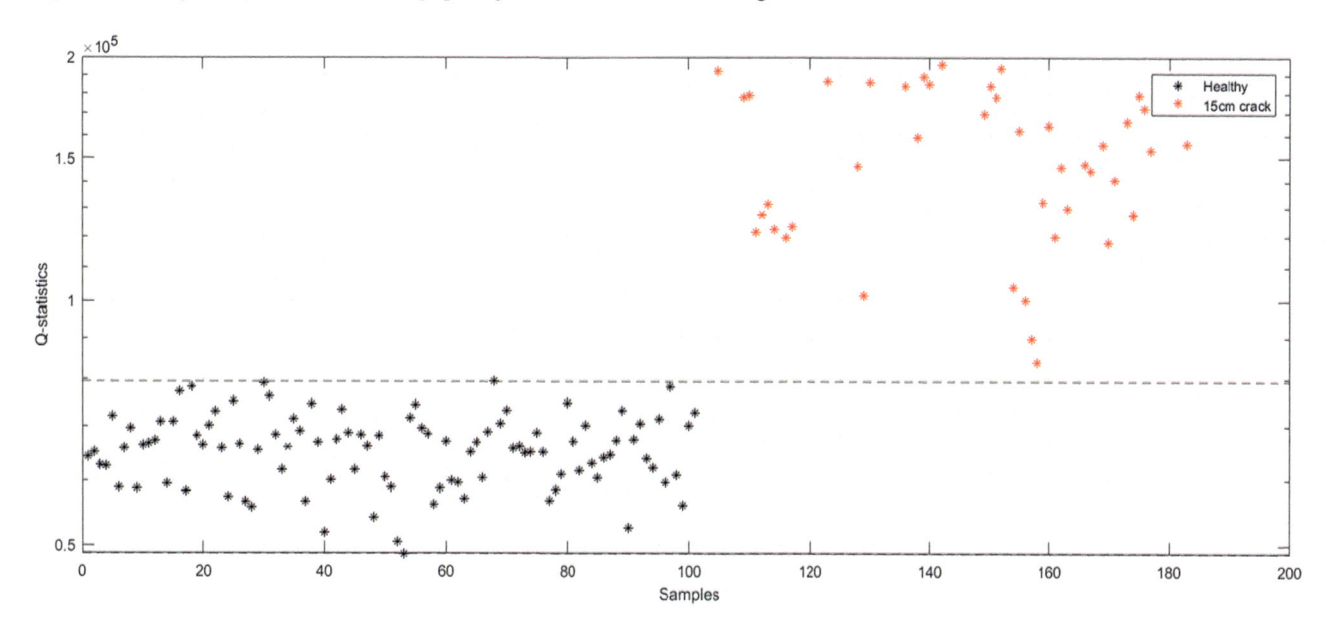

Fig. 23.7 Damage detection of 15 cm long opening with the Johanson cointegration

References

1. Gres, G., Ulriksen, M.D., Döhler, M., Johansen, R.J., Andersen, P., Damkilde, L., Nielsen, S.A.: Statistical methods for damage detection applied to civil structures. Procedia Eng. **199**, 1919–1924 (2017)
2. Mevel, L., Hermans, L., Van Der Auweraer, H.: Application of a subspace-based fault detection method to industrial structures. Mech. Syst. Signal Process. **13**(6), 207–224 (2014)
3. Qadri, B.A., Tcherniak, D., Ulriksen, M.D., Damkilde, L.: Damage detection in a reinforced concrete slab using outlier analysis. In: Proceedings of the 9th European Workshop on Structural Health Monitoring (2018)
4. Oliveira, G., Magalhães, F., Cunha, Á., Caetano, E.: Vibration-based damage detection in a wind turbine using 1 year of data. Struct. Control Health Monit. **25**(11), e2238 (2018)
5. Sohn, H.: Effects of environmental and operational variability on structural health monitoring. Philos Trans A Math Phys Eng Sci. **365**(1851), 539–560 (2007)
6. Farrar, C., Doebling, S., Cornwell, P., Straser, E.: Variability of modal parameters measured on the Alamosa Canyon Bridge: In: Proceedings of the 15th International Modal Analysis Conference, pp. 257–263 (1997)
7. Cross, E.J., Worden, K., Chen, Q.: Cointegration: A novel approach for the removal of environmental trends in structural health monitoring data. Philos Trans A Math Phys Eng Sci. **467**(2133), 2712–2732 (2011)

8. Dao, P.B.: Cointegration method for temperature effect removal in damage detection based on lamb waves. Diagnostyka. **14**(3), 61–67 (2013)
9. Shi, H., Worden, K., Cross, E.J.: A regime-switching cointegration approach for removing environmental and operational variations in structural health monitoring. Mech. Syst. Signal Process. **103**, 381–397 (2018)
10. Cross, E.J., Worden, K.: Cointegration and why it works for SHM. J. Phys. Conf. Ser. **382**, 012046 (2012)
11. Mujica, L.E., Rodellar, J., Fernandez, A., Güemes, A.: Q-statistic and T^2-statistics PCA-based measures for damage assessment in structures. Struct. Health Monit. **10**(5), 539–553 (2011)
12. Ulriksen, M.D., Tcherniak. D., Damkilde. L.: Damage detection in an operating vestas V27 wind turbine blade by use of outlier analysis. In: IEEE Workshop on Environmental, Energy and Structural Monitoring Systems (2015)
13. Tcherniak, D., Mølgaard, L.L.: Active vibration-based structural health monitoring system for wind turbine blade: Demonstration on an operating Vestas V27 wind turbine. Struct. Health Monit. **16**(5), 536–550 (2017)

Chapter 24
System Identification of a Five-Story Building Using Seismic Strong-Motion Data

Rodrigo Astroza, Francisco Hernández, Pablo Díaz, and Gonzalo Gutierrez

Abstract A full scale five-story reinforced concrete building was built and tested on the Large High Performance Outdoor Shake Table (LHPOST) at the University of California, San Diego in 2012. The main objective of the test program was to study the seismic response of the structure and the nonstructural components (NSCs) and their dynamic interaction at different levels of seismic excitation. The building specimen was first tested base-isolated and then fixed at its base. In the fixed-base configuration, a suite of six earthquake motions of various intensities was applied to the building to progressively increase the seismic demand. In this paper, the modal parameters of the fixed-base building are identified using the input-output dynamic data recorded during the seismic tests. The deterministic-stochastic subspace identification method (DSI) is employed to estimate the variations of the modal properties of the building during the seismic tests by employing a short-time windowing approach. The changes of the modal parameters during the seismic motions are tracked, analyzed, and compared to those previously obtained from ambient vibrations and low-amplitude white noise base excitation tests. The identified natural frequencies and equivalent damping ratios of the building changed with the intensity of the input motions and damage in the structure, while the mode shapes are found to be insensitive to them.

Keywords System identification · Damage · Shake table · Modal properties

24.1 Introduction

System identification is an active field of research aiming to characterize the dynamic properties of large and complex civil structures by using input-output or output-only vibration data recorded by sensors installed in the structure of interest. In particular, the results obtained from system identification have been used for vibration-based damage identification purposes (e.g., [1, 2]). The objective is to identify damage in the structure by analyzing the variation in the estimated dynamic characteristics from an initial (reference) state to a state after the structure has been subjected to potentially damage-inducing loading. The dynamic characterization of civil structures usually comprises the estimation of natural frequencies, damping ratios, and mode shapes from the recorded data.

Vibration data recorded during earthquake events on civil structures that have suffered damage is extremely scarce and large-scale shake tables have provided important data from structures tested at different states of damage and using different sources of dynamic excitation, including ambient vibrations (AV), white-noise (WN) base excitations, and seismic excitations (e.g., [3–5]). Most of previous studies have focused on the identification of the modal properties of structures using low-amplitude vibration data and a just a few researches have investigated the variations of the modal properties during strong motion earthquake excitations (e.g., [6, 7]). In this paper, the modal properties of a full-scale 5-story reinforced concrete (RC) building specimen, referred to as BNCS building hereafter, tested on the large-scale UC San Diego shake table [8–10] are identified using the acceleration response data recorded during seismic tests. The time-variant modal properties of the BNCS building are identified by using the deterministic-stochastic subspace identification method (DSI) applied to a moving short-time window of input-output acceleration data [6, 7, 11, 12]. These modal properties are compared to those previously identified before and after each seismic test from AV and WN base excitation test data.

R. Astroza (✉) · G. Gutierrez
Facultad de Ingeniería y Ciencias Aplicadas, Universidad de los Andes, Santiago, Chile
e-mail: rastroza@miuandes.cl

F. Hernández · P. Díaz
Department of Civil Engineering, Universidad de Chile, Santiago, Chile

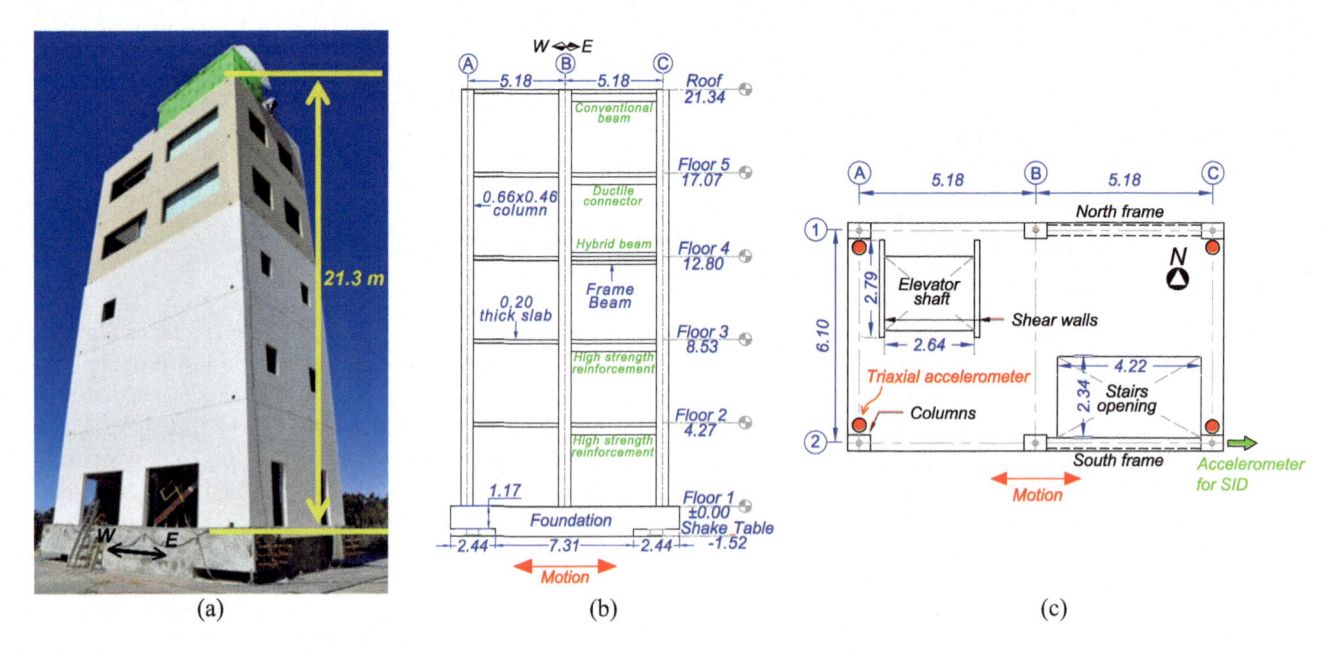

Fig. 24.1 Test structure: (**a**) general view of the building; (**b**) elevation view; (**c**) plan view (Dimensions in meters)

24.2 Building Description

The test specimen was a full-scale five-story RC structure fully furnished with nonstructural components and contents (Fig. 24.1a). The height of the building was 21.34 m, distributed in five stories of 4.27 m high (Fig. 24.1b). The building had plan dimensions of 6.6 × 11.0 m in the transverse and longitudinal directions (Fig. 24.1c), respectively, the latter coinciding with the direction of the movement of the uni-directional shaking table. In the longitudinal direction, the lateral resisting system of the building consisted of a pair of one-bay special moment resisting frames. The main structural components of the building included 0.30 × 0.71 m beams, 0.66 × 0.46 m columns, 0.20 m thick RC floor slabs, and two 0.15 m thick RC shear walls to accommodate an operating elevator. More information about the specimen can be found in [8, 9].

24.3 Instrumentation and Test Protocol

More than 550 sensors, including displacement transducers, strain gauges, and accelerometers, were installed in the structural and nonstructural components of the building. In particular, four triaxial accelerometers were installed on the corners of each floor slab (see Fig. 24.1c) and two triaxial accelerometers were installed on the shake table platen. The accelerometers were Episensor with a frequency bandwidth DC–200 Hz, dynamic range of 155dB, and full-scale range ±4g. A band-pass infinite impulse response Butterworth filter of order 4 with cut-off frequencies at 0.04 and 25.0 Hz was used to filter the recorded raw acceleration data.

A suite of six seismic records was applied to the BNCS building with the purpose of progressively increase the seismic demand of the structure. Spectrally-matched and actual motions were considered in the seismic test phase. Table 24.1 shows the strong motions employed and their corresponding peak input acceleration (PIA), peak input velocity (PIV) and peak input displacement (PID) achieved on the shake table platen. The acceleration time-histories and their acceleration and displacement elastic response spectra for a damping ratio of 5% are shown in Fig. 24.2.

Table 24.1 Description and nomenclature of seismic tests applied to the BI-BNCS building

Date of test	Motion	Name	PIA (g)	PIV (mm/s)	PID (mm)
May 7, 2012	Canoga Park—1994 Northridge earthquake	FB1-CNP100	0.21	235.0	87.8
May 9, 2012	LA City Terrace—1994 Northridge earthquake	FB2-LAC100	0.18	230.5	93.1
	ICA 50%—2007 Pisco (Peru) earthquake	FB3-ICA50	0.21	262.2	58.3
May 11, 2012	ICA 100%—2007 Pisco (Peru) earthquake	FB4-ICA100	0.26	284.9	73.2
May 15, 2012	TAPS Pump Station #9 67%—2002 Denali earthquake	FB5-DEN67	0.64	637.4	200.6
	TAPS Pump Station #9 100%—2002 Denali earthquake	FB6-DEN100	0.80	835.7	336.2

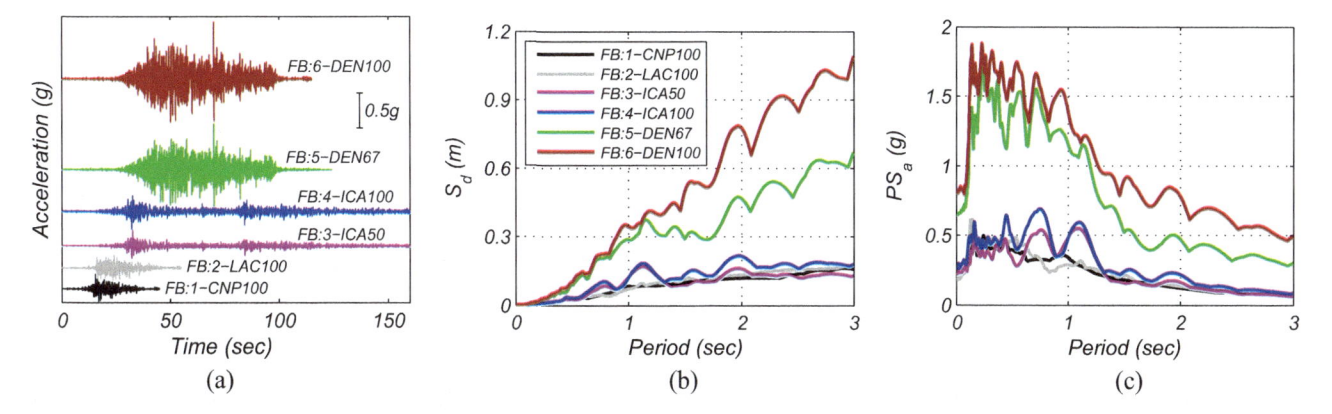

Fig. 24.2 Input ground motions achieved on the shake table: (**a**) acceleration time-histories; (**b**) elastic displacement response spectra ($\xi = 5\%$); (**c**) acceleration response spectra ($\xi = 5\%$)

24.4 System Identification

The modal properties of the BNCS building are identified using the DSI method with the recorded input-output acceleration data. A short-time windowing approach is employed in order to track the variation of the dynamic characteristics of the building during the seismic tests.

24.4.1 DSI Method

The following equations describe a discrete-time linear time-invariant (LTI) state-space model [13]:

$$\mathbf{x}_{k+1} = \mathbf{A_d}\mathbf{x}_k + \mathbf{B_d}\mathbf{u}_k + \mathbf{w}_k, \tag{24.1a}$$

$$\mathbf{y}_k = \mathbf{C_d}\mathbf{x}_k + \mathbf{D_d}\mathbf{u}_k + \mathbf{v}_k, \tag{24.1b}$$

with

$$E\left[\begin{pmatrix}\mathbf{w}_p\\\mathbf{v}_p\end{pmatrix}\begin{pmatrix}\mathbf{w}_q^T & \mathbf{v}_q^T\end{pmatrix}\right] = \begin{pmatrix}\mathbf{Q} & \mathbf{S}\\\mathbf{S}^T & \mathbf{R}\end{pmatrix}\delta_{pq} \geq 0, \tag{24.2}$$

where $\mathbf{x}_k$, $\mathbf{y}_k$, $\mathbf{u}_k$ = state, output, and input vectors, respectively; $\mathbf{A_d}$, $\mathbf{B_d}$, $\mathbf{C_d}$, $\mathbf{D_d}$ = state, input, output, and direct feed-through matrices, respectively; $\mathbf{w}_k$, $\mathbf{v}_k$ are the process and measurement noises with associated covariance matrices $\mathbf{Q}$, $\mathbf{R}$ and $\mathbf{S}$; δ_{pq} = Kronecker delta; and k = discrete time step.

To determine the modal frequencies (f_r), modal damping ratios (ξ_r), and mode shapes ($\boldsymbol{\phi}_r$) of the dynamic system, it is required to estimate the state and output matrices ($\mathbf{A_d}$ and $\mathbf{C_d}$) given s samples of the input and output data ($\mathbf{u}_j$ and $\mathbf{y}_j$ with $j = 0, \ldots, s-1$), because they are related trough the following expressions:

$$f_r = \frac{\sqrt{\lambda_r \lambda_r^*}}{2\pi}, \qquad (24.3)$$

$$\xi_r = \frac{-\operatorname{Re}(\lambda_r)}{|\lambda_r|}, \qquad (24.4)$$

$$\mathbf{\Phi} = \mathbf{C_d}\mathbf{\Psi} = [\boldsymbol{\phi}_1, \ldots, \boldsymbol{\phi}_r], \qquad (24.5)$$

where λ_r are the eigenvalues of the continuous-time state matrix $\mathbf{A_c}$ (with $\mathbf{A_d} = e^{\mathbf{A_c}\,\Delta t}$ and Δt = sampling time), $\mathbf{\Psi}$ are the eigenvectors of $\mathbf{A_d}$, and * and $|\cdot|$ denote complex conjugate and magnitude, respectively. In this article the DSI method is used to identify the modal properties of the BNCS building. Figure 24.3 presents a flowchart of the DSI method [13].

24.4.2 Identification of Modal Parameters from Seismic Experimental Data

The changes of the modal properties of the building during the seismic tests are estimated by using short-time windows of input-output data. The acceleration measured at the top of the foundation (floor 1) is considered as the input and the acceleration responses at the south-east corner of the upper floors as output (see Fig. 24.1c), i.e., $m = 1$ and $l = 5$ with m = number of inputs and l = number of outputs. A single acceleration per floor in the longitudinal direction of the building is considered for conducting the identification because the shake table imposed the excitation in that direction, therefore, small acceleration responses were measured in the transverse direction and the torsional effects were minor. The minimum length criteria (Eq. 24.6) was chosen to define the length of the windows of data [7].

$$s_{\min} = 2i\,(m + l + 1), \qquad (24.6)$$

In the case of the BNCS building, a number of block rows of the Hankel matrices equal to $i = 101$ was used based on preliminary analysis of identification results conducted with data from the beginning, strong motion, and final parts of the FB1-CNP100 seismic test. Then, the number of data samples on each short-time window is $s_{\min} = 2 \times 101(1 + 5 + 1) = 1414$. Because the sampling frequency was 200 Hz, the minimum window length corresponds to $1414/200 = 7.07$ s. The identification process was conducted considering an overlapping of 90% between consecutive time windows.

The physical modes were chosen as those satisfying the following criteria:

1. For a given short-time window, modes satisfying $|f^{(100)} - f^{(99)}|/f^{(100)} \leq 15\%$ and $\text{MAC}_{\phi^{(100)},\phi^{(99)}} \geq 75\%$, where $f^{(n)}$ = natural frequency for model order n, and $\text{MAC}_{\phi^{(n)},\phi^{(m)}}$ is the modal assurance criterion (MAC) [14] between modes identified for model orders n and m.
2. For consecutive short-time windows $(i - 1)$ and i, modes satisfying $|f_{(i-1)} - f_{(i)}|/f_{(i-1)} \leq 15\%$ and $\text{MAC}_{\phi_{(i-1)},\phi_{(i)}} \geq 90\%$, where $f_{(n)}$ = natural frequency for time window n, and $\text{MAC}_{\phi_{(n-1)},\phi_{(n)}}$ is the MAC value between corresponding modes identified for windows $(n - 1)$ and n.

Based on system identification analyses conducted for the BNCS building using AV and WN base excitation test data [15], modes with natural frequencies below 15 Hz were analyzed. It is noted that the frequency range from 0 to 15 Hz includes the first three longitudinal modes of the building (1-L, 2-L, and 3-L) [15]. Only longitudinal modes are identified with the seismic data because the shaking direction was imposed in that direction, therefore, the identification associated with transverse and torsional modes is not plausible because of their low contribution to the dynamic response of the building. The mode shapes identified for the first time window of the FB1-CNP100 test are shown in Fig. 24.4.

The bottom panel of Fig. 24.5 depicts the input acceleration time-histories measured on top of the foundation (floor 1) and upper panels show the natural frequencies of the first three longitudinal modes of the building identified using the short-time window approach described above. In these plots, natural frequencies of the corresponding modes identified with AV and WN base excitation data reported in [15] are marked with green and red dashes, respectively, at the beginning and at the end of each seismic test. In addition, for the fundamental mode (1-L), black dashes during the strong motion phases of tests FB1 to FB4 report the identification results obtained in Chen et al. [16] by using an optimization approach based on the roof displacement measured from collocated accelerometers and GPS antennas.

It is noted that at the beginning of each seismic test, the input motion exhibits low amplitudes and the identified natural frequencies match very well those identified using AV data (green dashes) and are higher than those identified using WN base

1. Define Hankel matrices:

$$\mathbf{U}_{0|2i-1} = \begin{pmatrix} \mathbf{U}_{0|i-1} \\ \mathbf{U}_{i|2i-1} \end{pmatrix} = \begin{pmatrix} \mathbf{U}_p \\ \mathbf{U}_f \end{pmatrix} \quad ; \quad \mathbf{Y}_{0|2i-1} = \begin{pmatrix} \mathbf{Y}_{0|i-1} \\ \mathbf{Y}_{i|2i-1} \end{pmatrix} = \begin{pmatrix} \mathbf{Y}_p \\ \mathbf{Y}_f \end{pmatrix}$$

2. Compute oblique projections:

$$\mathbf{O}_i = \mathbf{Y}_f /_{\mathbf{U}_f} \begin{pmatrix} \mathbf{U}_p \\ \mathbf{Y}_p \end{pmatrix} \quad ; \quad \mathbf{O}_{i+1} = \mathbf{Y}_f^- /_{\mathbf{U}_f^-} \begin{pmatrix} \mathbf{U}_p^+ \\ \mathbf{Y}_p^+ \end{pmatrix}$$

3. Determine order of the system using singular value decomposition:

$$\mathbf{W}_1 \mathbf{O}_i \mathbf{W}_2 = (\mathbf{U}_1 \quad \mathbf{U}_2) \begin{pmatrix} \mathbf{S}_1 & 0 \\ 0 & 0 \end{pmatrix} \begin{pmatrix} \mathbf{V}_1^T \\ \mathbf{V}_2^T \end{pmatrix} = \mathbf{U}_1 \mathbf{S}_1 \mathbf{V}_1^T$$

W_1, W_2: weighting matrices

4. Compute the extended observability matrices Γ_i and Γ_{i-1}:

$$\Gamma_i = \mathbf{W}_1^{-1} \mathbf{U}_1 \mathbf{S}_1^{1/2}$$

Γ_{i-1} is defined by removing the last l rows of Γ_i

5. Determine the state sequences $\tilde{\mathbf{X}}_i$ and $\tilde{\mathbf{X}}_{i+1}$:

$$\tilde{\mathbf{X}}_i = \Gamma_i^\dagger \mathbf{O}_i \quad ; \quad \tilde{\mathbf{X}}_{i+1} = \Gamma_{i-1}^\dagger \mathbf{O}_{i+1}$$

where $\bullet^\dagger$ denotes the Moore-Penrose pseudo-inverse of the matrix $\bullet$

6. Solve the set of linear equations for $\mathbf{A}_d, \mathbf{B}_d, \mathbf{C}_d$ and $\mathbf{D}_d$ using least-squares:

$$\begin{pmatrix} \tilde{\mathbf{X}}_{i+1} \\ \mathbf{Y}_{i|i} \end{pmatrix} = \begin{pmatrix} \mathbf{A}_d & \mathbf{B}_d \\ \mathbf{C}_d & \mathbf{D}_d \end{pmatrix} \begin{pmatrix} \tilde{\mathbf{X}}_i \\ \mathbf{U}_{i|i} \end{pmatrix} + \begin{pmatrix} \rho_w \\ \rho_v \end{pmatrix}$$

7. Compute eigenvalues μ and eigenvectors (Ψ) of the state matrix $\mathbf{A}_d$.

8. Determine modal properties of the system:

$$f_r = \frac{\sqrt{\lambda_r \lambda_r^*}}{2\pi}$$

$$\xi_r = \frac{-\text{Re}(\lambda_r)}{|\lambda_r|}$$

$$\Phi = \mathbf{C}_d \Psi$$

with

$$\lambda_r = \frac{\ln(\mu_r)}{\Delta t}$$

Fig. 24.3 Schematic overview of the DSI system identification method (adapted from [13])

excitation data (red dashes). As the amplitude of the base excitation increases, the identified natural frequencies decrease significantly, but they recover toward the end of each test, reaching a value lower than that identified at the beginning of the same seismic test but significantly higher than the lowest frequency values identified for the corresponding test. This implies that the equivalent lateral stiffness of the building decreases significantly during the strong motion phase (because of nonstructural and structural damage) but then it increases when the amplitude of the excitation decreases at the end of the seismic test. This behavior suggests that cracks in the concrete are opened during the strong motion phase and then they close after the amplitude of the excitation becomes small. However, some permanent degradation of the stiffness is observed, because the frequencies identified at the end of each seismic test are lower than those identified at the beginning of the same test. The same pattern is observed for the three longitudinal models. For model 1-L, at the beginning of test FB1-CNP100 a natural frequency 1.67 Hz is identified, progressively decreasing at the end of each of the seismic tests, and reaching a value of 0.73 Hz at the end of the final test FB6-DEN100. Similarly, modes 2-L and 3-L progress from 6.77 and 10.88 Hz at

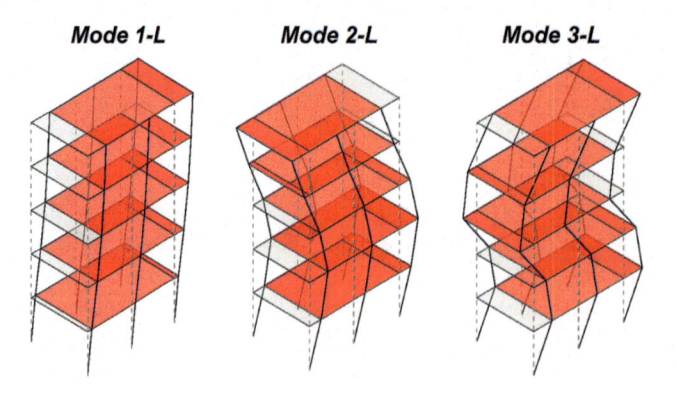

Fig. 24.4 Identified mode shapes of the building

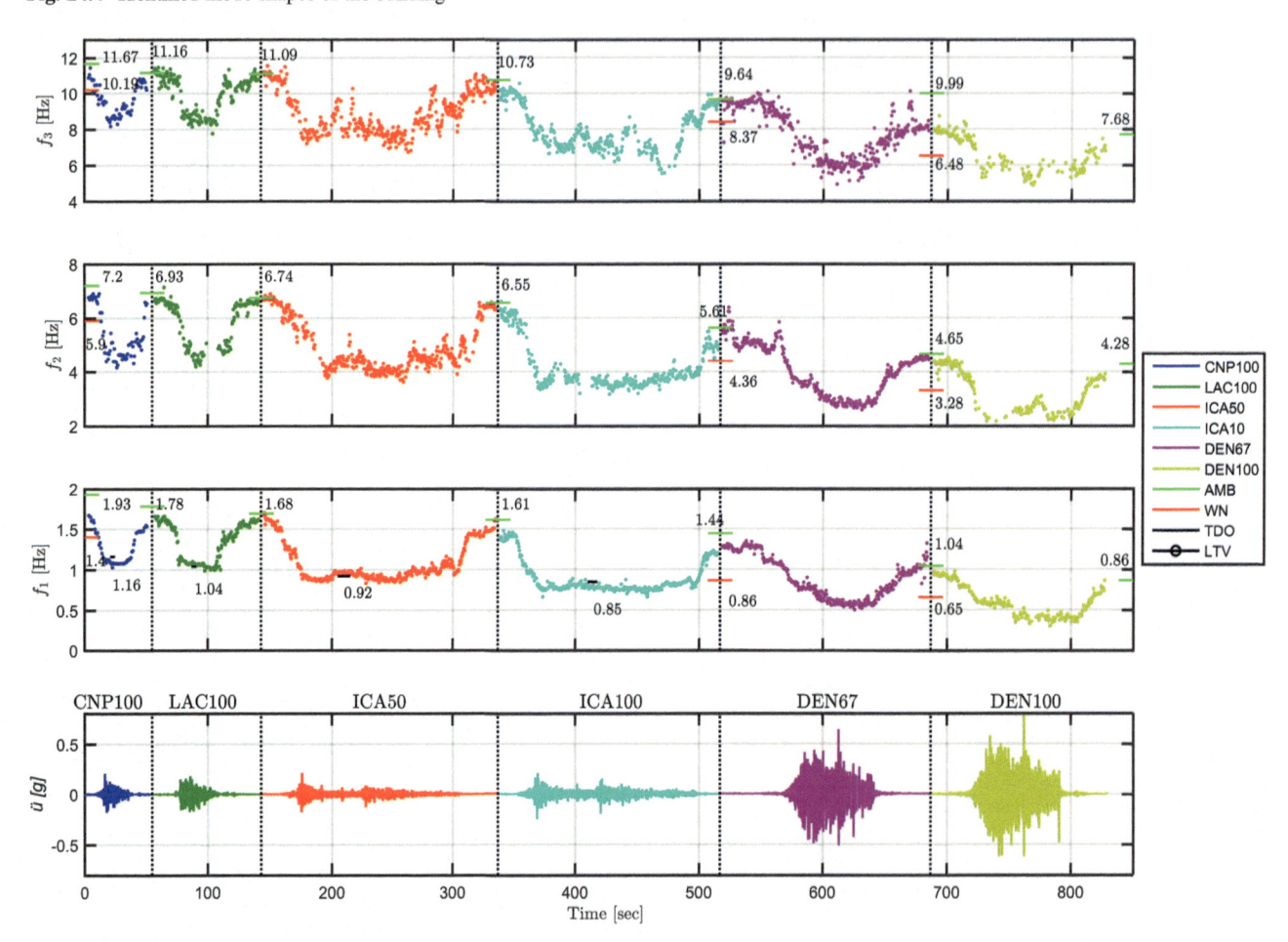

Fig. 24.5 Temporal evolution of the natural frequencies identified for first three longitudinal modes

the beginning of test FB1-CNP100 to 3.70 and 6.78 Hz at the end of test FB6-DEN100, respectively. For the first four tests (FB1-CNP100 to FB4-ICA100), the natural frequency of mode 1-L identified in this work in the strong motion phase and that obtained by Chen et al. [16] are in very good agreement.

Figure 24.6 shows the identification results for the equivalent viscous damping ratios. In the top three plots, the identified damping ratios for modes 1-L, 2-L, and 3-L are shown. Consistently with Fig. 24.5, identification results from AV and WN test data [15] and those reported in [16] for the strong motion parts of the first four tests are included in the plots with dashed lines. Significant variations of the identified damping ratios are observed. In general, they tend to increase during the strong motion phase for all modes. In addition, the values identified at the beginning and end of the seismic tests are relatively close to those identified when AV data was employed. It is noted that identification results obtained using AV data

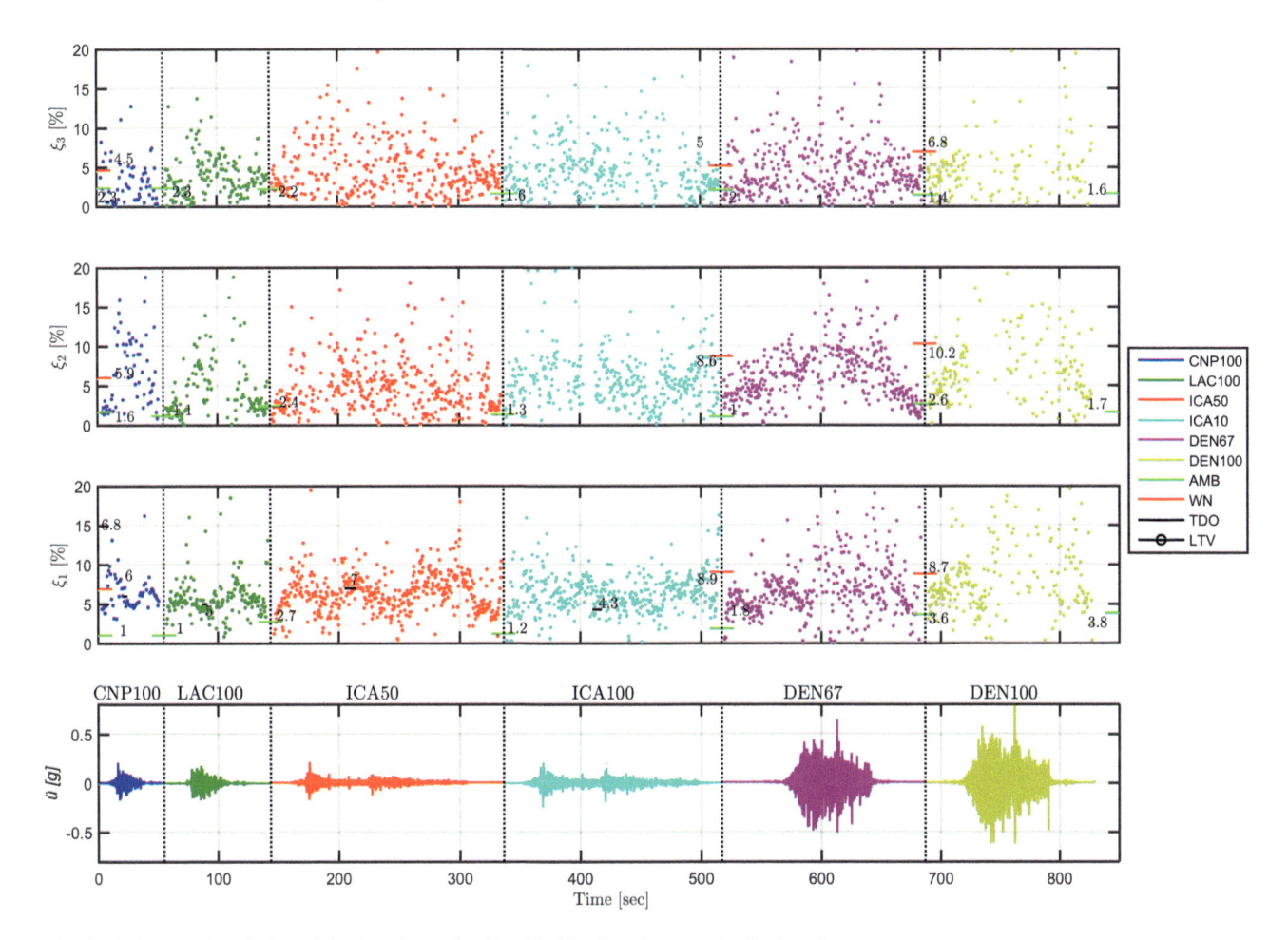

Fig. 24.6 Temporal evolution of the damping ratios identified for first three longitudinal modes

are affected by environmental conditions, and, in particular, the effects on damping ratios are significant [17]. As the seismic demand induced by the seismic excitation increases (from FB1-CNP100 to FB6-DEN100), the identified damping ratios also increase, suggesting that a larger amount of energy is dissipated when the excitation increases. It is worthy to note that the underlying mathematical model considered in the system identification method used, assumed that all the sources of energy dissipation are identified as equivalent viscous damping. For mode 1-L, the average value of the identified damping ratio in the strong motion part of the excitations is about 5–7% for the first four tests (FB1-CNP100 to FB4-ICA100) and increases to about 8% and 10% for tests FB5-DEN67 and FB6-DEN100, respectively. Similar values and trend are observed for the damping ratios identified for mode 2-L, while the damping ratio variation associated with mode 3-L is smaller.

The variation of the mode shapes during the seismic tests is also investigated in a similar fashion as the natural frequencies and damping ratios (Fig. 24.7). The MAC values are used to study the evolution of the mode shapes during the seismic test, taking the mode shapes identified at the beginning of test FB1-CNP100 as reference. For modes 1-L and 2-L, the MAC values only exhibit minor changes during the whole testing protocol (MAC > 90% for most values), suggesting that damage does not affect significantly the mode shapes. For mode 3-L, some deviations from the unity are observed along few specific time windows; however, no logical pattern is observed, implying that this can be generated by artifacts related to the identification process.

24.5 Conclusions

Experimental data collected from a series of six shake table tests conducted on a full-scale five-story reinforced concrete building specimen tested at the University of California, San Diego, was used to identified the modal properties of the building using seismic test data of varying intensities. The first three longitudinal modes of the building were identified using the

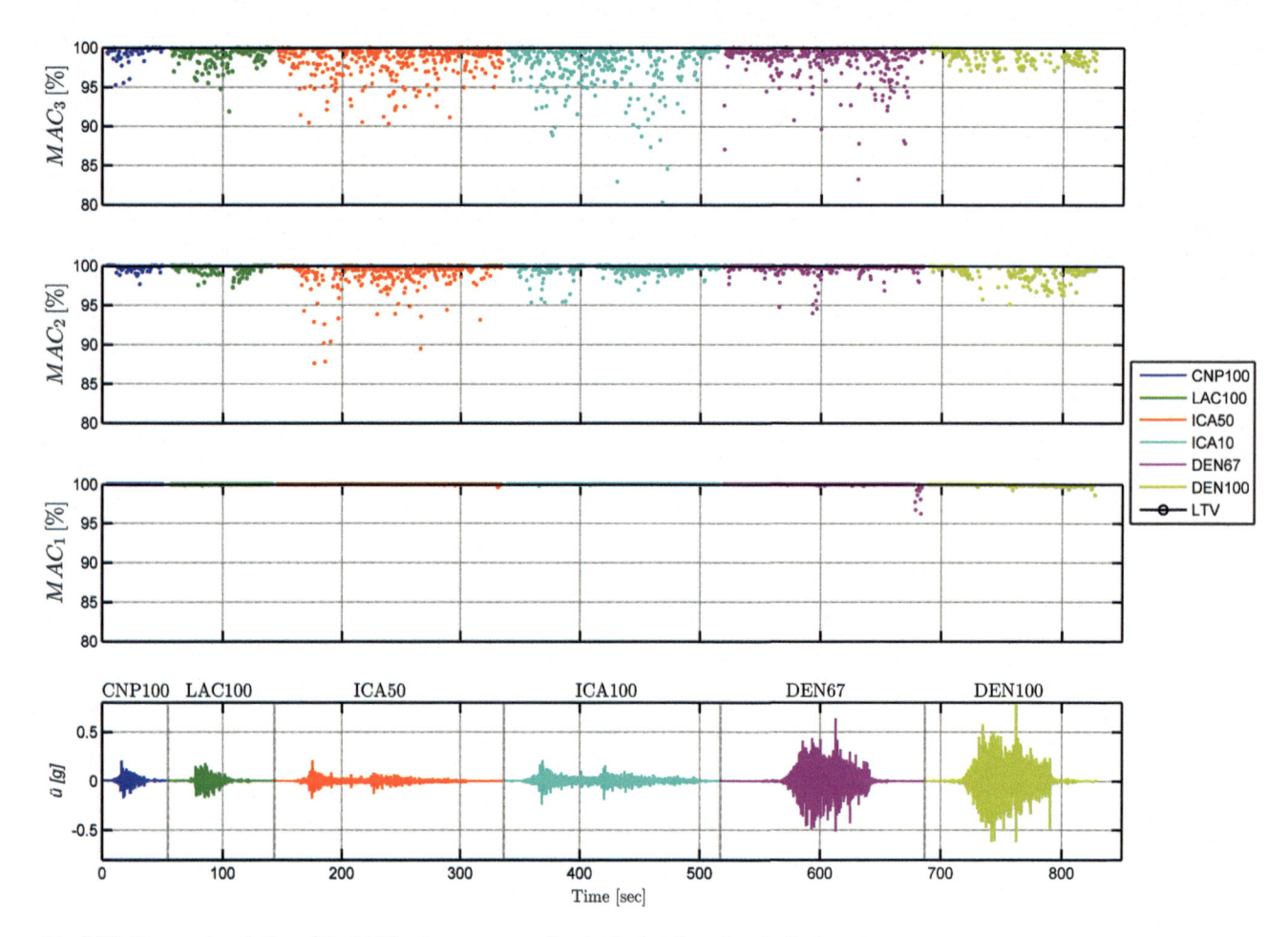

Fig. 24.7 Temporal evolution of the MAC values corresponding to the first three longitudinal modes

deterministic-stochastic subspace identification method with input-output acceleration data recorded during the seismic tests. The time evolution of the modal parameters was tracked by using a short-time windowing approach. The nonlinear behavior of the building was observed from the temporal variation of the identified natural frequencies and equivalent damping ratios. The stiffness reduction in the building due to nonstructural and structural damage implied a reduction in the identified natural frequencies and an increment in the identified damping ratios. Natural frequencies and damping ratios identified using the data at the beginning of each seismic test were found in good agreement with those identified from ambient vibration data. The variation of the mode shapes was analyzed using the MAC, taking the mode shapes identified at the beginning of the first seismic test as reference. It was found that the mode shapes did not change significantly during the seismic tests, i.e., they were not affected by the damage.

Acknowledgements The authors are very grateful to Professors Tara C. Hutchinson, Joel P. Conte, and José I. Restrepo from the University of California, San Diego for making these data available (DOI: 10.4231/D38W38349). R. Astroza acknowledges the financial support from the Chilean National Commission for Scientific and Technological Research (CONICYT), through FONDECYT research grant No. 11160009.

References

1. Doebling, S.W., Farrar, C.R., Prime, M.B., Shevit, D.W.: Damage identification and health monitoring of structural and mechanical systems from changes in their vibration characteristics: a literature review. Technical report LA-13070-MS, Los Alamos National Laboratory, Los Alamos (1996)
2. Fan, W., Qiao, P.Z.: Vibration-based damage identification methods: a review and comparative study. Struct. Health Monit. **10**(5), 83–111 (2011)

3. Ji, X., Fenves, G., Kajiwara, K., Nakashima, M.: Seismic damage detection of a full-scale shaking table test structure. ASCE J. Struct. Eng. **137**(1), 14–21 (2011)
4. Moaveni, B., He, X., Conte, J.P., Restrepo, J.I., Panagiotou, M.: System identification study of a seven-story full-scale building slice tested on the UCSD-NEES shake table. ASCE J. Struct. Eng. **137**(6), 705–717 (2011)
5. Astroza, R., Ebrahimian, H., Conte, J.P., Restrepo, J.I., Hutchinson, T.C.: Influence of the construction process and nonstructural components on the modal properties of a five-story building. Earthq. Eng. Struct. Dyn. **45**(7), 1063–1084 (2016)
6. Loh, C.-H., Weng, J.-H., Chen, C.-H., Lu, K.-C.: System identification of mid-story isolation building using both ambient and earthquake response data. Struct. Control Health Monit. **20**(2), 139–155 (2013)
7. Astroza, R., Gutierrez, G., Repenning, C., Hernández, F.: Time-variant modal parameters and response behavior of a base-isolated building tested on a shake table. Earthq. Spectra. **34**(1), 121–143 (2018)
8. Chen, M.C., Pantoli, E., Wang, X., Astroza, R., Ebrahimian, H., Hutchinson, T.C., Conte, J.P., Restrepo, J.I., Marin, C., Walsh, K., Bachman, R., Hoehler, M., Englekirk, R., Faghihi, M.: Full-scale structural and nonstructural building system performance during earthquakes: part I—SPECIMEN description, test protocol, and structural response. Earthq. Spectra. **32**(2), 737–770 (2016)
9. Pantoli, E., Chen, M.C., Wang, X., Astroza, R., Ebrahimian, H., Hutchinson, T.C., Conte, J.P., Restrepo, J.I., Marin, C., Walsh, K., Bachman, R., Hoehler, M., Englekirk, R., Faghihi, M.: Full-scale structural and nonstructural building system performance during earthquakes: part II—NCS damage states. Earthq. Spectra. **32**(2), 771–794 (2016)
10. Pantoli, E., Chen, M.C., Hutchinson, T.C., Astroza, R., Conte, J.P., Ebrahimian, H., Restrepo, J.I., Wang, X.: Landmark dataset from the building nonstructural components and systems (BNCS) project. Earthq. Spectra. **32**(2), 1239–1260 (2016)
11. Tobita, J., Izumi, M., Katukura, H.: Identification of vibration systems and nonlinear dynamics characteristics of structures under earthquake excitations. In: Proceedings of the 9st World Conference on Earthquake Engineering, August 2-9, 1988, Tokyo-Kyoto, Japan (1988)
12. Moaveni, B., Asgarieh, E.: Deterministic-stochastic subspace identification method for identification of nonlinear structures as time-varying linear systems. Mech. Syst. Signal Process. **31**, 40–55 (2012)
13. Van Overschee, P., De Moor, B.: Subspace Identification for Linear Systems: Theory, Implementation, Applications. Kluwer Academic Publishers, Dordrecht (1996)
14. Allemang, R.J., Brown, D.L.: A correlation coefficient for modal vector analysis. In: Proceedings of the 1st International Modal Analysis Conference (IMAC I), November 8–10, 1982, FL (1982)
15. Astroza, R., Ebrahimian, H., Conte, J.P., Restrepo, J.I., Hutchinson, T.C.: System identification of a full-scale five-story reinforced concrete building tested on the NEES-UCSD shake table. Struct. Control Health Monit. **23**(3), 535–559 (2016)
16. Chen, M.C., Astroza, R., Restrepo, J.I., Conte, J.P., Hutchinson, T.C., Bock, Y.: Predominant period and equivalent viscous damping ratio identification on a full-scale building shake table test. Earthq. Eng. Struct. Dyn. **46**(14), 2459–2477 (2017)
17. Astroza, R., Ebrahimian, H., Conte, J.P., Restrepo, J.I., Hutchinson, T.C.: Statistical analysis of the identified modal properties of a 5-story RC seismically damaged building specimen. In: Safety, Reliability, Risk and Life-cycle Performance of Structures and Infrastructures. CRC Press Taylor & Francis (2013)

Chapter 25
Structural Property Guided Gait Parameter Estimation Using Footstep-Induced Floor Vibrations

Jonathon Fagert, Mostafa Mirshekari, Shijia Pan, Pei Zhang, and Hae Young Noh

Abstract This paper presents an approach that estimates occupants temporal gait parameters (e.g. step time, stride time, and initial loading response time) through structural floor vibrations while adapting to varying structural conditions. Continuous monitoring of these temporal gait parameters is a key component for elderly fall risk assessment and diagnosing injuries and gait abnormalities in home healthcare. A primary research challenge with using floor vibration sensing is that real-world deployments are often conducted in existing structures where little is known about the underlying structural properties. As such, prior works utilizing floor vibrations combine data-driven approaches with key heuristic observations to achieve accurate results, but often fail to adapt to varying structural conditions without re-training. To overcome this challenge, we present our approach, which extracts dynamic properties of the structure from footstep-induced vibrations and uses those properties to update gait parameter estimation models. By passively learning the dynamic properties of the structure, our approach enables gait parameter estimation that is robust to structural changes. We evaluate our approach by conducting real-world walking experiments on a test bed at Carnegie Mellon University.

Keywords Structural vibrations · Wavelet decomposition · Gait parameters · Smart structures

25.1 Introduction

Temporal gait parameter estimation (e.g. step time, stride time, and initial loading response time) is essential for understanding health conditions such as Multiple Sclerosis, Parkinson's disease, and osteoarthritis as well as fall risk with elderly populations [1]. Existing methods for gait parameter estimation include direct observation by trained medical staff, camera/vision-based sensing, pressure sensor-based sensing, acoustic/sound-based sensing, and wearable devices [2–5]. These approaches are limited in non-clinical settings due to requirements of specialized staff (observation), line of sight (cameras), dense sensor deployment (pressure sensors), ambient/environmental noise (acoustic), and requiring users to carry or wear a device during monitoring (wearables).

To overcome these limitations, our prior work has utilized footstep-induced floor vibrations for occupant monitoring [6]. The primary challenge with these approaches is that real-world deployments are often conducted in existing structures where little prior information about the structure is available. As a result, a gait monitoring system trained in one structure is not effective in a different structure without re-training. In this work, we overcome this challenge by extracting the dynamic properties of the structure and use those properties to decompose the signal into components that correspond to the heel and toe components of the footstep strike. This enables estimation of temporal gait parameters such as cadence, step time, stride time, and initial loading response time that is robust to varying structural support conditions. We validate our approach by conducting real-world walking experiments on a wood floor structure using three different support conditions.

J. Fagert (✉) · M. Mirshekari · H. Y. Noh
Department of Civil and Environmental Engineering, Carnegie Mellon University, Pittsburgh, PA, USA
e-mail: jfagert@andrew.cmu.edu

S. Pan · P. Zhang
Department of Electrical and Computer Engineering, Carnegie Mellon University, Moffett Field, CA, USA

© Society for Experimental Mechanics, Inc. 2020
S. Pakzad (ed.), *Dynamics of Civil Structures, Volume 2*, Conference Proceedings
of the Society for Experimental Mechanics Series, https://doi.org/10.1007/978-3-030-12115-0_25

25.2 Temporal Gait Parameter Estimation Approach

Our temporal gait parameter estimation approach consists of three main modules: (1) a footstep detection module, (2) a structure-informed signal decomposition module, and (3) a gait-heuristic-based temporal parameter estimation module. An overview of our system can be found in Fig. 25.1.

25.2.1 Footstep Detection Module

In the first module of our temporal gait parameter estimation approach, our system isolates footstep-induced vibration responses from the vibration signal. In this work, vibration responses are collected using a SM-24 geophone sensor. To isolate the footstep responses from the signal, we utilize an chi squared-based anomaly detection algorithm based on our prior work [6]. In this way, we estimate an ambient noise variance based on the recorded signals when no footstep activity is present, and detect a footstep event when the windowed signal variance exceeds the ambient level.

25.2.2 Structure-Informed Signal Decomposition

In this module, our system extracts dynamic properties of the underlying structure and uses them to extract temporal gait information from the footstep-induced vibration response. To decompose the vibration signal, our method first transforms the signal to a spectral-temporal space using the continuous wavelet transform. We then determine the scale in the wavelet transform that has the highest energy for the entire footstep response and utilize its component of the signal for further analysis. This scale corresponds to the frequency of the structure that is most excited by the footstep response and contains the most information about the footstep excitation.

25.2.3 Temporal Gait Parameter Estimation

The final module of our method utilizes the decomposed signal to estimate temporal parameters of the subjects gait while walking in the sensing area. To accomplish this, we first determine the decomposed signal envelope by taking its absolute value and smoothing the curve. Next, we calculate the step time and stride time of each footstep in the walking trace (defined as several consecutive footsteps) by finding the start of the footstep signal (from our detection module) and comparing its time stamp to the next footstep (opposite foot: step time; same foot: stride time). Then, we use a peak detection algorithm to determine the peaks of the signal. To determine the loading response time, we find the first peak of the signal that is at least 8% into the total stride time for that footstep and subtract its time stamp from the start of the signal. This is based on the observations that this peak corresponds to the 'toe on' and that loading response time of the gait cycle is typically 10% of the total gait cycle time [7]. Figure 25.2 shows an example of a footstep response and the detected peaks.

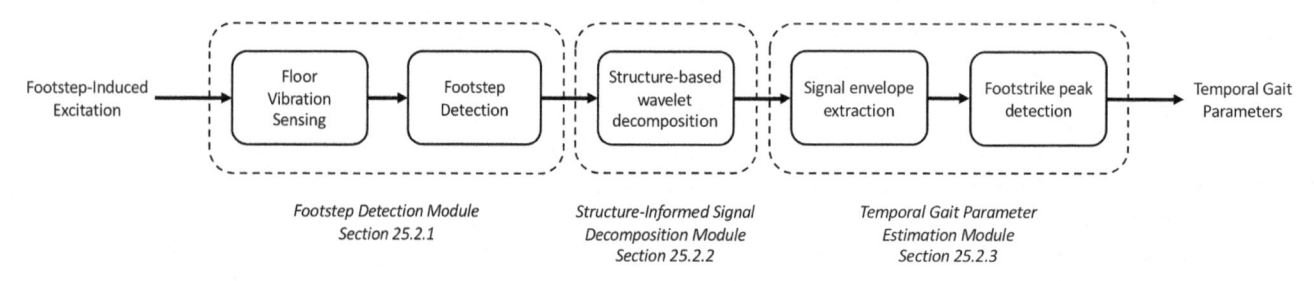

Fig. 25.1 Temporal gait parameter approach overview

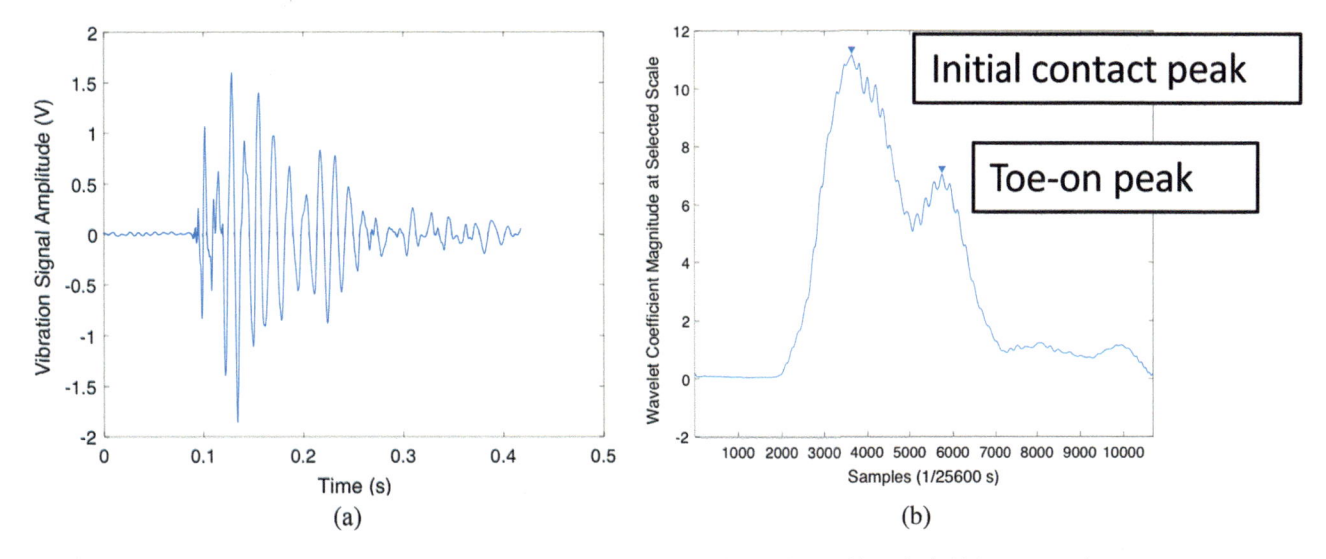

Fig. 25.2 (**a**) The raw footstep-induced vibration signal (**b**) After decomposition and smoothing, the initial contact and toe-on peaks are clearly visible in the signal

25.3 Evaluation

To evaluate the validity of our temporal gait parameter estimation approach, we conducted real-world walking experiments on a wood-framed testing structure at Carnegie Mellon University. To determine the robustness of our approach, we estimated stride time, step time, and loading response time for two participants walking at three step frequencies with three support conditions for the structure. Ground truth was collected using a camera with a frame rate of 60fps and step frequency was controlled with a metronome. A total of 540 footsteps were collected across the two participants and 3 structural configurations. To account for non-detected loading response peaks, we removed outlier estimations, which resulted in a total of 506 footsteps (94% detection rate). Our method achieves an average error of 0.015 s, 0.014 s, and 0.017 s for stride time, step time, and loading response time estimations, which corresponds to 99%, 98% and 86% accuracy, respectively and is on the same order of magnitude as natural variation in human walking [8]. Further, we note that each of the average errors for the three estimations is less than or equal to 1 frame in the ground truth video, which is the precision of the manual ground truth labeling.

25.4 Conclusions

In this paper, we introduce an approach for estimating temporal gait parameters using structural vibration sensing that is robust to changes in the structural support conditions. To overcome the limitation of little prior knowledge of structural conditions, our method extracts the peak energy scale in the time-frequency domain to decompose the signal into a component that contains the most information about the footstep-floor interaction. In this way, our approach achieves average errors of 0.014 s, 0.015 s, and 0.017 s for stride time, step time, and loading response time estimations, respectively, which corresponds to an average accuracy of 94% for temporal gait parameter estimation.

References

1. Moore, S.T., MacDougall, H.G., Ondo, W.G.: Ambulatory monitoring of freezing of gait in Parkinson's disease. J. Neurosci. Methods **167**(2), 340–348 (2008)
2. Kim, C.M., Eng, J.J.: Magnitude and pattern of 3d kinematic and kinetic gait profiles in persons with stroke: relationship to walking speed. Gait Posture **20**(2), 140–146 (2004)
3. Tarnita, D.: Wearable sensors used for human gait analysis. Romanian J. Morphol. Embryol. **57**, 08 (2016)

4. Wang, C., Wang, X., Long, Z., Yuan, J., Qian, Y., Li, J.: Estimation of temporal gait parameters using a wearable microphone-sensor-based system. Sensors **16**(12), 2167 (2016)
5. McDonough, L., et al.: The validity and reliability of the gaitrite system's measurements: a preliminary evaluation. Arch. Phys. Med. Rehabil. **82**(3), 419–425 (2001)
6. Mirshekari, M., Pan, S., Fagert, J., Schooler, E.M., Zhang, P., Noh, H.Y.: Occupant localization using footstep-induced structural vibration. Mech. Syst. Signal Process. **112**, 77–97 (2018)
7. Perry, J., et al.: Gait analysis: normal and pathological function. J. Pediatr. Orthop. **12**(6), 815 (1992)
8. Beauchet, O., et al.: Guidelines for assessment of gait and reference values for spatiotemporal gait parameters in older adults: the biomathics and Canadian gait consortiums initiative. Front. Hum. Neurosci. **11**, 353 (2017)

Chapter 26
Why Is My Coffee Cup Rattling: A Reassessment of the Office Vibration Criterion

Melissa W. Y. Wong and Michael J. Wesolowsky

Abstract In recent history, generic vibration limits have been developed which provide frequency-dependent sensitivities for wide classes of applications. These vibration criterion (VC) curves are internationally accepted as a basis for designing and evaluating the performance of structures. Within these VC curves, the Office (ISO) criterion has been used as a guideline for the design of office spaces. However, as offices gradually move away from paper-based offices to electronic ones, as well as from the traditional private office layouts to open concept offices, the Office (ISO) criterion may no longer be appropriate.

This paper provides three case studies of office floors that have been verified to meet the Office (ISO) criterion through measurements, but which have generated a number of objections from occupants due to excessive floor vibrations. Thus, these case studies suggest that a reassessment of the Office (ISO) criterion is necessary, to determine if a more stringent criterion is required for the design of office spaces.

Keywords Human-induced vibration · Vibration sensitive occupancies · Vibration measurements

26.1 Introduction

The main sources of vibration in buildings are due to human activity. In particular, footfall-induced vibrations from walking activities may easily transmit to nearby spaces if the floor structure is light and flexible. To minimize the floor vibrations, floor structures supporting office spaces are typically designed to meet the Office (ISO) criterion. However, in recent years, there have been many cases where floor structures have met the Office (ISO) criterion as proven by in-field measurements, but office occupants continue to raise concerns regarding the level of vibrations.

This paper presents three case studies of office buildings where vibration measurements were conducted in areas where the occupants had complained of excessive vibration levels and compares these measurements to the Office (ISO) criterion.

26.2 Vibration Criteria

For human comfort, vibration criteria are normally expressed as the root mean square (RMS) response of each one-third octave band from 1 to 80 Hz [1]. For sensitive equipment, the criteria may be expressed in one-third octave bands, or other formats, including power spectral densities, peak-to-peak levels, etc. Over the past 25 years, generic vibration limits have been developed which provide frequency-dependent sensitivities for wide classes of equipment, and are used extensively in the design of healthcare and research facilities [2]. These curves were originally based on the ISO 2631-2 (1989) [3] base curve for human response to whole body vibration, which is considered the threshold of human perception, but have since evolved. The ISO base curve is often referred to as the ISO-Operating Room criteria.

These vibration criterion (VC) curves are internationally accepted as a basis for designing and evaluating the performance of vibration sensitive equipment and the structures that support them. The VC curves range between Workshop (least stringent) through VC-G (most stringent). Most office areas target the Office (ISO) criterion when determining the suitability of the floor's performance for office use.

M. W. Y. Wong (✉) · M. J. Wesolowsky
Swallow Acoustic Consultants Ltd./Thornton Tomasetti, Mississauga, ON, Canada
e-mail: mwwong@thorntontomasetti.com

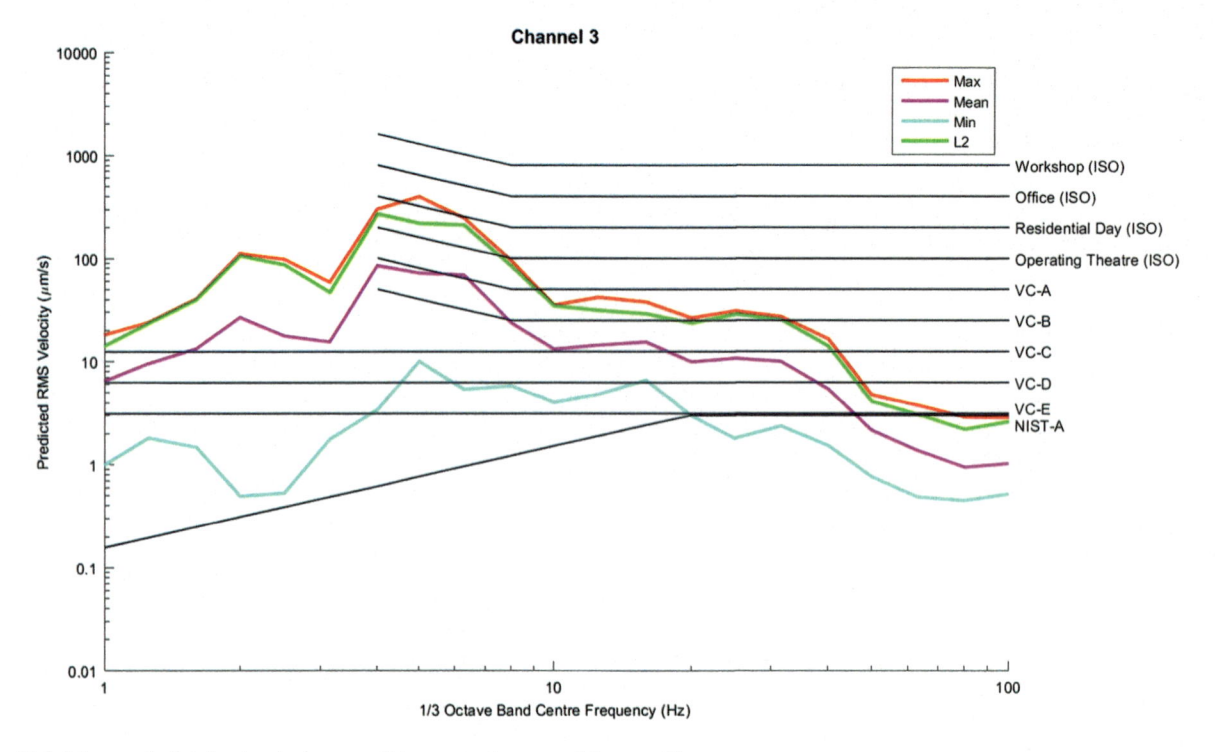

Fig. 26.1 Measured vibration levels due to walking excitation, complaint area #1

26.3 Case Study #1

The building used for Case Study #1 is a multi-storey building comprised of office spaces in an open concept layout. The building is a steel-framed structure with 89 mm thick concrete slab on a 38 mm thick steel deck supported on open web steel joists. The bay size is typically 9.15 m by 9.15 m. According to the office occupants, they could feel uncomfortably high vibrations whenever somebody walked by their desks. As such, vibration measurements were conducted at the main "complaint" areas to quantify the walking-induced vibrations.

To induce the worst-case response, the natural frequency of the floor structure was first determined using heel drops. Depending on the bay, the natural frequency ranged between 4.5 and 6.5 Hz. The higher frequencies were typically observed in areas that had partitions either above or below the floor. Once the natural frequency of the floor was determined, walking simulations were conducted at sub-harmonics of the floor's natural frequency. The walking simulations would follow the path of the existing corridors to induce realistic worst-case scenarios. In addition, all measured values were scaled to that of a walker weighing 76 kg, as 76 kg is defined as the weight of an 'average person' in standard literature [4]. This testing was repeated in all complaint areas. The vibration levels at each of the complaint areas are shown in Figs. 26.1, 26.2, and 26.3. Note that the L2 vibration levels are typically used when evaluating the measured response. In our opinion, the L2 is the most reasonable measure of performance as compared to using the maximum measured values, as it disregards extraneous measurements (i.e. outliers).

As can be seen from the above figures, all three areas measured velocity levels below the Office (ISO) criterion. However, these areas were where occupants had complained about the level of vibration. In addition, while on site, our staff subjectively also felt that the floor vibrations were objectionable.

26.4 Case Study #2

The next case study is again a multi-storey building where occupants on both the 15th and 16th floor complained about the vibration levels induced by people walking near their desks. On both floors, office occupants at the north and west side of the floor plate complained about the vibrations. The floor slab is comprised of an 89 mm thick lightweight concrete slab on

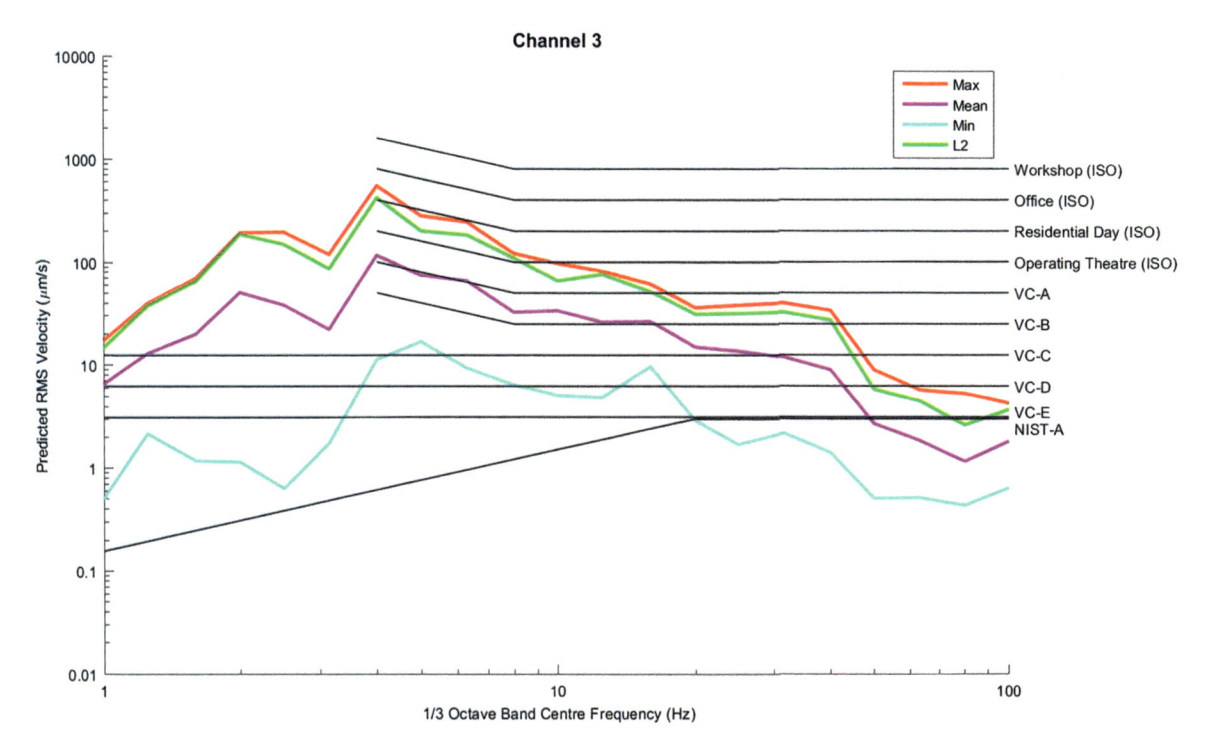

Fig. 26.2 Measured vibration levels due to walking excitation, complaint area #2

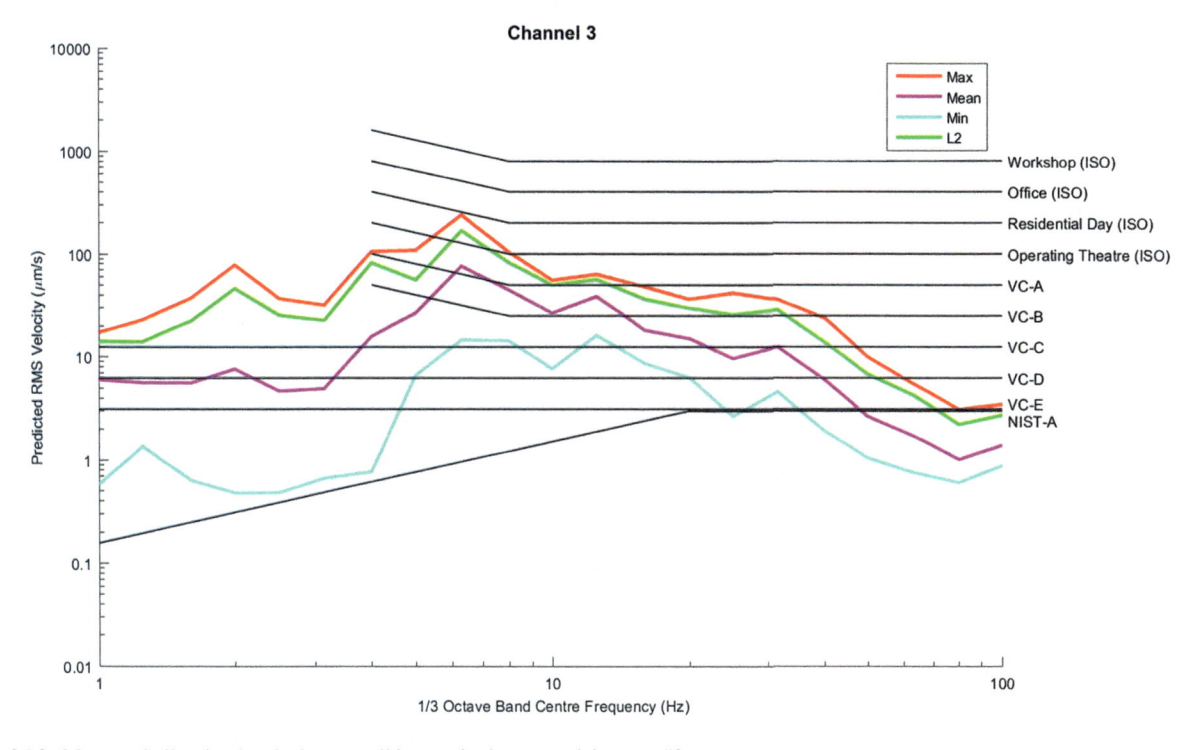

Fig. 26.3 Measured vibration levels due to walking excitation, complaint area #3

50 mm thick steel deck supported on steel I beams. The bay size of the north end of the floor is typically 12.2 m by 9.15 m, while the west end of the floor is typically 9.15 m by 9.15 m.

Similar to the first case study, walking simulations were conducted to induce the worst-case response of the floor. Based on heel drop tests, the natural frequency of the north end of the floor plate was approximately 4 Hz, while the west end of the floor plate measured between 5 and 6.3 Hz. Walking speeds at sub-harmonics of those determined floor natural frequencies

Table 26.1 Measured vibration levels due to walking excitation

Location		15th floor	16th floor
North end	Location 1	Office (ISO)	Office (ISO)
	Location 2	Office (ISO)	Residential Day (ISO)
	Location 3	Office (ISO)	Residential Day (ISO)
	Location 4	Office (ISO)	Office (ISO)
West end	Location 1	Residential Day (ISO)	Residential Day (ISO)
	Location 2	Operating Theatre (ISO)	Residential Day (ISO)
	Location 3	Operating Theatre (ISO)	Residential Day (ISO)
	Location 4	Residential Day (ISO)	Residential Day (ISO)

Fig. 26.4 Walking paths and measured vibration levels due to the walking paths, west section

were used for the simulations. Based on these measurements, the vibration levels measured in the complaint areas are shown in Table 26.1.

All measurements either met the Office (ISO) criterion or were significantly lower. Given that these values were seen at multiple areas along the floor plate, as well as on two different floors where multiple people were complaining about the vibration levels, this again suggests that the Office (ISO) criterion may not be a suitable design target for office areas. In addition, our staff noticed objectionably high vibration levels while conducting the floor vibration testing.

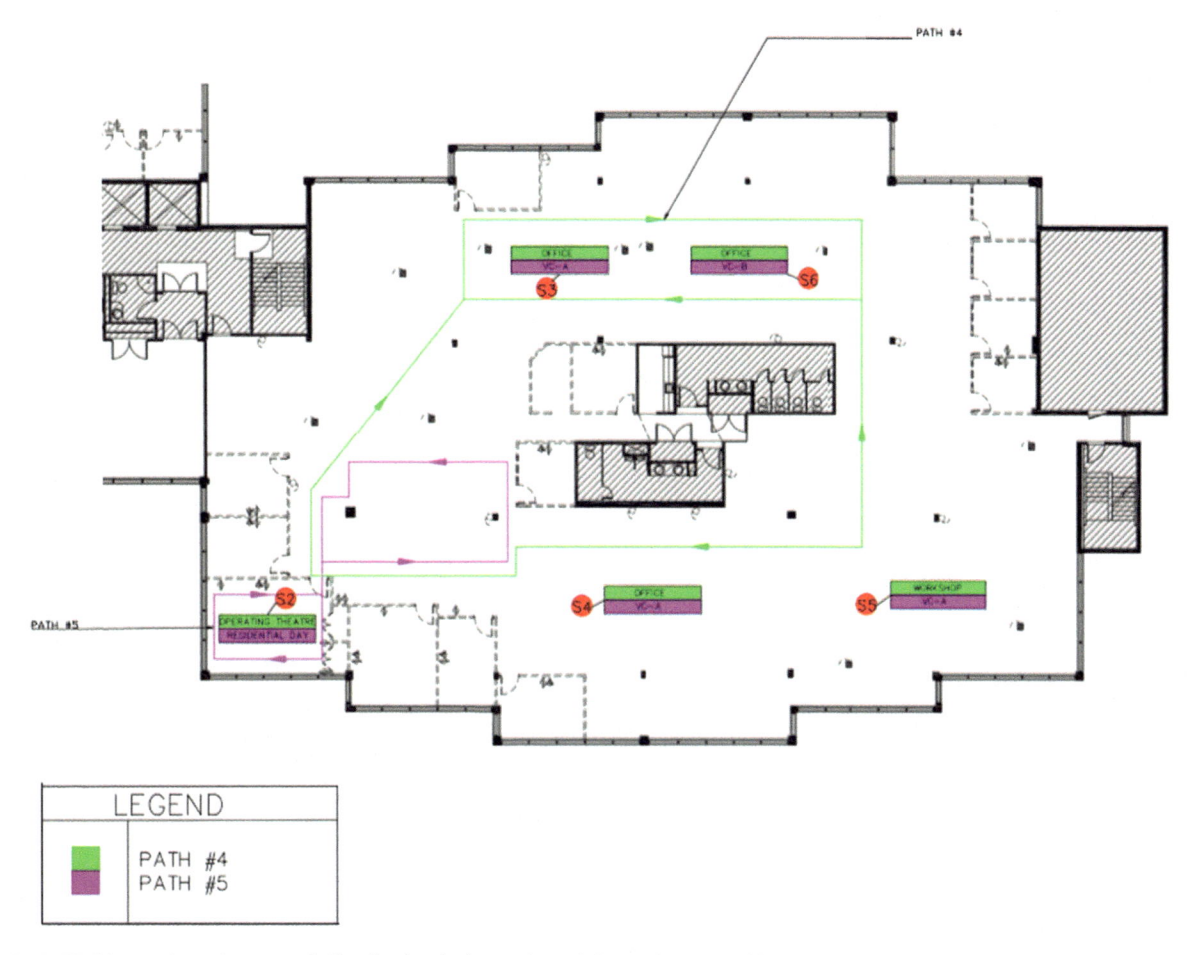

Fig. 26.5 Walking paths and measured vibration levels due to the walking paths, east section

26.5 Case Study #3

The last case study focuses on a multi-storey concrete building where the office occupants were complaining throughout the third floor. Note that unlike the first two case studies, the bay sizes of this floor plate were more irregular but typically measured around 8 m by 8 m. The natural frequency of the floor ranged between 5.7 and 6.3 Hz.

The walking paths and measured vibration levels as a result of the walking paths are shown in Figs. 26.4 and 26.5.

A total of five walking paths were conducted to induce the worst-case response at the accelerometer locations. It was found that during these walking simulations, only two sensor locations exceeded the Office (ISO) criterion. All other locations met the criterion. Although only two locations showed exceedances, occupants seated in areas all over the floor had filed complaints regarding the excessive vibration levels. Our staff also subjectively felt that the floor vibration levels were objectionably high, while conducting the floor vibration testing.

26.6 Conclusion

Three case studies have been presented to show that there have been multiple projects where the floor slab had met the Office (ISO) criterion, but the office occupants had still complained about excessive vibrations. The following conclusions have been made:

1. Using the Office (ISO) criterion may no longer be a suitable design target when designing structures for office use.
2. The vibration exceedances were mainly noticed in open office areas where the additional stiffness and weight of partitions were not available to reduce the floor motion.
3. Further research should be conducted to provide guidance to the design community such that a new appropriate criterion for office areas can be established.

References

1. BSI: Guide to Evaluation of Human Exposure to Vibration in Buildings (1 Hz to 80 Hz). BS 6472:1992. British Standards Institution, London (1992)
2. Amick, H., Gendreau, M., Busch, T., Gordon, C.: Evolving criteria for research facilities: I—vibration. In: Proceeding of SPIE Conference 5933: Buildings for Nanoscale Research and Beyond, San Diego, CA, July 31 to August 1, 2005 (2005)
3. ISO 2631-2: Evaluation of Human Exposure to Whole-Body Vibration—Part 2: Human Exposure to Continuous and Shock-Induced Vibrations in Buildings (1 to 80 Hz), ISO 2631-2. International Standard Organization, Geneva (1989)
4. Smith, A.L., Hicks, S.J., Devine, P.J.: Design of Floors for Vibration: A New Approach—Revised Edition (SCI P354). Steel Construction Institute, Ascot (2009)

Chapter 27
Response of a SDOF System with an Inerter-Based Tuned Mass Damper Subjected to Non-stationary Random Excitation

Abdollah Javidialesaadi and Nicholas E. Wierschem

Abstract Inerter-based tuned mass dampers (TMDs) have been developed recently with the goal of improving upon the performance of traditional TMDs. However, studies investigating the response of single-degree-of-freedom (SDOF) systems with inerter-based TMDs have been primarily limited to ones considering harmonic loads and stationary random excitation. Various relevant random loads have non-stationary characteristics (their frequency contents and/or amplitude change with time); therefore, these load types should be considered in the design of inerter-based TMDs. This paper presents an investigation to evaluate the mean squared response of SDOF systems with inerter-based TMDs that are subjected to a random non-stationary excitation. The non-stationary excitation considered is an evolutionary spectrum of the ground acceleration. The results of this study are used to determine the influence of the non-stationary excitation on the optimal damper properties in comparison to designs considering a stationary process.

Keywords Passive control · Non-stationary excitation · Rotational inertia damper · Tuned mass damper

27.1 Introduction

The dynamic response of civil and mechanical structures to extreme and service level loads can be mitigated using passive control devices. With passive control, the dynamic properties of the structure itself are modified by the addition of the control device. Tuned mass dampers (TMDs) are one of the most prevalently used devices for the passive control of structures. TMDs have been developed, designed, and utilized in structures as a passive control device for decades [1, 2]. When the properties of the TMD are properly tuned, a critical mode of vibration of the structure is replaced with modes that feature a response in which vibration is concentrated in the relative motion of the TMD. The effectiveness of TMDs is often limited by the size of the TMD mass used, with relatively large TMD masses capable of the most effective vibration control performance [3]; however, these large-mass TMDs are impractical in many cases. To address this, a new family of enhanced TMDs have been proposed and investigated recently [4, 5]. Multiple different types of inerter-based TMDs have been proposed and investigated recently [6]. The key innovation in this family of enhanced TMDs is the inclusion of a mechanical device called an "inerter" or a rotational inertia mass, which is capable of providing a large inertia mass by using a small physical mass [7]. In order to design the inerter-based TMDs, exact optimization procedures and numerical approaches have been used in previous studies [8, 9]. These recent investigations of inerter-based TMDs have been limited to considering stationary white noise and harmonic excitation. However, many different non-stationary loads exist, such as earthquake ground motions; therefore, this study focuses on response of SDOF with an inerter-based tuned mass damper subjected to a non-stationary excitation.

27.2 Inerter-Based Tuned Mass Damper

The rotational inertia double tuned mass damper (RIDTMD), which is one type of inerter-based TMD that has been studied under harmonic and white noise excitation [4, 8], is considered in this study. Figure 27.1 depicts the RIDTMD attached to the SDOF structure. In Fig. 27.1, k_s, m_s, c_s, and $\ddot{u}_g$ denote the stiffness, mass, damping, and ground acceleration of the structure, respectively. In the RIDTMD, the dashpot of the TMD is replaced with an inerter based-device, where k_2, c_2, and

A. Javidialesaadi · N. E. Wierschem (✉)
Department of Civil and Environmental Engineering, The University of Tennessee Knoxville, Knoxville, TN, USA
e-mail: nwiersch@utk.edu

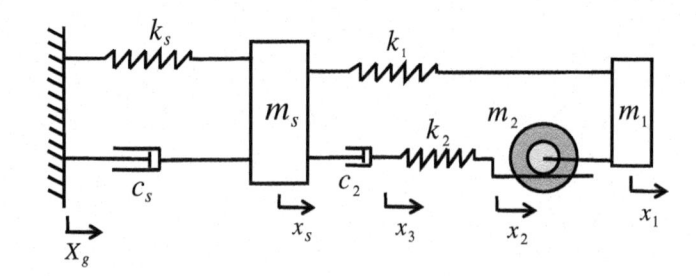

Fig. 27.1 A SDOF system with a rotational inertia double tuned mass damper (RIDTMD) attached

m_2 represent the stiffness, damping, and rotational inertia mass of the inerter-based device. This rotational inertia mass m_2 is an amplified effective mass and can be achieved with a flywheel with small physical mass and an inerter. For example, a device producing an amplified inertia mass equal to 350 kg by using a 2 kg physical mass has been reported [10].

27.3 Response to Non-stationary Excitation

The non-stationary excitation considered is an evolutionary spectrum of the ground acceleration, which can be calculated using Eq. (27.1) [11].

$$S_{\ddot{u}_g}(\omega, t) = \frac{\left(e^{-at} - e^{-bt}\right)}{\max\left(e^{-at} - e^{-bt}\right)} \left(\frac{\omega^4_g + 4\xi^2(t)_g \omega_g^2(t)\omega^2}{\left(\omega^2 - \omega^2_g(t)\right)^2 + 4\xi_g(t)\omega_g^2(t)\omega^2} \right)^{1/2} S_0 \qquad (27.1)$$

where, $a = 0.25 \text{ s}^{-1}$; $b = 0.5 \text{ s}^{-1}$; $S_0 = 1 \text{ cm}^2/\text{s}^3$, $\omega_g(t)$, and $\xi_g(t)$ are the spectrum characteristics. $\omega_g(t)$ and $\xi_g(t)$ are functions of time:

$$\xi_g(t) = \begin{cases} 0.64 \, (0 \leq t < 4.5) \\ 1.25(t - 4.5)^3 - 1.875(t - 4.5)^2 + 0.64 \, (4.5 \leq t < 5.5); \\ 0.015 \, (t < 5.5) \end{cases}$$

$$\omega_g(t) = \begin{cases} 15.56 \, (0 \leq t < 4.5) \\ 24.12(t - 4.5)^3 - 40.68(t - 4.5)^2 + 15.56 \, (4.5 \leq t < 5.5) \\ 2 \, (t < 5.5) \end{cases}$$

(27.2)

The TMD and RIDTMD, see Fig. 27.1, with mass ratios (m_1/m_s) equal to 0.01, 0.05 and 0.1 are considered therein. To evaluate and compare the performance of the TMD and the RIDTMD, the minimization of $E[u_s^2]$ (displacement mean square response function of the primary structure) is considered as the objective function and the optimum values of stiffness, damping, and rotational inertia mass obtained for TMDs and RIDTMD are obtained through numerical optimization for 10 s ($\Delta t = 0.1$). Unlike for a stationary input, the optimum values are changing during the time and the optimum design can be considered as the design that minimizes the maximum displacement mean square during the time. Utilizing the optimum values, the responses of the systems are obtained and presented in Fig. 27.2. Figure 27.2 shows that the displacement mean square is not constant for the TMD and RIDTMD but the performance of the RIDTMD is superior to the TMD during the entire time.

27.4 Conclusion

It has been previously shown that the RIDTMD is capable of superior performance, compared to the TMD, when attached to a structure that is subjected to a stationary excitation. In this case, the optimum design values and $E[u_s^2]$ are constant. The optimum design and performance of the RIDTMD in the non-stationary excitation case, where the input excitation

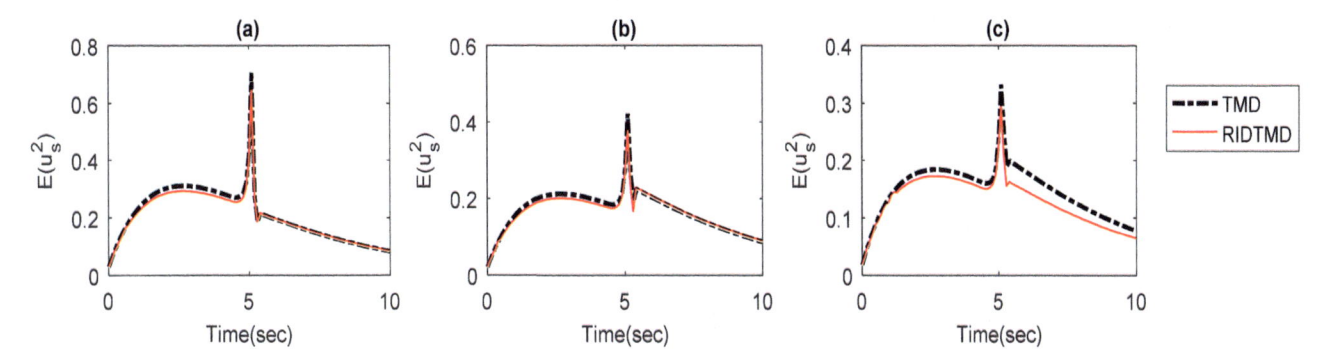

Fig. 27.2 Displacement mean square response of the primary structure vs. time: (**a**) mass ratio = 0.01; (**b**) mass ratio = 0.05; (**c**) mass ratio = 0.1

characteristics vary with time, are investigated in this paper. A traditional TMD and RIDTMD with three different mass ratios and a non-stationary excitation are studied to assess the performance of the inerter-based TMD, in comparison to the TMD. Results show the optimum values vary with time during the non-stationary load for both cases. For the mass ratios and non-stationary excitation considered, a 7–8% reduction of peak displacement mean square response of the primary structure was observed in the RIDTMD cases, in comparison to the TMD cases. While the trends here are expected to be repeated for other types of non-stationary loads, this specific value of peak reduction cannot be generalized.

References

1. Den Hartog, J.: Mechanical Vibrations, 4th edn. McGraw-Hill, New York (1956)
2. Warburton, G.B.: Optimum absorber parameters for minimizing vibration response. Earthq. Eng. Struct. Dyn. **9**(3), 251–262 (1981)
3. Sadek, F., Mohraz, B., Taylor, A.W., Chung, R.M.: A method of estimating the parameters of tuned mass dampers for seismic applications. Earthqu. Eng. Struct. Dyn. **26**(6), 617–636 (1997)
4. Garrido, H., Curadelli, O., Ambrosini, D.: Improvement of tuned mass damper by using rotational inertia through tuned viscous mass damper. Eng. Struct. **56**, 2149–2153 (2013)
5. Javidialesaadi, A., Wierschem, N.E.: Three-element vibration absorber–inerter for passive control of single-degree-of-freedom structures. J. Vib. Acoust. **140**(6), 11 (2018)
6. Hu, Y., Chen, M.Z.Q.: Performance evaluation for inerter-based dynamic vibration absorbers. Int. J. Mech. Sci. **99**, 297–307 (2015)
7. Smith, M.C.: Synthesis of mechanical networks: the inerter. IEEE Trans. Autom. Control. **47**(10), 1648–1662 (2002)
8. Javidialesaadi, A., Wierschem, N.E.: Optimal design of rotational inertial double tuned mass dampers under random excitation. Eng. Struct. **165**, 412–421 (2018)
9. Javidialesaadi, A., Wierschem, N.: Extending the fixed-points technique for optimum design of rotational inertial tuned mass dampers. Dyn. Civil Struct. **2**, 83–86 (2017)
10. Ikago, K., Saito, K., Inoue, N.: Seismic control of single-degree-of-freedom structure using tuned viscous mass damper: the tuned viscous mass damper. Earthqu. Eng. Struct. Dyn. **41**(3), 453–474 (2012)
11. Leung, A.Y.T., Zhang, H., Cheng, C.C., Lee, Y.Y.: Particle swarm optimization of TMD by non-stationary base excitation during earthquake. Earthqu. Eng. Struct. Dyn. **37**(9), 1223–1246 (2008)

Chapter 28
Experimental Study on Digital Image Correlation for Deep Learning-Based Damage Diagnostic

Nur Sila Gulgec, Martin Takáč, and Shamim N. Pakzad

Abstract Large quantities of data which contain detailed condition information over an extended period of time should be utilized to prioritize infrastructure repairs. As the temporal and spatial resolution of monitoring data drastically increase by advances in sensing technology, structural health monitoring applications reach the thresholds of big data. Deep neural networks are ideally suited to use large representative training datasets to learn complex damage features. In the previous study of authors, a real-time deep learning platform was developed to solve damage detection and localization challenge. The network was trained by using simulated structural connection mimicking the real test object with a variety of loading cases, damage scenarios, and measurement noise levels for successful and robust diagnosis of damage. In this study, the proposed damage diagnosis platform is validated by using temporally and spatially dense data collected by Digital Image Correlation (DIC) from the specimen. Laboratory testing of the specimen with induced damage condition is performed to evaluate the performance and efficiency of damage detection and localization approach.

Keywords Structural health monitoring · Digital image correlation · Convolutional neural networks · Damage detection

28.1 Introduction

It is important to establish lifetime safety of the infrastructure subjected to a wide range of environmental and operational conditions [1]. Providing timely damage assessment of these structures often requires long-term monitoring and dense instrumentation [2]. Sensor networks today provide an exciting set of opportunities and challenges to collect an enormous amount of data from any structure, which due to its nature is posing a big data problem [3]. Conventional approaches primarily focus on hand-crafting damage features and classifiers to interpret the health condition of the structures [4–7]. Although such methods are effective in identifying structural damage of a particular type, there are some constraints limiting these methods. The existing methods of analysis rely on estimating carefully crafted features that often are limited in what they can do and are not automated in nature, thus not appropriate for a broad range of big data applications [8, 9].

Deep Neural Networks (deep learning or DNN) are a state-of-the-art set of methods for taking advantage of the opportunities hidden in big data. They are designed such that they can learn from data, for this reason, deep learning is ideally suited to use large representative training datasets to learn complex features [10]. DNNs learn by training the network parameters by training which is then used to make data-driven predictions or decisions. One of the most widely used types of DNN is convolutional neural network (CNN) due to its ability to keep spatial features of the input and reduce memory requirements by using fewer parameters [11, 12].

The prior studies of authors [13, 14] addressed these issues by developing a CNN-based real-time damage identification and localization methodology that learns sophisticated damage features without hand-crafting them. The approach fed the network by using raw strain field measurements which are a direct indicator of stress, fatigue, and failure. The algorithm was trained and validated successfully on test cases created by FE simulations. In this paper, the performance of the trained algorithm is tested by using data collected by an optic-based technique called digital image correlation (DIC)[15]. Full-field measurement data obtained by DIC helps to understanding the local effects and material and component behavior.

N. S. Gulgec (✉) · S. N. Pakzad
Department of Civil and Environmental Engineering, Lehigh University, Imbt Labs, Bethlehem, PA, USA
e-mail: nsg214@lehigh.edu

M. Takáč
Department of Industrial and Systems Engineering, Lehigh University, Harold S. Mohler Laboratory, Bethlehem, PA, USA

The rest of the paper is organized as follows. First, a brief explanation of the CNN-based methodology is provided in Sect. 28.2; then, the test setup is described in Sect. 28.3. In Sect. 28.4, experimental validation of the proposed CNN architecture and the main findings are presented. Conclusions and future directions are given in Sect. 28.5.

28.2 CNN-Based Approach for Robust Structural Damage Diagnosis

In this section, the adopted technique [13] is briefly described. As presented in Fig. 28.1, a general map of the algorithm is composed of a training and testing phase. Training phase operates on the strain fields obtained from finite element simulations. After normalizing each strain field by its absolute maximum, the search mechanism finds a good set of hyperparameters. Once the network architecture is built by these hyperparameters, it is trained to determine the existence of damage (i.e. detection task) and estimate the boundaries of the damaged area (i.e. localization task). Both of the tasks share trainable parameters to extract local features which are common for them. This provides more efficient learning, shorter training time and lower computation cost. Trained parameters are saved to test the performance of strain field collected by DIC system. In this phase, raw strain fields are fed into the proposed architecture to estimate the labels for detection and localization tasks.

Training data is formed by using FE simulations of a connection shown in Fig. 28.3a. The connection consists of two 20 inch-long C8 × 11.5 channels welded to a steel plate with the dimension of 28 × 14 × 1/4 inches. Each channel member has 8-inch overlap with the main gusset plate. The material is modeled as elastic-perfectly plastic with a yield strength of 36 ksi. A total of 30,000 healthy and 30,000 damaged samples are generated with different loads, damage locations and measurement noise. Damage is modeled as 0.5 inch long cracks where coordinates are randomly selected from the area bounded by the two corners [$A(8.5, 1)$ and $B(19.5, 13)$]. The crack coordinates which are used to create training dataset are shown as black lines in the Fig. 28.3a. The location of the crack is stored as bounding box (a_1, b_1, a_2, b_2), where b_1 and b_2 indicate the coordinates of the tips of the crack; a_1 and a_2 are the y coordinate of the crack with 0.5 inch subtraction and addition, respectively. For the "healthy" samples, bounding box is set to [0, 0, 0, 0].

Varying loads between $\sim U[-100\,\text{kips (compression)}, 120\,\text{kips (tension)}]$ are applied to the end of the channels. Additive Gaussian noise $\sim N(0, \sigma^2)$ is added to noise-free samples where σ is the standard deviation of the measurement noise. Four different noise levels (i.e. the ratio between the standard deviation of measurement noise to actual strain values) are considered (2%, 5%, 10%, and 15%). Strain distribution in the direction of loading (ε_y) is represented as 28 × 56 × 1 tensors, then utilized to feed the CNN architecture.

The network found by hyperparameter search mechanism is shown in Fig. 28.2. It consists of three convolutional layers followed by two task-specific fully connected layers. The convolutional layers receive the input layer and pass them through a filter size of (3x3). The network forms 8, 16 and 32 feature maps after these convolutional layers. The max-pooling operation, which has the size of (2x2) with a stride of 2, performs right after the first and second convolution layer. The feature maps of

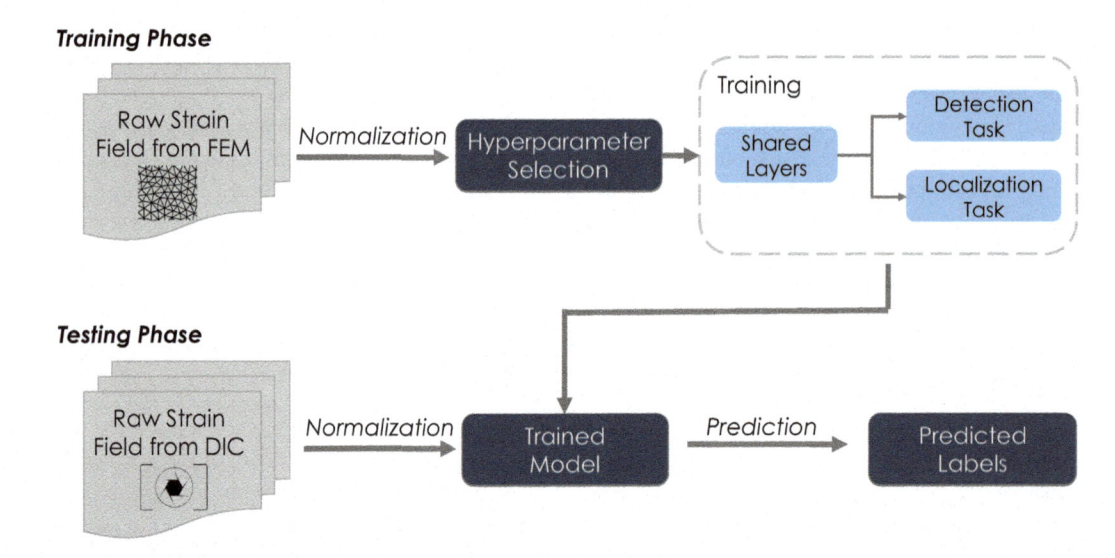

Fig. 28.1 Overview of the proposed methodology

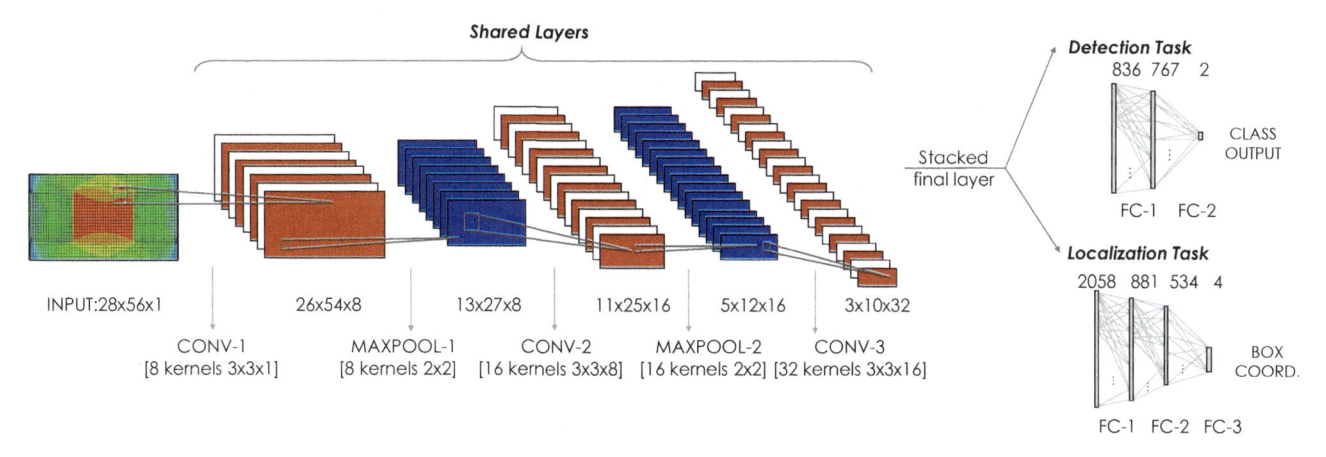

Fig. 28.2 Proposed CNN architecture

the last convolutional layer are stacked together in an array and employed as an input to the task-specific layers. The hidden layer sizes for the detection task are [836 − 767], whereas they are [1305 − 1191 − 406] for the localization task. The learning rate of $\eta_{det} = 0.0451$ and $\eta_{loc} = 0.0026$ are adopted for the detection and localization parts, respectively.

28.3 Experimental Validation

28.3.1 Test Setup

The principal behind DIC methodology is to determine the shift and/or rotation of elements of a reference image in an image taken under different conditions [16]. Figure 28.3 illustrates the test setup including the specimen, GOM Aramis 3D DIC system, and external light sources. A structural connection includes a stochastic pattern on the front surface of the specimen to track the changes in the gray-scale. In order to create the pattern, white paint is sprayed to generate the background, then black speckles are randomly spread with a rubber stamp. The closer look to the pattern is presented in Fig. 28.3b. The average speckle size of 1/8 inch is obtained. Damage is created as a grinded dent through the thickness from the coordinate (9, 17.5) to (10, 17.5) inches which was not used in the training dataset (Fig. 28.3a).

3D DIC system includes two high resolution (4000 × 3000 pixels) cameras with high precision lenses with a focal length of 24 mm. The cameras position 35 inches away from the specimen with 14.4-inch distance between the cameras to capture approximately 14 × 20 inch measurement volume. Polarize filters are attached to the lenses to remove glare and get consistent illumination. Calibration is done with coded panels to achieve good accuracy in both in-plane and out-of-plane measurements. The external light source is utilized to illuminate the specimen and balance the effect of the ambient light existing in the laboratory.

Dense strain measurements of the specimen are collected during different stages of the damage. The test is conducted within the linear elastic range of the material behavior. The plate is gradually loaded to 50 kips and unloaded to its zero-load position by using the SATEC 600 kip hydraulic testing system. The quasi-static loading scheme is presented in Fig. 28.3c. In every 5 kip load increment, the load is held for a minute to allow DIC to take pictures. For each constant load, 150 pictures are acquired with the sampling rate of 4 Hz.

28.3.2 Analysis of Data

ARAMIS Professional 2017 software [17] is used to achieve full-field analysis of the test object. The software evaluates high-resolution images recorded from the specimen during loading, then automatically computes 3D coordinates for all loading stages and derives strain results. The post-processing algorithm of software has a stage-wise analysis, in which each stage consists of one image. In this study, the reference image is selected as the first image taken under the load of 5 kips.

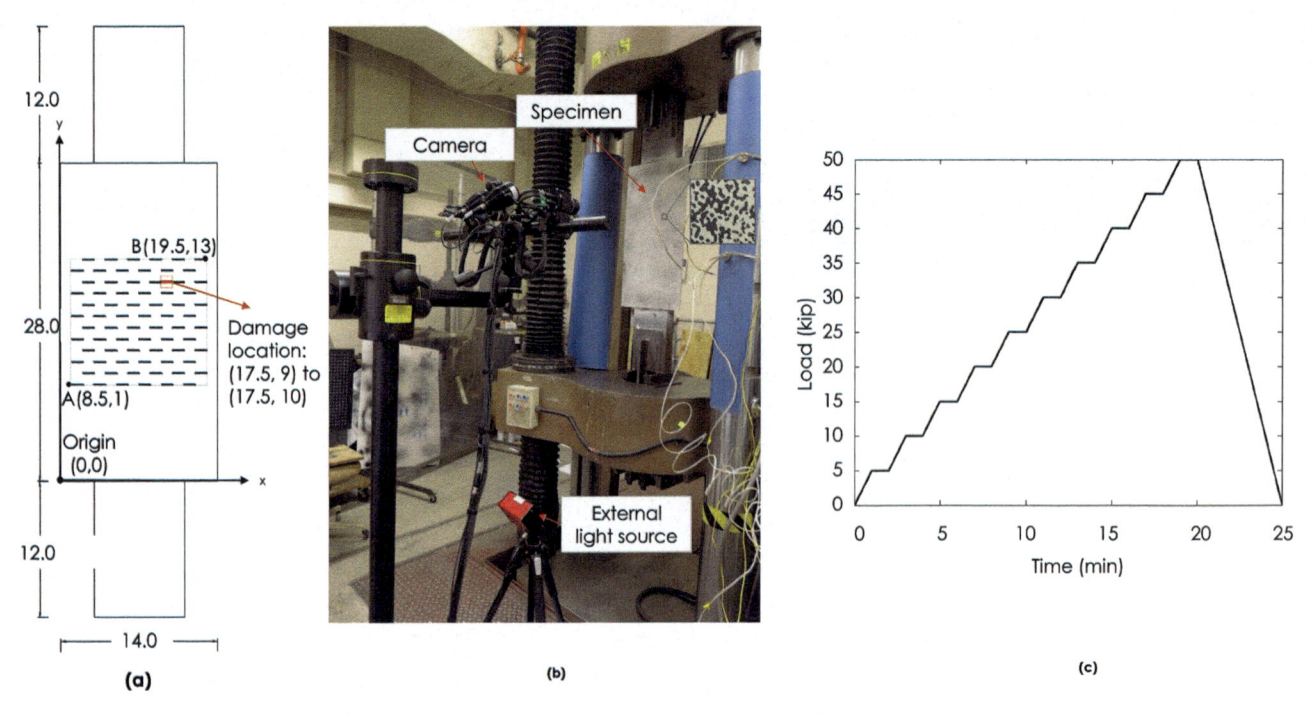

Fig. 28.3 (a) Specimen, (b) Test setup, (c) Loading scheme

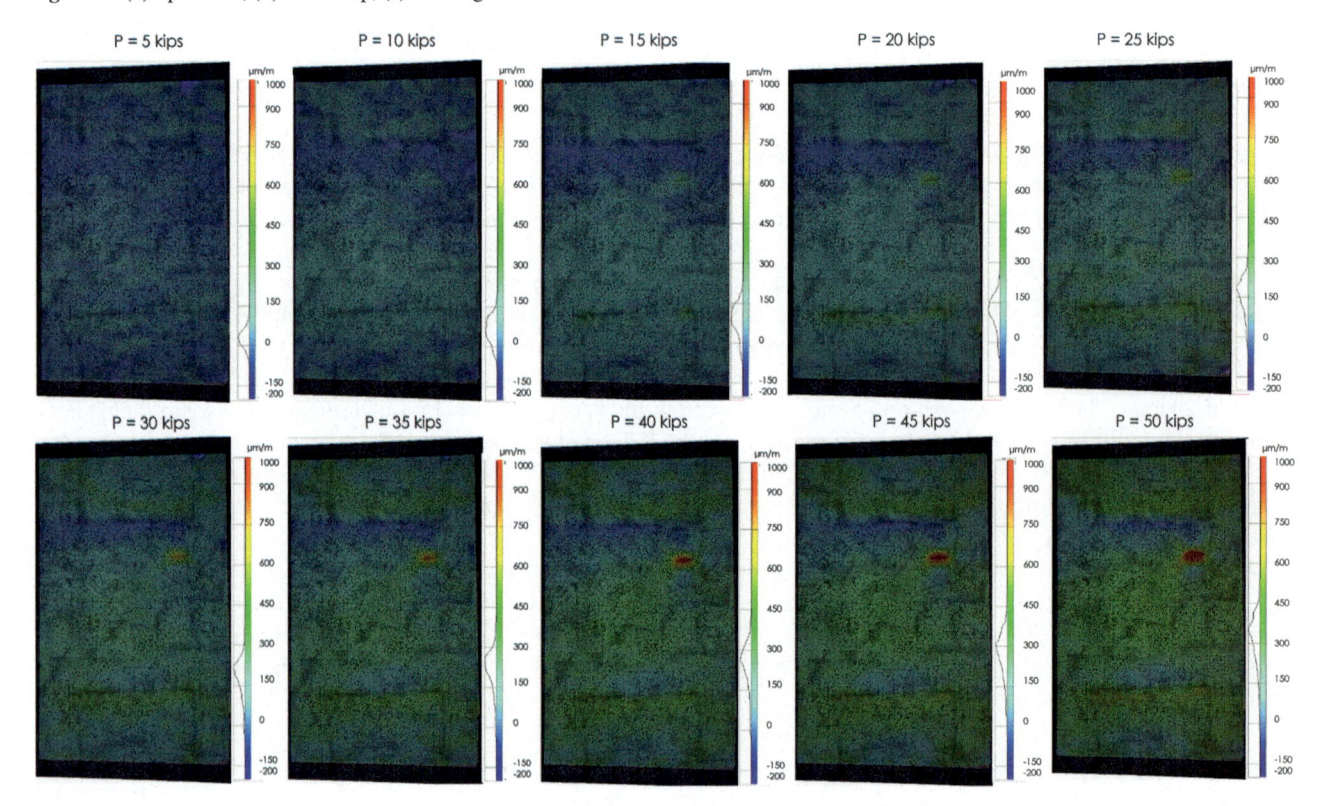

Fig. 28.4 Strain fields

Then, axial strain fields (ε_y) are computed as an input to trained network architecture. An example of strain field stages from each ten load levels {5 kip, 10 kip, ..., 50 kip} are shown in Fig. 28.4. The axis of the colorbar is set to $[-200, 1000]$ μm/m for all stages to illustrate the effect of the crack on the axial strain. The histogram of the strain values is also shown next to the legend. It is noticeable that strain gradients start to occur near crack tips when the load is greater than 35 kips.

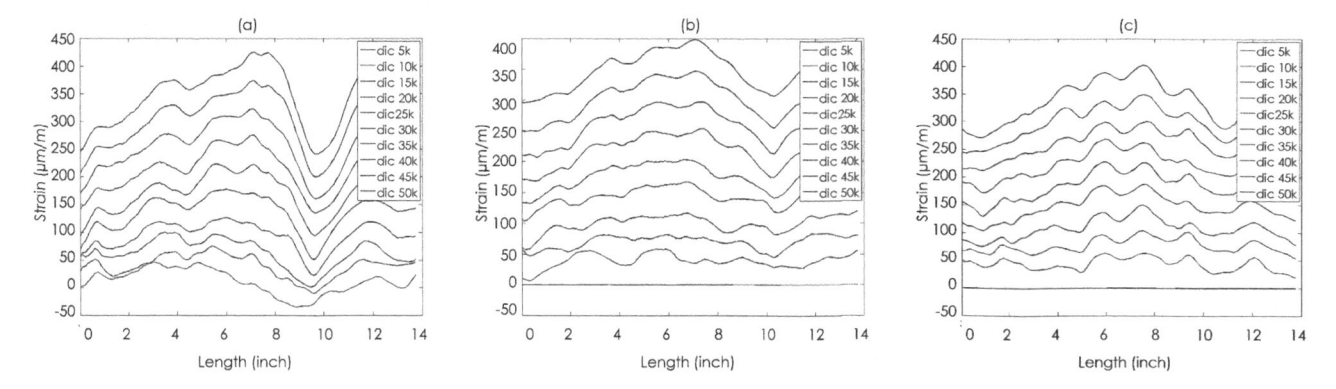

Fig. 28.5 Strain values of the cross section at (**a**) Plane Y = 12.5 inch, (**b**) Plane Y = 13 inch, (**c**) Plane X = 15.5 inch

In order to better analyze the measurements, the strain measurement obtained from several cross sections (i.e. Plane Y = 12.5, 13, 15.5 inches) are plotted for all load stages (Fig. 28.5). In the figure, each line represents 150 stages that have been averaged together and each color indicates the strain measurements at a specific load. The figure shows that strains increase linearly in a similar trend with each consecutive loading. In other words, the difference of the measurements at 50 kips and 45 kips are identical with the measurements at 45 kips and 40 kips. The only exception is observed for loads smaller than 15 kips where the maximum strain is less than 100 μm/m. This level of strain measurement is accepted as noise floor which is very difficult to measure for the size of the specimen.

28.4 Damage Detection and Localization Results

This section evaluates the performance of the trained network on data collected from DIC for detection and localization tasks. The data is sampled from every 0.5 inch to adapt the mesh size in FE model. Since cameras capture 14 × 20 inch measurement volume, the areas that are not covered by the cameras are filled by padding zeros. In the end, strain fields of 28 × 56 × 1 tensors are normalized by its absolute maximum are tested by the saved model parameters. Detection accuracy is defined as the correct prediction of a sample being damaged or healthy. Detection performance of the network is visualized in Fig. 28.6a. All samples from different loading stages are correctly identified. It is worthwhile to mention that high detection accuracy is accomplished although there were strain gradients caused by other imperfections rather than the crack. High stress concentrations are also observable near the welds.

Further analysis is achieved by performing the localization task since the data from healthy samples are not available. In this task, the accuracy is defined as predicted values being within the boundaries pre-defined by the threshold values (i.e. for the threshold value of 0.5, bounding box expands from the sides of 0.5 inch). Three different thresholds are used such as thr = 0.5 inch, thr = 1 inch, thr = 2 inch. According to the Fig. 28.6b, the proposed architecture localizes the crack with 100% accuracy when the threshold value is 0.5 inch and the load is greater than 30 kips. Although the accuracy seems to decrease for the small loads for thr = 0.5 inch, the accuracy reaches 80% when the crack location is searched in the larger area by increasing the threshold.

28.5 Conclusion

In this study, the proposed real-time damage diagnosis platform is validated by using spatially dense data collected by Digital Image Correlation. Laboratory testing of the specimen with an induced damage condition is performed to evaluate the performance and efficiency of damage detection and localization approach.

Deep learning achieves remarkable generalization when it is designed carefully such that it can perform successfully even with unseen cases. In this study, designed architecture diagnoses damages on samples collected by DIC with high accuracy although training dataset only includes finite element simulations. Moreover, this generalization is observable for the localization task. The location of the crack is predicted successfully although the crack location was not given as input during training. Therefore, the proposed methodology is promising for automatizing the real-time structural damage diagnosis.

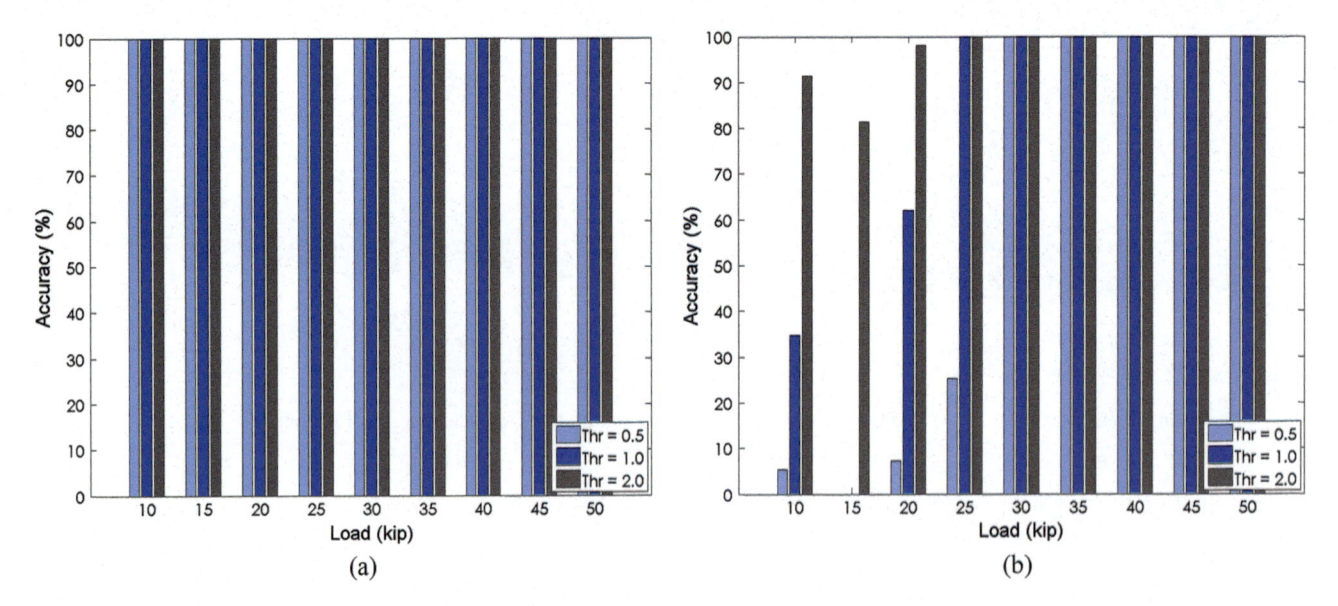

Fig. 28.6 Performance of the network for (**a**) detection and (**b**) localization tasks

Acknowledgements Research funding is partially provided by the National Science Foundation through Grant No. CMMI-1351537 by Hazard Mitigation and Structural Engineering program, and by a grant from the Commonwealth of Pennsylvania, Department of Community and Economic Development, through the Pennsylvania Infrastructure Technology Alliance (PITA). Martin Takáč was supported by National Science Foundation grant CCF-1618717 and CMMI-1663256.

References

1. Sohn, H., Farrar, C.R.: Damage diagnosis using time series analysis of vibration signals. Smart Mater. Struct. **10**(3),446 (2001)
2. Fang, X., Luo, H., Tang, J.: Structural damage detection using neural network with learning rate improvement. Commun. Strateg. **83**(25), 2150–2161 (2005)
3. Gulgec, N.S., Shahidi, G.S., Matarazzo, T.J., Pakzad, S.N.: Current challenges with bigdata analytics in structural health monitoring. In: Structural Health Monitoring & Damage Detection, vol. 7, pp. 79–84. Springer, Berlin (2017)
4. Nair, K.K., Kiremidjian, A.S., Law, K.H.: Time series-based damage detection and localization algorithm with application to the ASCE benchmark structure. J. Sound Vib. **291**(1), 349–368 (2006)
5. Fujimaki, R., Yairi, T., Machida, K.: An approach to spacecraft anomaly detection problem using kernel feature space. In: Proceedings of the Eleventh ACM SIGKDD International Conference on Knowledge Discovery in Data Mining, pp. 401–410. ACM, New York (2005)
6. Shi, A., Yu, X.-H.: Structural damage detection using artificial neural networks and wavelet transform. In: 2012 IEEE International Conference on Computational Intelligence for Measurement Systems and Applications (CIMSA) Proceedings, pp. 7–11. IEEE, Tianjin (2012)
7. Valeti, B., Pakzad, S.: Automated detection of corrosion damage in power transmission lattice towers using image processing. In: Structures Congress, pp. 474–482. American Society of Civil Engineers, Reston (2017)
8. Shahidi, S.G., Gulgec, N.S., Pakzad, S.N.: Compressive sensing strategies for multiple damage detection and localization. In: Dynamics of Civil Structures, vol. 2, pp. 17–22. Springer, Cham (2016)
9. Gulgec, N.S., Shahidi, S.G., Pakzad, S.N.: A comparative study of compressive sensing approaches for a structural damage diagnosis. In: Geotechnical and Structural Engineering Congress, pp. 1910–1919. American Society of Civil Engineers, Reston (2016)
10. LeCun, Y., Bengio, Y., Hinton, G.: Deep learning. Nature **521**(7553), 436–444 (2015)
11. LeCun, Y., Bengio, Y.: Convolutional Networks for Images, Speech, and Time-Series. MIT Press, Cambridge (1995)
12. He, K., Zhang, X., Ren, S., Sun, J.: Deep residual learning for image recognition. arXiv preprint arXiv:1512.03385 (2015)
13. Gulgec, N.S., Takac, M., Pakzad, S.N.: Convolutional neural network approach for robust structural damage detection and localization. J. Comput. Civ. Eng. **30**(3), 04015038 (2018)
14. Gulgec, N.S., Takáč, M., Pakzad, S.N.: Structural damage detection using convolutional neural networks. In: Model Validation and Uncertainty Quantification, vol. 3, pp. 331–337. Springer, Berlin (2017)
15. Pan, B., Qian, K., Xie, H., Asundi, A.: Two-dimensional digital image correlation for in-plane displacement and strain measurement: a review. Meas. Sci. Technol. **20**(6), 062001 (2009)
16. Yoneyama, S., Kitagawa, A., Iwata, S., Tani, K., Kikuta, H.: Bridge deflection measurement using digital image correlation. Exp. Tech. **31**(1), 34–40 (2007)
17. GOM. Aramis User Manual–Software (2013)

Chapter 29
Dynamic Response of the Suspended on a Single Cable Footbridge

Mikolaj Miskiewicz, Lukasz Pyrzowski, and Krzysztof Wilde

Abstract The article presents numerical simulations, dynamic in situ load tests and a structural health monitoring (SHM) system installed in a suspended on a single cable footbridge. Numerical simulations performed prior to construction indicated the possibility of structural dynamics problems, finally confirmed in the course of dynamic test loading. In the dynamic load course the bridge deck developed vibrations displaying accelerations up to 4.5 m/s^2. Such footbridge behavior causes unacceptable discomfort to the users and risk of structural damage. The tests brought about the need for repair works and the use of a monitoring system to increase operational safety of the object.

Keywords Suspended footbridge · Dynamic testing · Dynamic analysis · SHM

29.1 Introduction

Suspension bridges have been erected throughout the entire bridge engineering history, even in the Middle Ages. Nowadays this structural type is still successfully and intentionally applied, since a cable suspension allows crossing large obstacles without intermediate supports. A suspension takes a significant load amount making it possible to apply more slender decks. Unfortunately, it leads to problems with a dynamic performance of such structures due to wind or human-induced excitations. The well-known examples of excessive vibration occurrence are the cases of the Tacoma Narrows Bridge [1] and the London Millennium Bridge [2]. Most suspension bridges include at least two main cables. Structural cases including a deck suspended to a single cable are very rare anywhere in the world. Such solution may lead to a successive dynamic problem—a swing effect, i.e. torsional-flexural deck vibrations with simultaneous oscillation of the cable in the horizontal plane. Such a problem may occur especially in footbridges due to characteristics of excitations induced by pedestrians [3–6]. Nevertheless, such a structure was designed and erected for pedestrian-bicycle traffic in Radom (Poland), presented in Fig. 29.1. Unfortunately, a dynamic load test within the course of acceptance structural tests has confirmed the occurrence of excessive vibrations. The article presents a sequential problem solution. It includes description of dynamic in situ load tests, numerical analysis of a solution proposed in the repair project and structural health monitoring (SHM) system, installed in order to increase operational safety of the structure.

29.2 Structure Description

The structure is a suspension footbridge with a composite steel-concrete deck suspended by hangers to a single cable stretched between two towers, see Fig. 29.1. The concrete pylons are further stabilized by double guy ropes. The steel grid of the footbridge consists of three main HEB500 beam girders and welded transverse bars, see Fig. 29.2. The concrete slab shows a variable thickness of 16–20 cm. The cantilevers with elements to provide active hanger anchoring are located on the extension of the transverse bars. Abutments are erected in the massive form, the support foundations are situated indirectly on the piles. The deck girders are supported by elastomeric bearings. Characteristic structural dimensions are: tower height 19.6 m, span length 40.2 m, span width 4.5 m and usable deck width 4.0 m. The footbridge was designed for a pedestrian crowd load of $q = 4$ kN/m^2 according to [7].

M. Miskiewicz (✉) · L. Pyrzowski · K. Wilde
Department of Mechanics of Materials and Structures, Faculty of Civil and Environmental Engineering, Gdansk University of Technology, Gdańsk, Poland
e-mail: mikolaj.miskiewicz@pg.edu.pl

Fig. 29.1 View on the single-cable suspension footbridge in Radom (Poland)

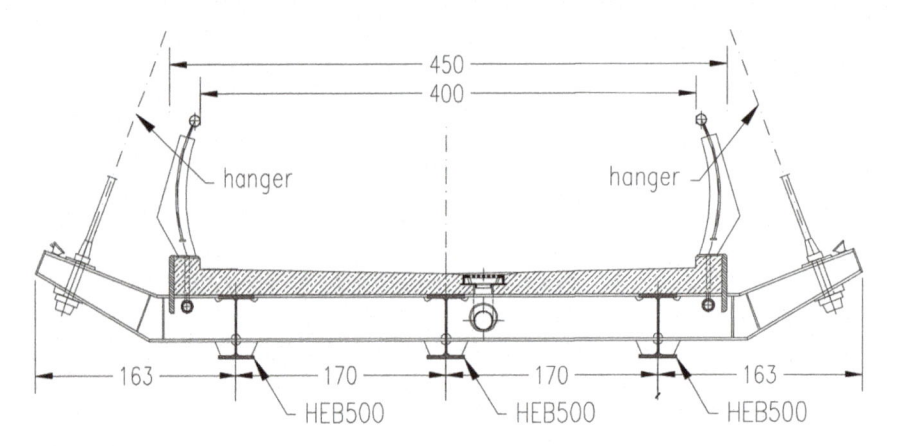

Fig. 29.2 Cross-section of the footbridge

29.3 Dynamic Load Tests

A thorough description of the performed acceptance tests is included in [8]. In the presented case dynamic tests are considered only. These tests were performed in two stages. The first stage concerned structural excitation by pedestrians and measurements performed by inductive displacement sensors (deck) and accelerometers (deck and main cable). The localization of measurement points is shown in Fig. 29.3. The tests program included dynamic loadings reflecting normal operating conditions (Fig. 29.4a), as well as vandal loading tests, i.e. synchronous marching, running, squatting and jumping (Fig. 29.4b). An entirety of 40 dynamic tests was carried out. The second stage was focused on the analysis of bridge free vibrations only. Impact tests were performed in order to identify natural frequencies and modal shapes of the footbridge superstructure. Dynamic pulse load was induced by a modal hammer at selected points of the tested structure. In the test course the accelerometers were placed on the deck, each cross-section was attached two units with a suspended transverse beam.

During the standard operating condition tests, no excitements of excessive vibrations were observed. The situation changed in the case of vandal loading. Excessive vibrations of the footbridge emerged during rhythmic jumping involving a group of 12 people with a frequency approximately equal 2.6 Hz. The registered deck accelerations in the vertical direction

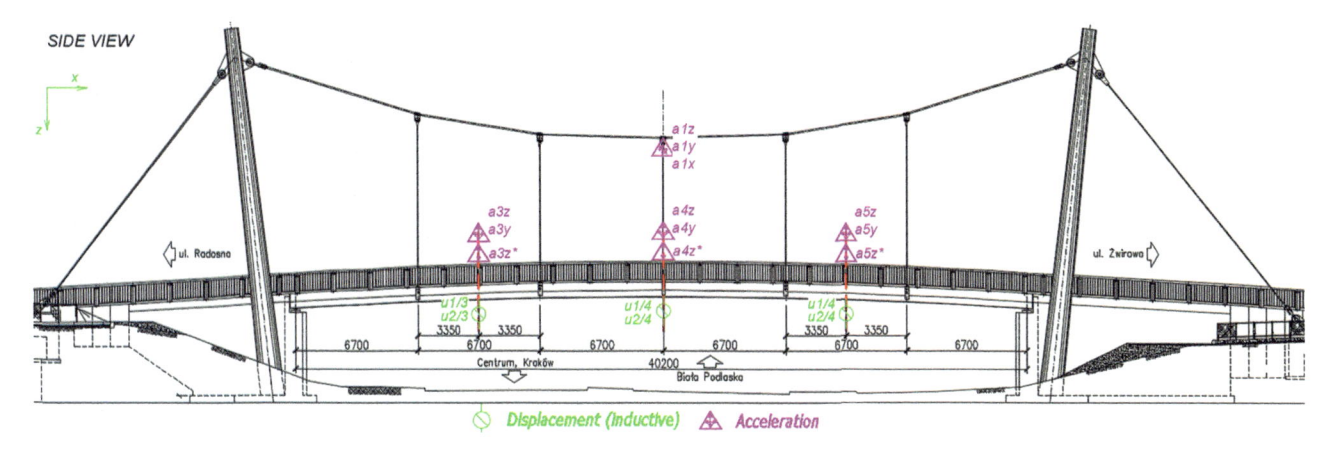

Fig. 29.3 Localization of measurement points during dynamic tests based on pedestrian excitation

Fig. 29.4 Dynamic loadings: (**a**) standard operational test—non-synchronous marching, (**b**) vandal test—jumping

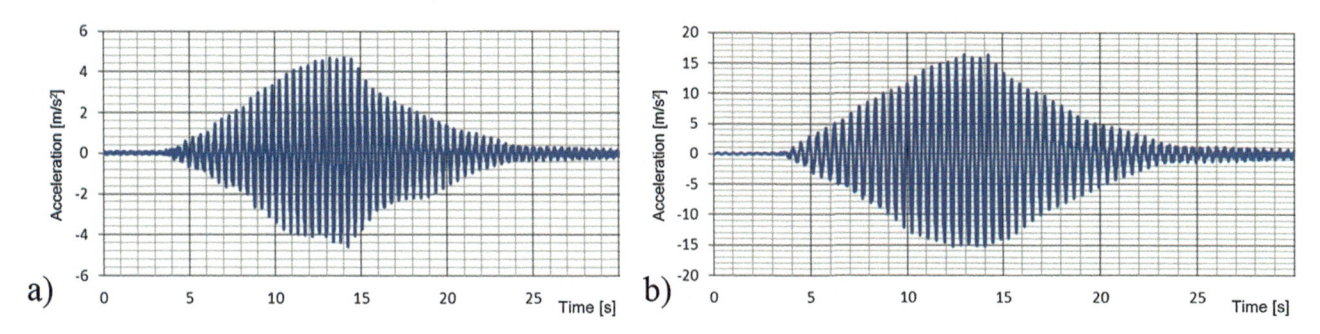

Fig. 29.5 Time history of accelerations registered during a group of 12 people jumping: (**a**) point a4z (deck), (**b**) point a1y (cable)

were up to 4.5 m/s^2, while cable horizontal accelerations were up to 16.4 m/s^2, see Fig. 29.5. The amplitude of deck displacements reached 15.4 mm. Such behavior, regarding comfort and serviceability conditions, is classified unacceptable by [9], it may lead to the possible damage of the structure, therefore concerning ultimate limit states. Furthermore, the induced vibrations increased without stabilization, therefore load application was terminated after approximately 14 s. The persons involved in the testing clearly felt the presence of a dangerous vibration level.

29.4 Repair Project and Numerical Analysis of Its Effectiveness

The repair project assumed installation of two 350 kg tuned mass dampers (TMD) to external girders of the deck in the middle of its span. A more detailed information regarding TMD applications is included in [10, 11]. Numerical analysis,

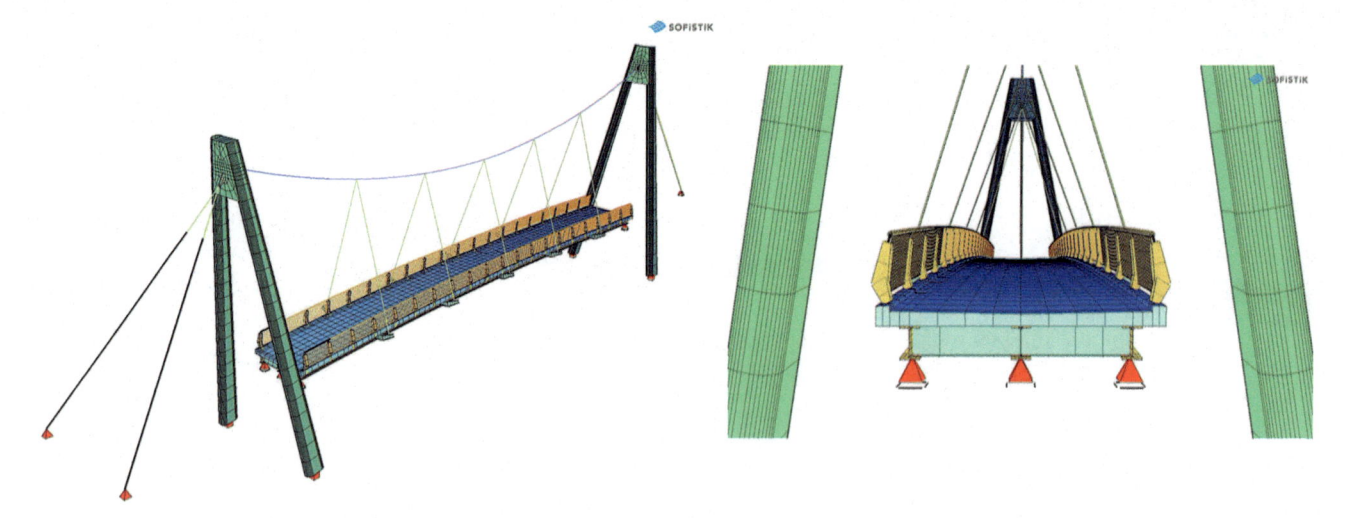

Fig. 29.6 Visualization of the model after validation procedure

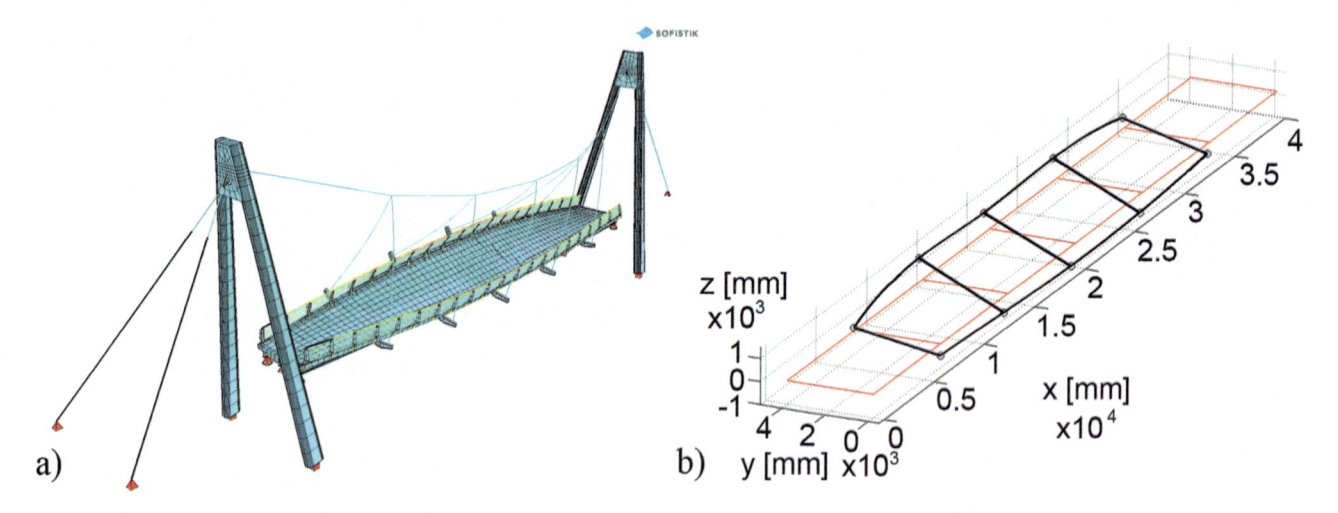

Fig. 29.7 Second mode shape, the results of (**a**) numerical analysis, (**b**) identification

conducted by the FEM-based Sofistik system, required validation of the model previously applied for the assembly design. The procedure was aimed at obtaining similar dynamic characteristics, priorly identified on the basis of dynamic test results. This aim was completed taking into account handrail stiffness, elastomeric bearings stiffness, an increased concrete elasticity modulus and actual damping ratio, i.e. 0.54% (Fig. 29.6). The differences between calculated and identified natural frequencies have been reduced to 1%. As an example, due to the impact of pedestrians, the most dominant second mode shape corresponds to the calculated frequency of 2.54 Hz, while the identified value is 2.55 Hz (Fig. 29.7).

Dynamic analysis was performed in a geometrically nonlinear range, it also included an assembly history in order to obtain real force values in the cable and hangers. Excitation in the form of synchronous jumps of 12 people, 70 kg per person, was applied using a sine function. As a reference result the same vertical accelerations were obtained in the model without TMDs, as in the real test (Fig. 29.8a). After introducing the assumed TMDs to the model the deck vertical accelerations dropped to the level of 1.4 m/s^2 (Fig. 29.8b), the cable horizontal accelerations were 5.5 m/s^2. These results confirmed correctness of the repair project. The analyzed TMDs were considered sufficient for the task of structural vibration control.

29.5 SHM System of the Structure

Due to problematic experiences with dynamic structural behavior, a potential risk of damage during its subsequent use and in order to diagnose the effects of environmental loads, the footbridge was fitted with a SHM. A detailed description of the

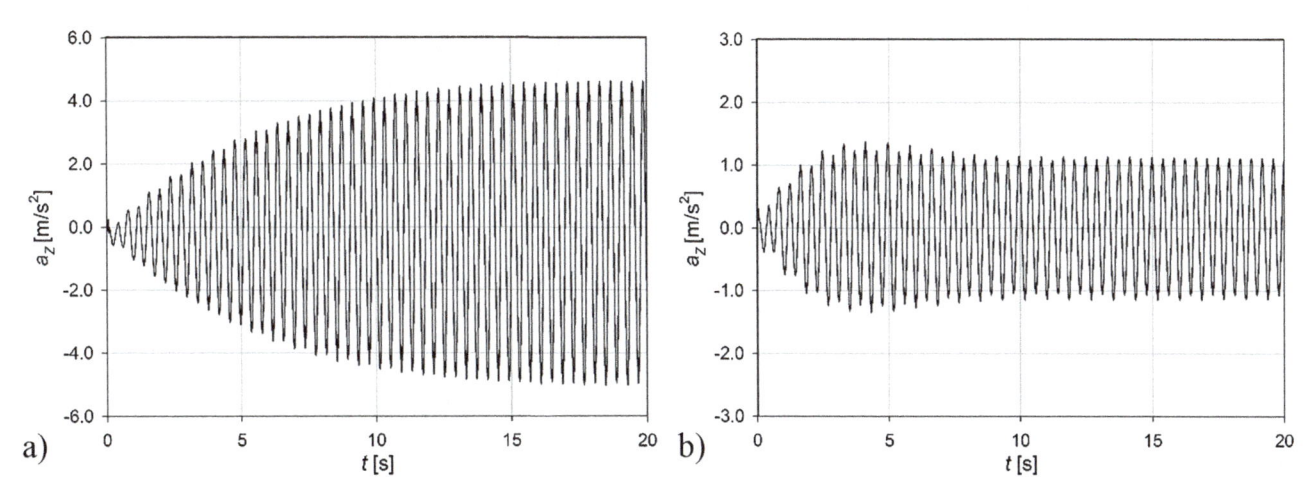

Fig. 29.8 Time history of deck vertical accelerations (point a4z) obtained in numerical analysis (**a**) without TMDs, (**b**) with TMDs

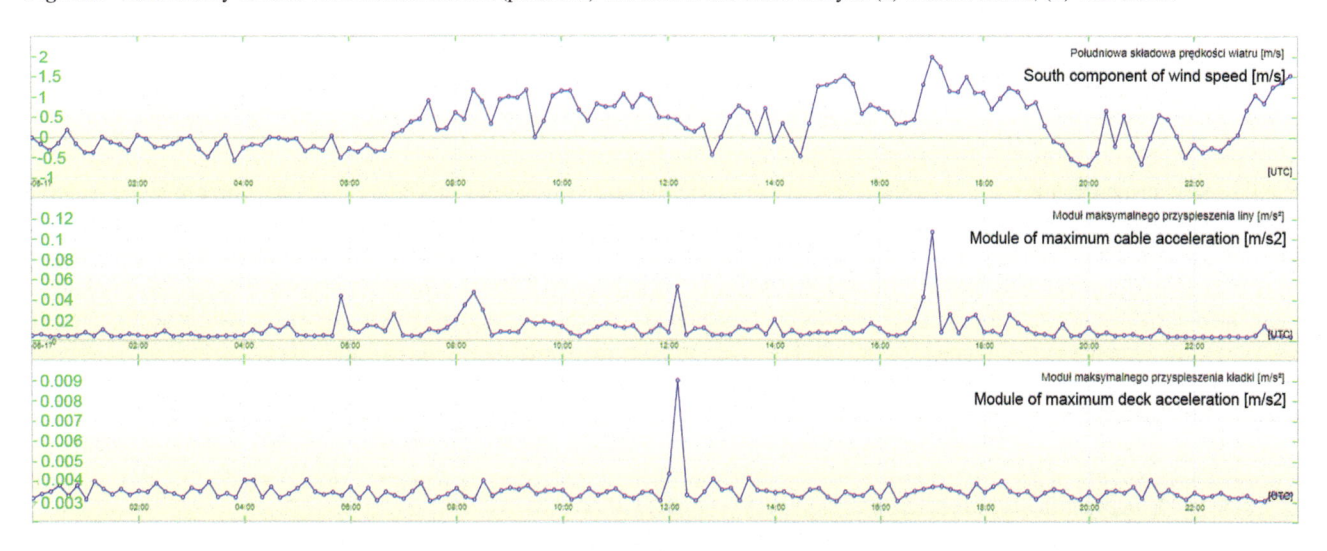

Fig. 29.9 Values of wind speed and maximum accelerations of cable and deck registered by the SHM system in June 17, 2018

system is included in [12]. It consists of three functional modules: measuring, expert and notification. The main hardware component of the system is the central management unit which controls its function and collects the measurement data. The first module performs the following measurements: wind force and direction, air temperature, atmospheric pressure, cable and deck accelerations, changes in the angle (recalculated to displacements) at specific points of the structure. The tasks of the expert module include: analysis of environmental and operational forces in real time, an expert algorithm for determining the index of the structural condition and generating information signals. The notification module is used to present the measurement data via a dedicated website and to communicate the current state of the structure. The system has been designed based on the authors' own experience and on the works of other authors [13–22].

The SHM system has been operating since February 2017. The registered acceleration values of acceleration are mostly of a low range. The example results registered on June 17, 2018 presented in Fig. 29.9 are the ordinary ones. The vibrations of the cable are mostly wind-excited, the accelerations do not exceed 0.1 m/s^2. Few instants occur of the operational time to observe higher vibrations. Such an extraordinary case happened in November 27, 2017 when accelerations of the cable, probably induced by pedestrians reached almost 4 m/s^2 (Fig. 29.10). This case shows that the deck vibrations are sufficiently controlled by the TMDs. Its accelerations are not higher than 0.02 m/s^2.

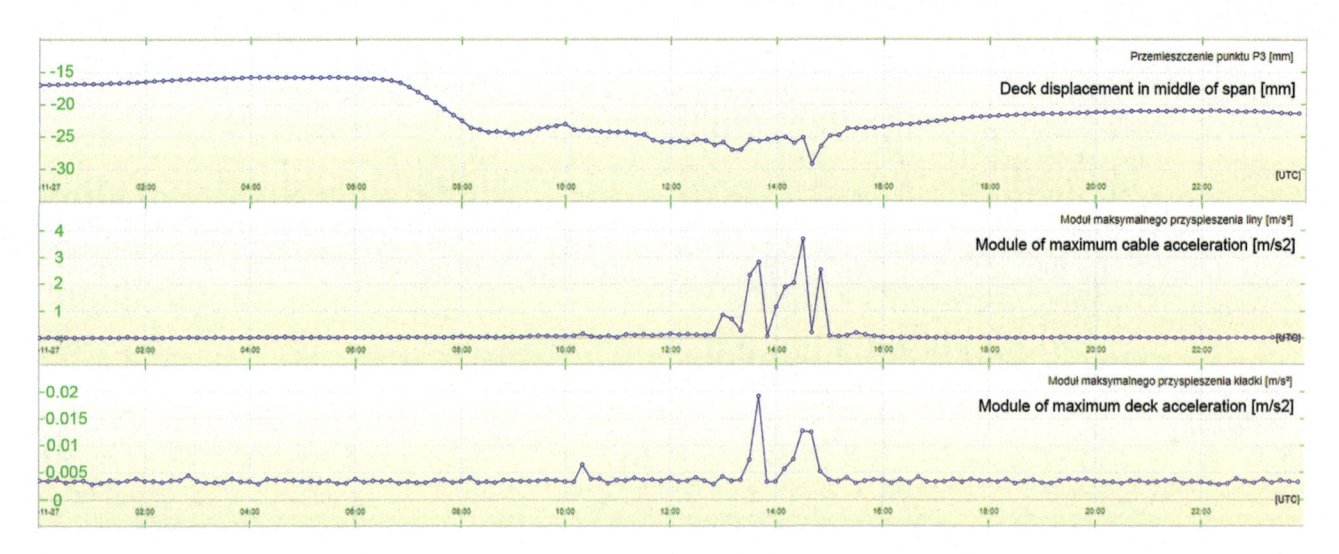

Fig. 29.10 Values of deck displacement and maximum accelerations of cable and deck registered by the SHM system in November 27, 2017

29.6 Conclusion

Single-cable suspension bridges are sensitive to asymmetric dynamic excitations causing vibrations which may be called a swing effect. Such case occurred in the newly built footbridge in Radom (Poland). The dynamic acceptance test revealed the problem, that excluded opening the footbridge for pedestrian-bicycle traffic after termination of construction works. It was necessary to perform the repair works. The project assumed installing two TMD dampers in the deck. Numerical simulations confirmed correctness of the presented approach. After the repair and repeated dynamic tests, this time with a positive result, the footbridge has been finally open for public. However, to provide operational safety of the object, as well as to provide data on the structural behavior during operation and information on its general technical condition, the SHM system has been permanently installed on the bridge.

Acknowledgements The authors send their words of gratitude to the following members of the research team: M. Rucka, R. Kędra, B. Meronk, M. Groth, M. Zieliński and the Accredited Research Laboratory ASPEKT: J. Kałuża.

References

1. Arioli, G., Gazzola, F.: Torsional instability in suspension bridges: the Tacoma Narrows Bridge Case. Commun. Nonlinear Sci. Numer. Simul. **42**, 342–357 (2017)
2. Pavic, A., Armitage, T., Reynolds, P., Wright, J.: Methodology for modal testing of the Millennium Bridge, London. Proc. Inst. Civil Eng. Struct. Build. **152**(2), 111–121 (2002)
3. Huang, M.H., Thambiratnam, D., Perera, N.: Dynamic performance of slender suspension footbridges under eccentric walking dynamic loads. J. Sound Vib. **303**(1–2), 239–254 (2007)
4. Altin, S., Kaptan, K., Tezcan, S.S.: Dynamic analysis of suspension bridges and full scale testing. Open J. Civil Eng. **2**, 58–67 (2012)
5. Bruno, L., Venuti, F., Nascé, V.: Pedestrian-induced torsional vibrations of suspended footbridges: proposal and evaluation of vibration countermeasures. Eng. Struct. **36**, 228–238 (2012)
6. Blachowski, B., Gutkowski, W., Wisniewski, P.: Dynamic substructuring approach for human induced vibration of a suspension footbridge. In: Proceedings PCM-CMM-2015—3rd Polish Congress of Mechanics & 21st Computer Methods in Mechanics, pp. 307–308 (2015)
7. PN-85/S-10030 Bridge structures. Loads (in Polish)
8. Miskiewicz, M., Pyrzowski, Ł., Wilde, K., Chroscielewski, J., Kaluza, J.: Load Testing of a Suspended Footbridge in Radom (Poland). Shell Structures: Theory and Applications, vol. 4, pp. 437–440. CRC Press/Balkema, Leiden (2018)
9. SÉTRA: The Technical Department for Transport, Roads and Bridges Engineering and Road Safety. Assessment of vibrational behavior of footbridges under pedestrian loading. SÉTRA, Paris (2006)
10. Li, H.N., Ni, X.L.: Optimization of non-uniformly distributed multiple tuned mass damper. J. Sound Vib. **308**, 80–97 (2007)
11. Poovarodom, N., Kanchanosot, S., Warnitchai, P.: Application of non-linear multiple tuned mass dampers to suppress man-induced vibrations of a pedestrian bridge. Earthq. Eng. Struct. Dyn. **32**, 1117–1131 (2003)
12. Miskiewicz, M., Pyrzowski, L., Wilde, K.: Structural Health Monitoring System for suspension footbridge. In: 2017 Baltic Geodetic Congress (BGC Geomatics), June 2017

13. Miskiewicz, M., Pyrzowski, Ł., Chroscielewski, J., Wilde, K.: Structural health monitoring of composite shell footbridge for its design validation. In: Proceedings 2016 Baltic Geodetic Congress (Geomatics), pp. 228–233 (2016)
14. Wilde, K., Miskiewicz, M., Chroscielewski, J.: SHM system of the roof structure of sports arena "Olivia". In: Chang, F.K. (ed.) Proceedings of the 9th International Workshop on Structural Health Monitoring (IWSHM 2013), 10–12 Sep 2013, Stanford, CA, pp. 1745–1752 (2013)
15. Clemente, P., De Stefano, A.: Novel methods in SHM and monitoring of bridges: foreword. J. Civil Struct. Health Monit. **6**(3), 317–318 (2016)
16. Mariak, A., Miskiewicz, M., Meronk, B., Pyrzowski, L., Wilde, K.: Reference FEM model for SHM system of cable-stayed bridge in Rzeszów. In: Kleiber, M., et al. (eds.) Advances in Mechanics: Theoretical, Computational and Interdisciplinary Issues, pp. 383–387. Taylor & Francis Group, London (2016)
17. Hou, J., Jankowski, L., Ou, J.: An online substructure identification method for local structural health monitoring. Smart Mater. Struct. **22**, 095017 (2013)
18. Miskiewicz, M., Pyrzowski, L., Wilde, K., Mitrosz, O.: Technical monitoring system for a new part of Gdansk Deepwater Container Terminal. In: Polish Maritime Research Special Issue 2017 S1 (93), vol. 24, pp. 149–155 (2017)
19. Klikowicz, P., Salamak, M., Poprawa, G.: Structural health monitoring of urban structures. Procedia Eng. **161**, 958–962 (2016)
20. Siwowski, T., Kaleta, D., Rajchel, M.: Structural behaviour of an all-composite road bridge. Compos. Struct. **192**, 555–567 (2018)
21. Tysiac, P., Wojtowicz, A., Szulwic, J.: Coastal cliffs monitoring and prediction of displacements using terrestrial laser scanning. In: 2016 Baltic Geodetic Congress (BGC Geomatics), pp. 61–66 (2016)
22. Miskiewicz, M., Meronk, B., Brzozowski, T., Wilde, K.: Monitoring system of the road embankment. Baltic J. Roads Bridge Eng. **12**(4), 218–224

Chapter 30
Event Detection and Localization Using Machine Learning on a Staircase

Blake Feichtl, Caleb Thompson, Tyler Liboro, Saad Siddiqui, V. V. N. Sriram Malladi, Tim Devine, and Pablo A. Tarazaga

Abstract Recent years have seen a push towards smart buildings that are energy efficient and proactive in decision making by detecting building events. Instrumentation of structures with sensors such as accelerometers or thermocouples is an essential element for providing the building with the necessary capabilities to enhance the occupant's comfort, safety and overall quality-of-life. As a result, vast amounts of data are collected that if correctly parsed can produce meaningful information for this purpose. One of the much-needed information about a building's activities is event localization. In many situations, event localization, based on traditional wave propagation techniques associated with vibrations, is a challenging task in an active environment as there is little control over the noise concurrent with the event. Determining ways to process sensor data efficiently and effectively will make user interaction with the building more intuitive and enhance user experience.

The present work pursues and evaluates an in-situ machine learning based approach for detecting and localizing footsteps on an instrumented staircase. The first part of the algorithm takes in live data from three accelerometers on a staircase and identifies footsteps based on a spike in the signal-to-noise ratio based on power spectral densities. The second part of the algorithm is the localization of the footstep once it is detected. Additionally, the performance of various features extracted from the time data (collected through controlled experiments) to generate an accurate machine learning model is also part of the current work. A nested tree algorithm is developed which yields 87% accuracy, showing potential for future stand-alone applications.

Keywords Machine learning · Localization · Detection · Footstep · Feature extraction

30.1 Introduction

The use of smart buildings is a concept with several applications yet to be fully explored by the wider community. Potential benefits of smart buildings can lie in detecting structural abnormalities. If the building is vibrating at a different frequency and exciting slightly different modes, then there may be structural changes in a building that causes the damping of the building to decrease. Similar to detecting structural health conditions, detection of occupants within the building can also enhance the capability of a building. HVAC and lighting can be controlled by detecting the location of occupants, saving energy and lowering operating costs of the building. Occupant detection can also be used for the benefit of security, with the detection of any abnormal activity. In this case, if a building is entered and the appropriate method of entry (keypad, swipe entry) is not detected, then the alert system of the building can be activated. Another aspect of occupant detection can be used in the medical industry. If an occupant falls in a hospital or nursing home, an alert system can be activated, pinpointing the location of the fallen occupant. This will enable faster medical attention to the patient, potentially saving many lives.

While the present work aims to add to current research benefiting smart buildings, several studies have been conducted which relate footstep detection and machine learning. One such project involved the collection of biometric footstep data, which was reliably collected in time and space, without the drawbacks of standard sensors [1]. Many studies have used sensors from smartphones as a means for data collection of footsteps [2, 3] as an alternative to embedded sensors in a building such as [4]. Analyzing footstep data can be partitioned into two main sections, detection and localization. Finding reliable ways to detect footsteps has been addressed using energy and periodicity of footsteps in [5]. A study based in Virginia Tech's Goodwin Hall was performed using vibration-based event localization which used the K-Nearest Neighbor (KNN)

B. Feichtl · C. Thompson · T. Liboro · S. Siddiqui · V. V. N. S. Malladi · T. Devine (✉) · P. A. Tarazaga
Vibration, Adaptive Structures and Testing Lab (VAST), Department of Mechanical Engineering, Virginia Polytechnic Institute and State University, Blacksburg, VA, USA
e-mail: timd@vt.edu

technique for location-based classification [6]. Other approaches that have been tested utilize decision tree classifiers for localization with promising results can be seen in [7]. The present work seeks to create a robust machine learning algorithm for footstep detection and localization.

30.2 Background

To determine the detection of a footstep, both mathematical and physics-based approaches were evaluated. Initial testing used the L2 norm of the voltage data, however this approach did not account for varying levels of noise. Consequently, by calculating energy from the Power Spectral Density (PSD) for a band of 50–400 Hz, where only the energy of footsteps is contained, as determined experimentally. Utilizing this energy of the waveform, the signal-to-noise ratio (SNR) is used to detect an event and keep track of a noise floor. The noise floor is based on the lowest energy recorded as calculated from the PSD. Once the SNR passes the set threshold, a footstep is flagged as detected. Once the event is detected, the sensor with the maximum SNR is identified and the window of the signal with the highest peak is saved. The location that the largest spike occurs within that sensor's frame is saved. Note that events which might contain similar energy and frequency content cannot be discriminated against in this method. For example, dropping an object can be misinterpreted as a footstep. Future work will look into more tighter restrictions on footstep analysis and classification.

In anticipation of a machine learning algorithm to predict footstep localization, a database of correctly identified footsteps needed to be collected. A National Instruments DAQ with a sampling rate of 900 Hz controlled by MATLAB was used to collect footsteps on the desired staircase. Several experiments were performed, resulting in 120 trials of subjects traversing the staircase. Through analysis of each sample, the individual footsteps could be extracted and classified by their step number on the staircase (steps 1–10). An example of a single footstep and its representation in the frequency domain as a power spectral density plot is included below in Fig. 30.1. This formed the database of samples to which the machine learning algorithm would be trained. Different footstep window sizes were evaluated, and a size of 200 points was selected, which on average contains the complete footstep data.

Once all footsteps were extracted from the data set, features of each footstep signal were calculated and several machine-learning algorithms were investigated to determine the best-fitting algorithm for the data. Throughout the study of various algorithms, a k-fold cross validation of 10 folds is used.

30.2.1 Implementation

For the hardware of the in-situ device, a relatively cost effective data acquisition system and computer combination was used, consisting of a LabJack T7 DAQ and Raspberry Pi Model 3 computer. The sensors where three PCB393B04 accelerometers

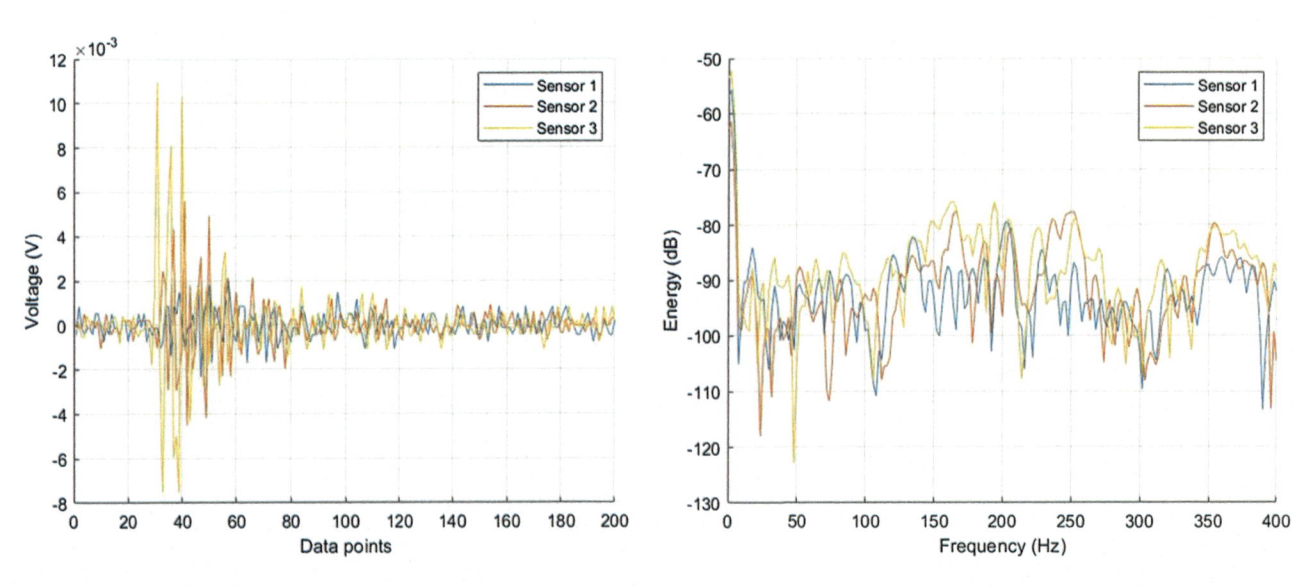

Fig. 30.1 Example footstep data and corresponding PSD

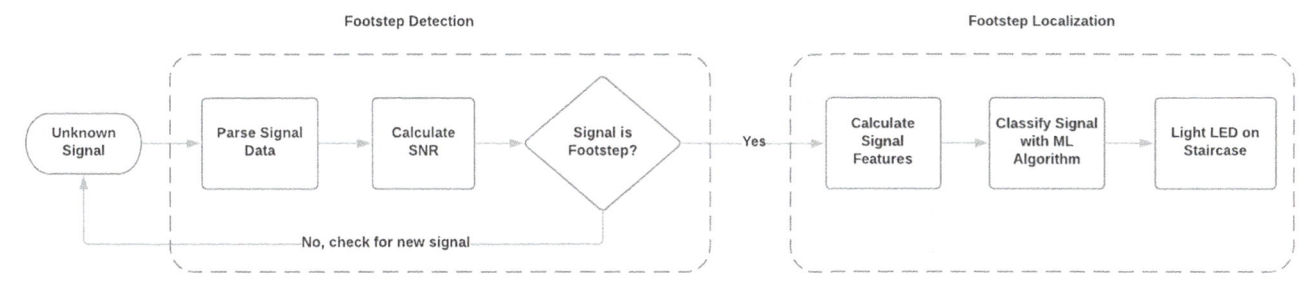

Fig. 30.2 Footstep detection and localization process

Fig. 30.3 Hardware implementation on staircase

and a signal conditioner set to condition the signal before the data acquisition. The combination of these components allow the implementation of a Python code and an overall small form factor. The LabJack DAQ reads data from the three PCB accelerometers at a sample rate of 500 Hz and monitors the SNR of the sample. Once the SNR exceeds a predetermined level, the calculation of all features is executed on that sample. These metrics are then fed to the machine learning classifier and a predicted stair step is returned. This stair step is then sent to an LED light strip which lights up the associated LEDs with the predicted step. The footstep detection/localization process and setup of the device on the staircase are shown below in Figs. 30.2 and 30.3.

30.3 Analysis

A list of ten main features were deemed appropriate for classification of a footstep. These 10 features were calculated across the samples read in from all 3 sensors, for a total of 30 metrics. Then the ratio of each feature across all sensors was also calculated and added for an increased accuracy rating (1–2, 2–3, 1–3; for each feature). This resulted in a total of 60 metrics to classify a footstep.

The ten main features to classify the footstep signal include standard deviation of the voltages, energy from the PSD between 50 and 400 Hz, peak amplitude, Root Mean Squared (RMS) value, L2 norm, kurtosis, maximum value of the Discrete Fourier Transform (DFT), crest factor, and Root Sum of Squares (RSSQ) value. The integral of 7 evenly spaced quadrants of the PSD (50–400 Hz) was calculated for the footstep, and the maximum value is used as a feature. The PSD is found using *pwelch* in MATLAB, with no overlap for the footstep signal for each sensor. Other features of the footstep signal

Table 30.1 Investigation of various algorithms

Algorithm	Accuracy (%)
KNN	~47
Single tree	~50
SVM	~54
Bagged tree	~54
Nested tree	~87

that were tested, but not included with the final algorithm were level crossings, sensor with the highest SNR, and the sum of PSD values across specified frequency bands. The difference in index location of the peak across all sensors was also used as an additional feature, bringing the total number of metrics to 63.

A variety of algorithms were tested to determine the localization of footsteps. The results of several different algorithms are presented below in Table 30.1. The algorithms are evaluated on their ability to predict the correct step. The single tree algorithm generally yielded an accuracy of 50%. Since limited accuracy was reached with a tree-based algorithm, investigation was made into a nested tree algorithm. This algorithm makes an initial guess at the footstep location, then splits the data into a smaller subset to analyze the data with another tree algorithm which retains the initial guess as a feature. Yet another smaller subset group is formed after the second guess, then a final prediction of step location is made which incorporates the previous two guesses of step location as additional features to the algorithm. With the staircase labeled from steps 1–10, an initial guess of step 1 will put the footstep in the subset algorithm for steps 1–5. Data is split into steps 1–5, 4–8, and 7–10, with overlapping of the subset groups added in case the first guess is off by a step. The data is then further split into step groups 1–3, 3–5, 4–6, 6–8, 7–9, and 8–10 based on a second location guess, which also incorporates the initial guess as a feature. Once split into these groups, it will make the final guess for the footstep location, incorporating the previous two sensor location guesses for the final prediction. The nested tree algorithm's accuracy was estimated from an average accuracy for each final sub-group, which resulted in an accuracy of 87%.

30.4 Conclusion

This work evaluates the use of a footstep detection and localization algorithm for use in an in-situ device located on a staircase of a building. Data processing methods were used to develop a diverse set of participant data that could make generalizations and predictions for a larger data set of the general population. Resulting hardware, that could be applied remotely within a building, was implemented and used for algorithm testing. A method of nesting algorithms for localization that proves promising in the application of footsteps on a staircase was developed with an 87% accuracy rating.

The main limitations in this approach are the processing power and device latency in hardware extensions of the device used. Microcontroller/computer selection could be explored in order to optimize processing power. Future applications of this project could focus on improving the accuracy of the machine learning algorithm and exploring additional options for creating the in-situ device. Future research can also aid in improving the device's application in other locations like hallways and other staircases. This can be done by further developing the comprehensive feature list suited to the desired application.

Acknowledgements C. Thompson and B. Feichtl would like to thank the Luther and Alice Hamlett Undergraduate Research Support program in the Academy of Integrated Science as well as the Department of Mechanical Engineering at Virginia Tech for supporting travel and conference expenses. The authors would also like to thank the Student Engineers Council at Virginia Tech for partially funding this project. C. Thompson and B. Feichtl would also like to thank the other members of the VAST Lab for their continued support of the project. Dr. Tarazaga would like to acknowledge the support provided by the John R. Jones III Faculty Fellowship.

References

1. Vera-Rodriguez, R., Mason, J.S., Fierrez, J., Ortega-Garcia, J.: Comparative analysis and fusion of spatio-temporal information for footstep recognition. IEEE Trans. Pattern Anal. Mach. Intell. **35**(4), 823–834 (2013)
2. Bayat, A., Bayat, A.H., Sina, A.: Classifying human walking patterns using accelerometer data from smartphone. Int. J. Comput. Sci. Mobile Comput. **3**, 1–6 (2017)
3. Bayat, A., Pomplun, M., Tran, D.: A study on human activity recognition using accelerometer data from smartphones. In: The 11th International Conference on Mobile Systems and Pervasive Computing, vol. 2014, pp. 450–457 (2014)

4. Poston, J.D., Schloemann, J., Buehrer, R.M., Malladi, V.V.N.S.A., Woolard, G., Tarazaga, P.A.: Towards indoor localization of pedestrians via smart building vibration sensing. In: 2015 International Conference on Location and GNSS (ICL-GNSS), Gothenburg, pp. 1–6 (2015)
5. Mirshekari, M., Zhang, P., Noh, H.Y. Calibration-free footstep frequency estimation using structural vibration. In: IMAC XXXV A Conference and Exposition on Structural Dynamics. SEM (2017)
6. Woolard, A.G.: Supplementing localization algorithms for indoor footsteps. Ph.D. dissertation, Virginia Polytechnic Institute and State University, Blacksburg (2017)
7. Fan, L., Wang, Z., Wang, H.: Human activity recognition model based on decision tree. In: 2013 International Conference on Advanced Cloud and Big Data, Nanjing, pp. 64–68 (2013)

Chapter 31
Footbridge Vibrations and Their Sensitivity to Pedestrian Load Modelling

Lars Pedersen and Christian Frier

Abstract Pedestrians may cause vibrations in footbridges, and these vibrations may potentially be problematic from a footbridge serviceability point-of-view. Foreseeing (already at the design stage) unfit conditions is useful, and the present paper employs a probability-based methodology for predicting vibrational performance of a bridge. The methodology and the walking load model employed for calculation of bridge response accounts for the stochastic nature of the walking parameters of pedestrians (step frequency, step length etc.) and the end result is central statistical parameters of bridge response (quantiles of bridge acceleration) to the action of a pedestrian. The paper explores the impact that selected decisions made by the engineer in charge of computations have on the statistical parameters of the dynamic response of the bridge. The investigations involve Monte Carlo simulation runs as walking parameters are modelled as random variables and not as deterministic properties. Single-person pedestrian traffic is the load scenario considered for the investigations of the paper and numerical simulations of bridge accelerations are made for artificial but realistic footbridges.

Keywords Footbridge vibrations · Walking loads · Walking parameters · Stochastic load models · Serviceability-limit-state

Nomenclature

a	Bridge acceleration
f_1	Bridge fundamental frequency
f_s	Step frequency
i	Integer
l_s	Step length
m_1	Bridge modal mass
t	Time
v	Pacing speed
F	Walking load
L	Bridge length
Q	Modal load
W	Weight of pedestrian
α	Dynamic load factor
ζ_1	Bridge damping ratio
μ	Mean value
σ	Standard deviation
Θ	Phase
Φ	Mode shape

L. Pedersen (✉) · C. Frier
Department of Civil Engineering, Aalborg University, Aalborg, Denmark
e-mail: lp@civil.aau.dk

31.1 Introduction

Footbridges may become so slender that excessive vibrations may occur as a result of loads on the bridge generated by pedestrians. The Millennium Bridge in London [1] is an example of a footbridge for which the engineers in charge of design had overlooked that the bridge would be unfit for its intended use, namely carrying people in locomotion.

Maybe based on tradition, walking load models specified in some codes of practice (such as [2, 3]) are deterministic in the sense that they do not account for intersubject variability in walking forces. Variability is known to exist (for instance variability in step frequency, step length and dynamic load factors). Variability in these parameters is documented in [4–7].

It is considered useful to employ a load model that accounts for uncertainty related to these parameters. The load model used in this paper (developed and described in [8]) models parameters such as step frequency, step length and dynamic load factors as random variables. This has the effect that the bridge response to the action of walking is best described by the probability of various vibration levels.

When applying any model there will be input parameters to select. Hence, this is also the case for the walking load model employed in the paper. Input parameters are the stochastic nature of walking parameters (such as the statistical distribution of step frequency, step length and dynamic load factors). The aim of the paper is to investigate how sensitive the stochastic nature of bridge response (the result obtained by the calculations) is to decisions about the stochastic nature of step frequencies.

Specifically, the stochastic nature of bridge response will be the stochastic nature of vertical bridge accelerations obtained using Monte Carlo simulation and Newmark-time-integration. One way of representing the stochastic nature is by the probability distribution function (cumulative distribution function) of bridge acceleration. Instead of displaying the entire function, selected acceleration quantiles can be sampled from the probability distribution function. This is what is done in the paper, which hence is concerned with investigating how sensitive selected acceleration quantiles are to decisions made about the stochastic nature of step frequency of pedestrians.

A set of artificial footbridges are used for the investigations. They are all single-span pin-supported bridges, and only the response of the first bending mode of the bridge will be considered and it will be single-person pedestrian traffic that causes bridge vibrations.

Section 31.2 introduces the walking load modal, and Sect. 31.3 outlines the methodology. Section 31.4 presents the results in terms of acceleration quantiles of bridge vibrations obtained under different assumptions with respect to the stochastic nature of step frequency of pedestrians, and Sect. 31.5 provides conclusions.

31.2 Modelling of Walking Loads

On a footbridge, the vertical modal load, $Q(t)$, may be calculated using Eq. (31.1):

$$Q(t) = \Phi(t)F(t) \quad (31.1)$$

where $\Phi(t)$ is the mode shape function and $F(t)$ is the vertical load generated by the pedestrian as he crosses the bridge.

The mode shape function for the first bending mode of the bridge can be determined using Eq. (31.2):

$$\Phi(t) = \sin(\pi vt/L) \quad (31.2)$$

provided that the pedestrian walks with a constant velocity v, which will be assumed. In the equation, L is the length of the bridge between pin supports.

The walking velocity is derived using Eq. (31.3):

$$v = f_s l_s \quad (31.3)$$

where l_s is the step length and f_s is the step frequency of the pedestrian.

In terms of the load $F(t)$ the following expressions will be applied:

$$F(t) = \sum_{i=1}^{5} F_i(t) + \sum_{i=1}^{5} F_i^S(t) \quad (31.4)$$

$$F_i(t) = W\alpha_i \sum_{\overline{f}_j=i-0.25}^{i+0.25} \overline{\alpha}_i\left(\overline{f}_j\right) \cos\left(2\pi \overline{f}_j f_s t + \theta\left(\overline{f}_j\right)\right) \qquad (31.5)$$

$$F_i^S(t) = W\alpha_i^S \sum_{\overline{f}_j^S=i-0.75}^{i-0.25} \overline{\alpha}_j^S\left(\overline{f}_j^S\right) \cos\left(2\pi \overline{f}_j^S f_s t + \theta\left(\overline{f}_j^S\right)\right) \qquad (31.6)$$

They represent the load model suggested in [8] in which a detailed description of the meaning of the many parameters is given. Here it suffices to mention that the model accounts for the fact that idealised periodic action of a pedestrian is not possible.

An idealised periodic action would result in excitation frequencies at the step frequency and at integer multipliers of this frequency like the models proposed in [9–11]. The present model accounts for the fact that excitation in between multipliers of the step frequency will also occur and that some degree of leakage of energy prevails around the excitation frequencies.

Central parameters in the load model are the dynamic load factors and their stochastic nature.

Equation (31.7) provides the expressions employed for the mean value (μ) and standard deviation (σ) for the first dynamic load factor, α_1.

$$\mu = -0.2649 f_S^3 + 1.3206 f_S^2 - 1.7597 f_S + 0.7613; \quad \sigma = 0.16\mu \qquad (31.7)$$

The first and the following dynamic load factors are modelled using Gaussian distributions. Table 31.1 presents the mean values and standard deviations for the other dynamic load factors accounted for in the load model.

In the model also phases, Θ, between load harmonics need to be defined and to this end a uniform distribution in the range $[-\pi,\pi]$ is employed.

The value of W (the weight of the pedestrian) is set to 750 N corresponding to the value used in [8], and l_s (step frequency) is assumed to follow a Gaussian distribution with a mean value of 0.71 m and a standard deviation of 0.071 m [8].

31.3 Methodology

A complete stochastic load model will also involve a stochastic model for the step frequency of pedestrians. Such model was not introduced in Sect. 31.2.

In terms of the stochastic nature of step frequency, different proposals are available in existing literature. Based on findings from experiments, different authors have fitted a Gaussian distribution to their experimental data. Table 31.2 lists three proposals in terms of the mean value, μ, and the standard deviation, σ.

Bridge vibrations will be calculated with model I, II, and III, respectively, as input for the load model.

For the calculations, the response of a single bridge could be investigated. However, it is chosen to investigate the response of three bridges in order to establish a broader basis for understanding the mechanisms.

For the investigations, it is chosen to focus on bridges that have natural frequencies close to or coinciding with mean values of the probability density functions for step frequency introduced in Table 31.2.

Modal properties of the bridges considered for the studies of the paper are shown in Table 31.3, in which f_1 represents the natural frequency of the bridge, ζ_1 the damping ratio, m_1 the modal mass, and L the assumed length of the bridges.

Table 31.1 Mean values and standard deviations [8]

–	α_2	α_3	α_4	α_5
μ	0.07	0.05	0.05	0.03
σ	0.030	0.020	0.020	0.015

Table 31.2 Mean values and standard deviations

Model	μ [Hz]	σ [Hz]	Reference
I	1.87	0.186	[5]
II	1.99	0.173	[4]
III	2.20	0.300	[12]

Beside from that fact that the bridges have natural frequencies close to the mean values of the step frequency models introduced in Table 31.2, it can be seen that the bridges are assumed lightly damped. The modal masses and bridge lengths decrease with increase in bridge frequency which is considered meaningful and realistic.

The bridge response in focus is the acceleration response of the bridges encountered at bridge midspan. This response will vary while a pedestrian crosses the bridge, but it will only be the peak value of acceleration that is selected for further processing of results.

Realizations of time histories of bridge response are generated using standard Newmark-time-integration procedures and Monte Carlo simulations. For each pedestrian, walking parameters are sampled from the statistical distributions described in Sects. 31.2 and 31.3. From computed probability distribution functions of peak acceleration, the acceleration quantiles a_{95}, a_{90} and a_{75} were extracted (the 95%, the 90%, and the 75% quantile of bridge acceleration, respectively). It is three quantiles in the upper end of the probability distribution that are selected for extraction as it is believed that it would be relatively high quantiles that would be of interest to the operator of the bridge.

It proved sufficient to employ 100,000 numerical simulation runs (bridge crossings) to obtain converged results.

31.4 Results

This section is concerned with presentation of results in terms of acceleration quantiles derived from calculations.

Table 31.4 presents results for a_{95} (the 95% quantile of bridge acceleration) obtained for the three bridges (A, B, and C) employing the three different stochastic models for step frequency of pedestrians (I, II and III), respectively.

For the three bridges it is not the same stochastic model for step frequency that results in the highest value of a_{95}. For bridge A it is model I that provides the highest value of a_{95}, for bridge B it is model II, and for bridge C it is model III.

This to some extent has to do with the fact that the natural frequency of bridge A (1.85 Hz) is very close to the mean value of stochastic model I (1.87 Hz), the natural frequency of bridge B is close to 1.99 Hz, and the natural frequency of bridge C is equal to 2.2 Hz.

Table 31.5 presents normalised acceleration quantiles (for a_{95}) in the sense that for every bridge, the acceleration quantiles are normalised by the acceleration quantile found using stochastic model II.

In this representation of results it is fairly simple to see that the smallest relative difference between the three values of a_{95} (calculated for a specific bridge) is at seen bridge B. The highest relative difference is seen at bridge C.

Overall, the results suggests that the value of a_{95} is rather sensitive to which stochastic model for step frequencies of pedestrians is applied for the calculation of bridge vibrations. This is useful to know.

Table 31.3 Modal properties of bridges and bridge length

Bridge	f_1 [Hz]	ζ_1 [%]	m_1 [10^3 kg]	L [m]
A	1.85	0.5	46.2	46.5
B	2.00	0.5	39.5	43.0
C	2.20	0.5	32.6	39.1

Table 31.4 Acceleration quantile a_{95}

–	Stochastic model for step frequency and mean value of the step frequency (in bracket)		
Bridge	I (1.87 Hz)	II (1.99 Hz)	III (2.20 Hz)
A (1.85 Hz-bridge)	0.3093 m/s^2	0.2879 m/s^2	0.1750 m/s^2
B (2.00 Hz-bridge)	0.3904 m/s^2	0.4256 m/s^2	0.3442 m/s^2
C (2.20 Hz-bridge)	0.2743 m/s^2	0.4715 m/s^2	0.5033 m/s^2

Table 31.5 Normalised acceleration quantile for a_{95}

–	Stochastic model for step frequency and mean value of the step frequency (in bracket)		
Bridge	I (1.87 Hz)	II (1.99 Hz)	III (2.20 Hz)
A (1.85 Hz-bridge)	1.07	1.00	0.61
B (2.00 Hz-bridge)	0.92	1.00	0.81
C (2.20 Hz-bridge)	0.58	1.00	1.07

Table 31.6 Acceleration quantiles a_{90} and a_{75}

Bridge	Quantile	Stochastic model for step frequency and mean value of the step frequency (in bracket)		
		I (1.87 Hz)	II (1.99 Hz)	III (2.20 Hz)
A (1.85 Hz-bridge)	a_{90}	0.2399 m/s^2	0.2092 m/s^2	0.0967 m/s^2
	a_{75}	0.1167 m/s^2	0.0971 m/s^2	0.0493 m/s^2
B (2.00 Hz-bridge)	a_{90}	0.2941 m/s^2	0.3438 m/s^2	0.2110 m/s^2
	a_{75}	0.1307 m/s^2	0.1799 m/s^2	0.0901 m/s^2
C (2.20 Hz-bridge)	a_{90}	0.1389 m/s^2	0.3210 m/s^2	0.3719 m/s^2
	a_{75}	0.0631 m/s^2	0.1274 m/s^2	0.1598 m/s^2

Table 31.6 lists results for a_{90} (the 90% quantile of bridge acceleration) and a_{75} (the 75% quantile of bridge acceleration) obtained for the three bridges (A, B, and C) employing the three different stochastic models for step frequency of pedestrians (I, II and III), respectively.

It is noted, and as expected, that for a specific stochastic model for step frequency, the values of a_{90} are smaller than values of a_{95}. Similarly, the values of a_{75} are smaller than values of a_{90}.

Else the tendencies already mentioned when commenting on the results for a_{95} is also seem to prevail for a_{90} and a_{75}.

For the three bridges it is not the same stochastic model for step frequency that results in the highest value of a_{90}. For bridge A it is model I that provides the highest value of a_{90}, for bridge B it is model II, and for bridge C it is model III. Similar observation for a_{75}.

Again it is for bridge B that you find the smallest deviations between values of a_{90} (and values of a_{90}).

31.5 Conclusion and Discussion

With offset in a probability-based methodology for predicting vibration generated by pedestrians, the paper examined the sensitivity of bridge acceleration quantiles to decisions related to the modelling of the stochastic nature of step frequencies of pedestrians. Three models for the stochastic nature of step frequencies were found in existing literature, and these were assumed (one by one) as input in the walking load model used in the context of prediction the stochastic nature of bridge acceleration response.

The results suggest that for the three bridges for which response is calculated in the paper, the acceleration quantiles (95%, 90% and 75%) are rather sensitive to which model for the stochastic nature of step frequencies of pedestrians that is employed for calculations. On one hand this is useful information. On the other hand, it leaves the engineer in charge of a design-stage vibration serviceability check of a bridge with some challenges.

There would be different strategies that may be applied to handle the fact that the acceleration quantiles are sensitive to the stochastic model for step frequency. A rather conservative approach would be to perform calculation with a set of different stochastic models for step frequency—and then consider only the result for the model that results in the highest vibration levels.

Another strategy could be to obtain as much relevant information as possible about the stochastic nature of step frequencies for the population of pedestrians/people which is expected to be using the bridge in the future. This may be difficult but it is an option that may be considered.

There would also be the strategy of evaluating the quality of the different stochastic models found in literature. Probably the quality of the data sets used for fitting a stochastic model differs—instance in terms of the sample size or because of differences in the experimental procedures.

When it comes to the stochastic models for step frequency employed for the calculations of this paper, it is noticeable that one of the models has a standard deviation almost twice as high as those in the other models. Also the mean value of that model is relatively high—and of a size that suggests that maybe quite fast walking have been exercised by people tested in the experiments forming the basis for fitting a stochastic model to the dataset.

Acknowledgements This research was carried out in the framework of the project "Urban Tranquility" under the Interreg V program and the authors of this work gratefully acknowledge the European Regional Development Fund for the financial support.

References

1. Dallard, P., Fitzpatrick, A.J., Flint, A., Le Bourva, S., Low, A., Ridsdill-Smith, R.M., Wilford, M.: The London Millennium Bridge. Struct. Eng. **79**, 17–33 (2001)
2. Ontario Highway Bridge Design Code, Highway Engineering Division; Ministry of Transportation and Communication, Ontario, Canada (1983)
3. British Standard Institution: Steel, concrete and composite bridges. Specification for loads, BS 5400: Part 2 (1978)
4. Matsumoto, Y., Nishioka, T., Shiojiri, H., Matsuzaki, K.: Dynamic design of footbridges, In: IABSE Proceedings, No. P-17/78, pp. 1–15 (1978)
5. Živanovic, S.: Probability-based estimation of vibration for pedestrian structures due to walking. PhD Thesis, Department of Civil and Structural Engineering, University of Sheffield (2006)
6. Kerr, S.C., Bishop, N.W.M.: Human induced loading on flexible staircases. Eng. Struct. **23**, 37–45 (2001)
7. Pedersen, L., Frier, C.: Sensitivity of footbridge vibrations to stochastic walking parameters. J. Sound Vib. (2009). https://doi.org/10.1016/j.jsv.2009.12.022
8. Živanovic, S., Pavic, A., Reynolds, P.: Probability-based prediction of multi-mode vibration response to walking excitation. Eng. Struct. **29**, 942–954 (2007). https://doi.org/10.1016/j.engstruct.2006.07.004
9. Ellis, B.R.: On the response of long-span floors to walking loads generated by individuals and crowds. Struct. Eng. **78**, 1–25 (2000)
10. Bachmann, H., Ammann, W.: Vibrations in Structures—Induced by Man and Machines, IABSE Structural Engineering Documents 3e, Zürich, Switzerland (1987)
11. Rainer, J.H., Pernica, G., Allen, D.E.: Dynamic loading and response of footbridges. Can. J. Civ. Eng. **15**, 66–78 (1998)
12. Kramer, H., Kebe, H.W.: Man-induced structural vibrations (in German). Der Bauingeniuer. **54**(5), 195–199 (1979)

Chapter 32
Recreating Periodic Events: Characterizing Footsteps in a Continuous Walking Signal

Ellis Kessler, Pablo Tarazaga, and Serkan Gugercin

Abstract Multiple fields present a need to characterize what a footstep response looks like, or detect when footsteps are occurring in a periodic walking signal: localization, classification, and gait analysis among them. At the heart of this problem is a periodic signal encoded with biomechanical information of the person walking. In this work, the periodic nature of gait is used to recreate a template of the floor acceleration response to a single step for a particular person. Underfloor accelerometers positioned to measure vertical acceleration will be used. The floor acceleration response is expected to take the form of a sum of decaying sinusoidal responses, and from this assumed response the expected analytical correlation can be found. Correlation functions of the actual response over the course of multiple steps will be used to exploit the periodic nature of walking to give a clean correlation signal which represents the frequency content of each step. Next, these correlation functions are used to fit coefficients which are in turn used to recreate a template of a single step response. After recreating a template of the footstep event, the template could be used as a direct characterization of the person's gait or to further process the response data such as footstep detection.

Keywords Event detection · Localization · Footstep · Smart infrastructure · Gait

32.1 Introduction

There has been increasing interest in using structurally mounted sensing to analyze human gait with examples including occupancy tracking for sustainable power consumption [1, 13], localization of individuals [2, 15], and classification of individuals [3]. Data collected from these sensors (such as accelerometers or geophones) contains periodic repetitions of a similar response occurring at each footstep event. Some previous works overlook or undervalue the challenge of actually detecting the steps within a signal and defining exactly when the footstep occurred. To that end, it would be helpful in many use cases to have a time domain template of what a footstep response is expected to look like. This template could be assumed from past experience with a particular setup, but a quantitative method for deriving such a template would be more generally useful. In this work, floor acceleration data measured by underfloor accelerometers mounted in Virginia Tech's Goodwin Hall will be used to recreate a template for a person's footstep response. Since one end goal could be to enable a better method for detecting the beginning of a step event, the exact timing of each footstep is assumed to be unknown. For this reason, no previously documented time-domain averaging ([16]) techniques will be used because it would require a way to align each step before being averaged. Whatever method is used for this alignment would then have an effect on the template itself. Instead, information needed to create the template will be derived from the autocorrelation of the periodic response. Fitting the correlation function has direct parallels to the Geophysics community, where waves travelling through a dispersive media are also studied and green functions are constructed via measurements, see, e.g., [4, 5, 11].

E. Kessler (✉) · P. Tarazaga
Virginia Polytechnic Institute and State University, Virginia Tech's Smart Infrastructure Laboratory, Department of Mechanical Engineering, Blacksburg, VA, USA
e-mail: ellisk1@vt.edu

S. Gugercin
Virginia Polytechnic Institute and State University, Department of Mathematics, Blacksburg, VA, USA

32.2 Modeling Framework and Methodology

The goal of this work is to recreate the floor acceleration response, $x(t)$, of a single footstep event. The entire acceleration signal $A(t)$ can be represented as a series of N_s footstep responses with different amplitudes translated in time by an amount of time defined by the frequency at which steps are taken, ω_s:

$$A(t) = \sum_{i=0}^{N_s} A_i x\left(t - \frac{2\pi i}{\omega_s}\right). \tag{32.1}$$

An example of an acceleration response containing three steps is shown in Fig. 32.1. A single instance of the step response $x(t)$ will include noise, and there will be some variability between each step response. It is therefore desirable to average together multiple step responses to get a representative template of the response. However, if the footstep responses are averaged directly in the time domain, the time where each step begins (the $t = 0$ point for each step) must be chosen in order to align multiple steps for averaging. Figure 32.2 shows that it is difficult to define the beginning of a step response. Any method to determine the beginning of the step would also have an effect on the final template, and will therefore be avoided.

Instead of using the step response $x(t)$ directly, the autocorrelation of a series of steps, $R_{xx}(\tau)$ will be considered. The autocorrelation is defined by the equation

$$R_{xx}(\tau) = E[x(t)x(t+\tau)], \tag{32.2}$$

where τ is the time lag between signals. The advantage of using this autocorrelation is that the effects of multiple steps can be averaged together without explicitly defining the beginning of each step. If the autocorrelation can be understood, then a

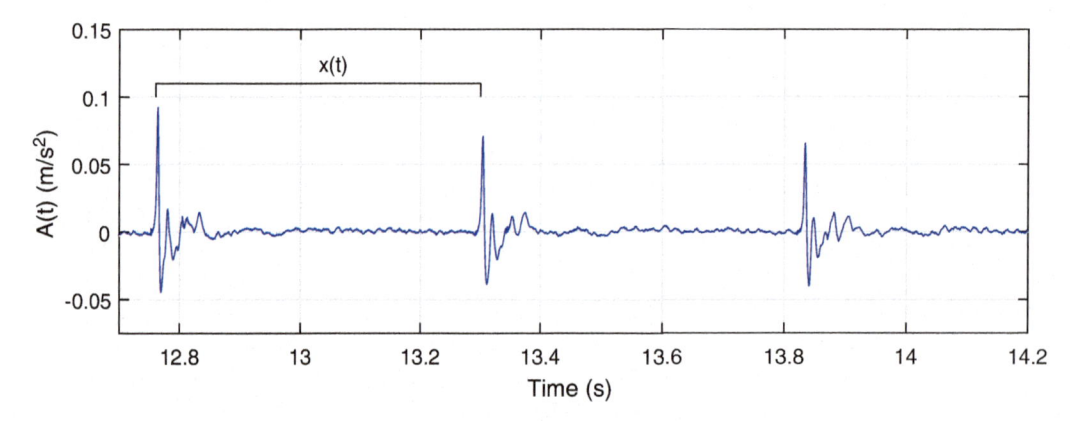

Fig. 32.1 An example of an acceleration signal from a person walking, containing three footstep responses

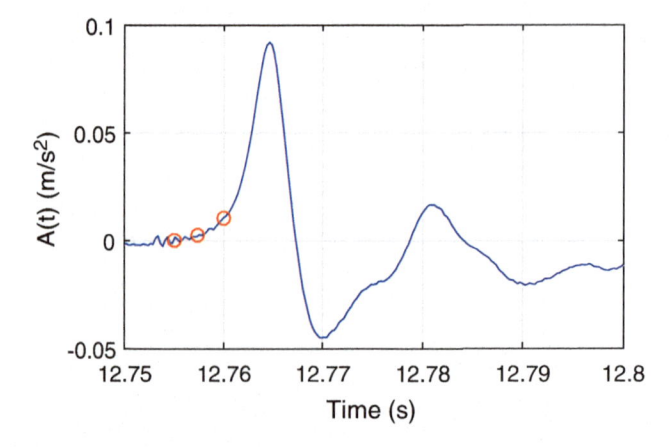

Fig. 32.2 A zoomed region of the beginning of a footstep response. Possible definitions of the $t = 0$ condition are shown in red, highlighting the difficulty of choosing a starting point for alignment in order to average

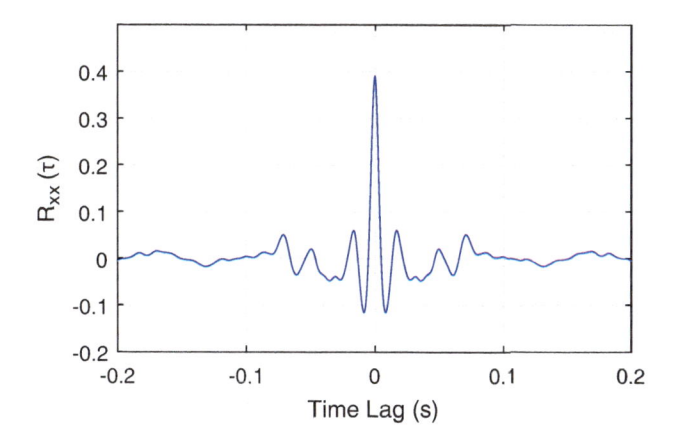

Fig. 32.3 An example of an autocorrelation created using the acceleration signal in Fig. 32.1

time domain template $x(t)$ that is representative of all steps that went into $R_{xx}(\tau)$ can be recreated. Figure 32.3 shows an example of an autocorrelation created from the three steps shown in Fig. 32.1.

In order to recreate a template of $x(t)$, an analytical formulation for $x(t)$ will be assumed. Equation (32.2) will then be used to get an analytical form of $R_{xx}(\tau)$, which can be fitted with a non-linear least squares solver. The coefficients which are used to fit $R_{xx}(\tau)$ can then be used to recreate the template for $x(t)$. As seen from Fig. 32.1 the step responses qualitatively look similar to a summation of damped sine waves. Previous research into modelling footstep induced vibrations also backs up the assumption that the response can be predicted by treating the input force into the floor as an impulse [17]. If the input from a foot to the floor is similar to an impulse, then the floor response will be similar to the impulse response. Therefore the response is assumed to be the sum of decaying sine waves, and the equation for $x(t)$ can be represented by the following equation:

$$x(t) = \sum_{n=1}^{\infty} a_n e^{-b_n t} \sin(n\omega_s t), \tag{32.3}$$

where ω_s is the frequency at which the person is taking steps, a_n are the amplitude coefficients, b_n are the damping coefficients, and $t \geq 0$. This summation is similar to a generalized fourier series over the period of the person's walking frequency, with the addition of an exponential decay term to each term in the summation. The analytical autocorrelation of a single step can therefore be derived from $x(t)$ using Eq. (32.2). The process is not stationary, and therefore $R_{xx}(\tau)$ will decay with t as well as τ. By letting $t = 0$ the autocorrelation is found to be:

$$R_{xx}(\tau) = \sum_{n=1}^{\infty} \frac{1}{2} a_n^2 e^{-b_n |\tau|} \cos(n\omega_s \tau). \tag{32.4}$$

In practice it is not possible to apply Eq. (32.2) since the acceleration signal is discrete. Therefore, the autocorrelation $R_{xx}(\tau)$ will be approximated from experimental data using the formula

$$R_{xx}(i) = \frac{1}{L} \sum_{j=1}^{L-1} x_i x_{i+j}, \tag{32.5}$$

where L is the length of the signal $x(t)$. Once the autocorrelation of experimental data is found as shown in Fig. 32.3, non-linear least squares fitting can be used to find the various coefficients ($a_1 - a_n$ and $b_1 - b_n$) describing the autocorrelation. These same coefficients can be then be used to recreate the original response, $x(t)$.

A Remark on Implementation The methods described above present a few challenges in implementation. First is that the autocorrelation of a series of footsteps will not decay quickly to zero, but instead will have echoes at multiples of the period of each footstep frequency. This hurdle can be overcome by padding the acceleration signal with zeros between each footstep. This echo can also be leveraged to get a value for ω_s instead of adding it as another parameter in the non-linear least squares problem. A second challenge is the difficulty in fitting a large number of coefficients in a single function without getting stuck in a sub-optimal local minimum. In this work this was overcome by iteratively fitting more frequencies in increasing order. For example when going from ℓ summed terms to $\ell + 1$ summed terms, the original solution is used as the starting point for all coefficients except $a_{\ell+1}$ and $b_{\ell+1}$. By initializing the least squares fitting algorithm using a previous solution, issues with local minimums and convergence time were mitigated. Although the summations for $x(t)$ and $R_{xx}(\tau)$ theoretically go up to infinity, only 40 terms are used in this work. In all examples investigated, 40 terms was enough to capture the dynamics of the response.

Effective computational tools for nonlinear exponential least-squares fitting are well-studied; for example, see the Ph.D. Thesis [8] and the recent work [9] for an extensive literature survey and new analytical and computationally effective approaches we plan to incorporate in our future work together with other tools from systems theory such as [6, 7, 10, 12, 14].

32.3 Results

The results of the non-linear least squares fitting for an autocorrelation found from a series of steps is shown in Fig. 32.4a. The autocorrelation fit with 40 terms results in an absolute RMSE of 10^{-2}. With 40 terms the RMSE had also stabilized and further terms did not improve fit accuracy. Figure 32.4b also shows the recreation of a time-domain template from the coefficients used in the autocorrelation fit. This template is shown overlaid with the 5 most prominent steps in the walking signal used. More walking trials were fitted using the same method, showing RMSE errors in the autocorrelation fit, and similar quality of the time-domain template.

32.4 Discussion

By inspecting Fig. 32.4 it is clear that although the autocorrelation was fit well, the time-domain template captures major trends of the step response but could be improved. The template would never perfectly match the step responses because the force from the foot to the floor is not an ideal impulse. The step response $x(t)$ therefore cannot be composed exactly of decaying sinusoidal responses all starting at the same time. For example, some acceleration will be induced after the heel-strike event, creating acceleration responses not captured by the assumed form of $x(t)$. The template, however, does appear to be useful. The main dynamics in the response are captured: a large initial spike with subsequent oscillations occurring at the correct intervals. Since the beginning of the time domain template matches well, this is encouraging for use in detecting when a footstep occurs. Another use of this method could be in using the coefficients found in fitting the autocorrelation as a feature to represent an individual themselves. Since the footstep response is a combination of the floor transfer function

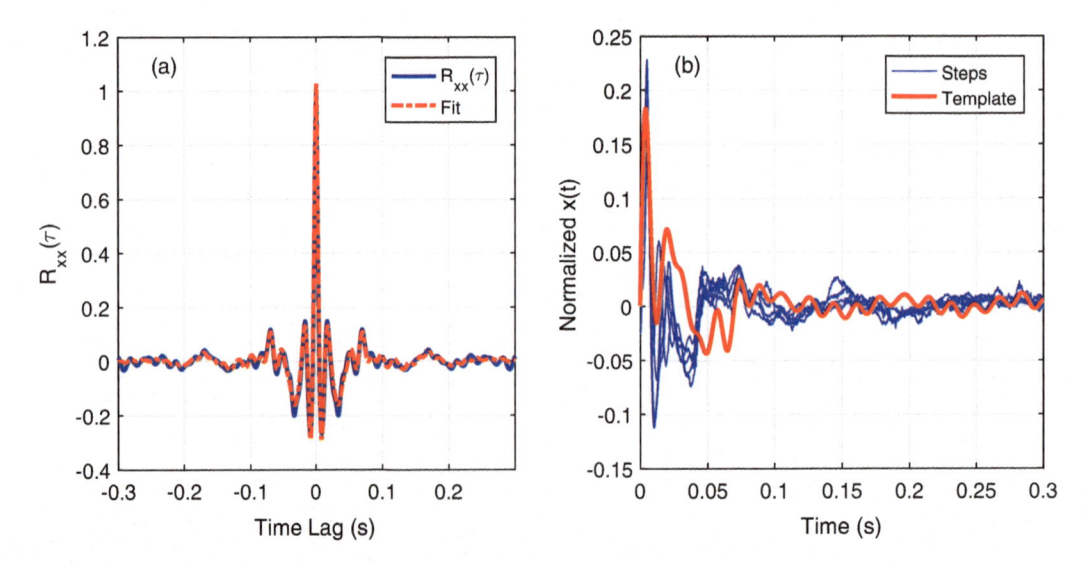

Fig. 32.4 Autocorrelation from an experimental trial, along with the fitted autocorrelation (**a**). The five most prevalent steps in the signal, along with the time-domain template assembled using the same coefficients from the autocorrelation fit (**b**)

frequency content of the force input into the floor from the person's foot, there is innately information about how a particular person walks encoded in the template. In this work only the autocorrelation of a single sensor was considered to create the time-domain template. Future work will consider cross correlations between multiple sensors, as well possible benefits from fitting measurements in the frequency domain.

Acknowledgements Dr. Tarazaga would like to acknowledge the financial support of the John R. Jones Faculty Fellowship.

The authors wish to acknowledge the support as well as the collaborative efforts provided by our sponsors, VTI Instruments, PCB Piezotronics, Inc.; Dytran Instruments, Inc.; and Oregano Systems. The authors are particularly appreciative for the support provided by the College of Engineering at Virginia Tech through Dean Richard Benson and Associate Dean Ed Nelson as well as VT Capital Project Manager, Todd Shelton, and VT University Building Official, William Hinson. The authors would also like to acknowledge Gilbane, Inc. and in particular, David Childress and Eric Hotek. We are especially thankful to the Student Engineering Council (SEC) at Virginia Tech and their financial commitment to this project. The work was conducted under the patronage of the Virginia Tech Smart Infrastructure Laboratory and its members.

The work is supported in part by the National Science Foundation via grant no. DGE-1545362, UrbComp (Urban Computing): Data Science for Modeling, Understanding, and Advancing Urban Populations. Any opinions, findings, and conclusions or recommendations expressed in this material are those of the authors and do not necessarily reflect the views of the National Science Foundation.

References

1. Agarwal, Y., Balaji, B., Gupta, R., Lyles, J., Wei, M., Weng, T.: Occupancy-driven energy management for smart building automation. In: Proceedings of the 2Nd ACM Workshop on Embedded Sensing Systems for Energy-Efficiency in Building, BuildSys '10, pp. 1–6. ACM, New York (2010)
2. Alajlouni, S., Albakri, M., Tarazaga, P.: Impact localization in dispersive waveguides based on energy-attenuation of waves with the traveled distance. Mech. Syst. Signal Process. **105**, 361–376 (2018)
3. Bales, D., Tarazaga, P.A., Kasarda, M., Batra, D., Woolard, A., Poston, J.D., Malladi, V.S.: Gender classification of walkers via underfloor accelerometer measurements. IEEE Internet Things J. **3**(6), 1259–1266 (2016)
4. Bardos, C., Garnier, J., Papanicolaou, G.: Identification of Green's functions singularities by cross correlation of noisy signals. Inverse Prob. **24**(1), 015011 (2008)
5. Claerbout, J.: Fundamentals of Geophysics Data Processing: With Applications to Petroleum Prospecting. 06 (1985)
6. Drmac, Z., Gugercin, S., Beattie, C.: Quadrature-based vector fitting for discretized h_2 approximation. SIAM J. Sci. Comput. **37**(2), A625–A652 (2015)
7. Gustavsen, B., Semlyen, A.: Rational approximation of frequency domain responses by vector fitting. IEEE Trans. Power Delivery **14**(3), 1052–1061 (1999)
8. Hokanson, J.: Numerically Stable and Statistically Efficient Algorithms for Large Scale Exponential Fitting, PhD thesis. Rice University, Houston (2013)
9. Hokanson, J.M.: Projected nonlinear least squares for exponential fitting. SIAM J. Sci. Comput. **39**(6), A3107–A3128 (2017)
10. Ljung, L.: System Identification. Prentice Hall, New Jersey (1987)
11. Malcolm, A., Scales, J., Tiggelen, B.: Extracting the green function from diffuse, equipartitioned waves. Phys. Rev. E Stat. Nonlin. Soft. Matter. Phys. **70**, 015601 (2004)
12. Mayo, A., Antoulas, A.: A framework for the solution of the generalized realization problem. Linear Algebra Appl. **425**(2–3), 634–662 (2007)
13. Pan, S., Bonde, A., Jing, J., Zhang, L., Zhang, P., Noh, H.Y.: BOES: building occupancy estimation system using sparse ambient vibration monitoring. In: Proceedings of SPIE–The International Society for Optical Engineering, vol. 9061, p. 90611O. International Society for Optics and Photonics (2014)
14. Peherstorfer, B., Gugercin, S., Willcox, K.: Data-driven reduced model construction with time-domain loewner models. SIAM J. Sci. Comput. **39**(5), A2152–A2178 (2017)
15. Poston, J.D., Buehrer, R.M., Tarazaga, P.A.: Indoor footstep localization from structural dynamics instrumentation. Mech. Syst. Signal Process. **88**, 224–239 (2017)
16. Racic, V., Chen, J., Pavic, A.: Advanced fourier-based model of bouncing loads. In: Pakzad, S. (ed.) Dynamics of Civil Structures, vol. 2, pp. 367–376. Springer International Publishing, Cham (2019)
17. Willford, M., Young, P., Field, C.: Improved Methodologies for the Prediction of Footfall-Induced Vibration. vol. 5933, 03 (2006)

Chapter 33
On Wave Propagation in Smart Buildings

Mauro S. Maza, Mohammad I. Albakri, V. V. N. S. Malladi, and Pablo A. Tarazaga

Abstract In this work, the problem of wave propagation in a smart building, Virginia Tech's Goodwin Hall, is investigated. Goodwin Hall is a five-story, L-shaped engineering building that is instrumented with 225 accelerometers. The accelerometers are permanently mounted on the building's steel columns and girders allowing for a continuous monitoring of vibration activities. Using the building's accelerometers, the propagation of elastic waves emitted from a series of floor impact excitations is studied. Wave propagation within a given floor as well as floor-column interaction are investigated in this study. Experimental results suggest that waves of the first anti-symmetric mode are induced in the floor due to impact excitation. Wave mode conversion takes place at the floor-column interface, and the waves propagating along the column are found to be of the first symmetric mode. Time-of-arrival-based calculations are implemented to obtain an estimate of wave speed in the building's components and the limitations of this approach are discussed.

Keywords Smart buildings · Wave propagation · Floor impact · Floor-column interaction · Wave mode conversion

33.1 Introduction

Structural health monitoring, energy efficiency, and security requirements have been driving the spread of instrumented smart civil structures. Such applications deal with analyzing data from multiple sensors detecting impact-like events such as footsteps, door slams, and human falls [1–3]. Central to these applications is the understanding of elastic wave propagation characteristics in the structures. The heterogeneous nature of steel reinforced concrete along with the interaction between concrete slabs, horizontal girders, and vertical columns add to the complexity of this problem [4]. For instance, it has been observed that signal time-of-arrival and distortion is not only dependent on the distance from the source but also on the structural component to which sensors are attached. Therefore, there is a need to investigate the nature of wave propagation in such complex environment.

The problem of elastic wave propagation in one and two dimensional structures has been thoroughly investigated over the last few decades [5, 6]. Such theories have been applied to study wave propagation in smart buildings, where concrete slabs are approximated by anisotropic heterogeneous plates. Impacts due to events taking place in residential/commercial buildings, such as footsteps, door slams, or falling objects, are modelled as seismic disturbances propagating in concrete slabs as a combination of symmetric and antisymmetric wave modes. In the frequency range excited by such events, >1 kHz, antisymmetric waves are strongly dispersive, resulting in waveform distortion and spread as it propagates through the structure. Symmetric waves, on the other hand, are weakly dispersive over this frequency range, and hence, minimal waveform distortion is expected to occur. However, wave transmission/reflection at the interfaces between different structural components strongly affect the measured waveforms. Previous studies in the literature addressed this problem numerically [7]. Experimental investigations are limited to component level where isolated structural components, such as concrete slabs, are investigated. Experimental investigation of wave propagation in steel-reinforced concrete structures has not been fully addressed in the literature.

M. S. Maza
Department of Mechanical Engineering, Universidad Nacional de Córdoba, Córdoba, Argentina

Virginia Tech Smart Infrastructure Laboratory (VTSIL), Department of Mechanical Engineering, Virginia Tech, Blacksburg, VA, USA

M. I. Albakri (✉) · V. V. N. S. Malladi · P. A. Tarazaga
Virginia Tech Smart Infrastructure Laboratory (VTSIL), Department of Mechanical Engineering, Virginia Tech, Blacksburg, VA, USA
e-mail: malbakri@vt.edu

The present work is the first step towards investigating wave propagation characteristics in multi-story smart buildings. For this purpose, experiments have been conducted in Virginia Tech's Goodwin Hall, a five-story building permanently instrumented with 225 accelerometers. Using the building's accelerometers, the propagation of elastic waves emitted from a series of impact excitations is studied. Wave propagation along the building's concrete floors and vertical columns is investigated. The interaction between these structural components is also investigated, and its impact on wave speed and wave mode conversion is explored.

33.2 Description of Experiments

The experiments presented and discussed in this work were conducted at Virginia Tech's Goodwin Hall. Goodwin Hall, an engineering building opened Fall 2014, is the most instrumented public building in the world for high-fidelity vibration studies. The building is instrumented with 225 highly-sensitive uniaxial accelerometers. These sensors are permanently mounted on the building's structural beams. In a number of locations, sets of three sensors are used, allowing multi-axis measurements, along the x-, y-, and z-directions, to be made. The sensors are located throughout the structure, including the building's major corners, corridors, labs and office floors, allowing the characterization of activities throughout the building.

Figure 33.1 shows sensor deployment on the fourth floor where numbers indicate that sensors measure along 1-, 2-, or 3-axes. The black square represents the data acquisition hub on that floor. The inset shows the way an accelerometer is mounted to an I-beam. The figure also highlights the subset of sensors that are used for the two experiments discussed in this work, the floor wave propagation experiment and the floor-column interaction experiment.

For the experiments presented in this study, a small subset of building's sensors has been used. The first experiment presented here is designed to study wave propagation in a given floor. For this experiment, a subset of five sensors located on the fourth floor is used, these sensors are highlighted in Fig. 33.1. Figure 33.2 shows the exact location of these sensors. The event studied in this experiment is a hammer impact taking place 1.5 m from sensor S1. The characteristics of the elastic wave generated by this event is studied by tracking its propagation in the floor using sensors S1–S5.

The second experiment presented in this study investigates the wave propagation from floors to vertical columns. For this experiment a subset of four sensors is used. These sensors are mounted on the middle-northern column of the building. This column is highlighted in Fig. 33.1, while the exact location of these sensors is shown in Fig. 33.3. The event studied in this experiment is a hammer impact on the third floor, 5 m from the column. To study floor-column interaction, the propagation of the elastic waves generated by this event along the column is tracked using sensors S1–S4. For all experiments, a sampling rate of 32 kHz is used for data acquisition.

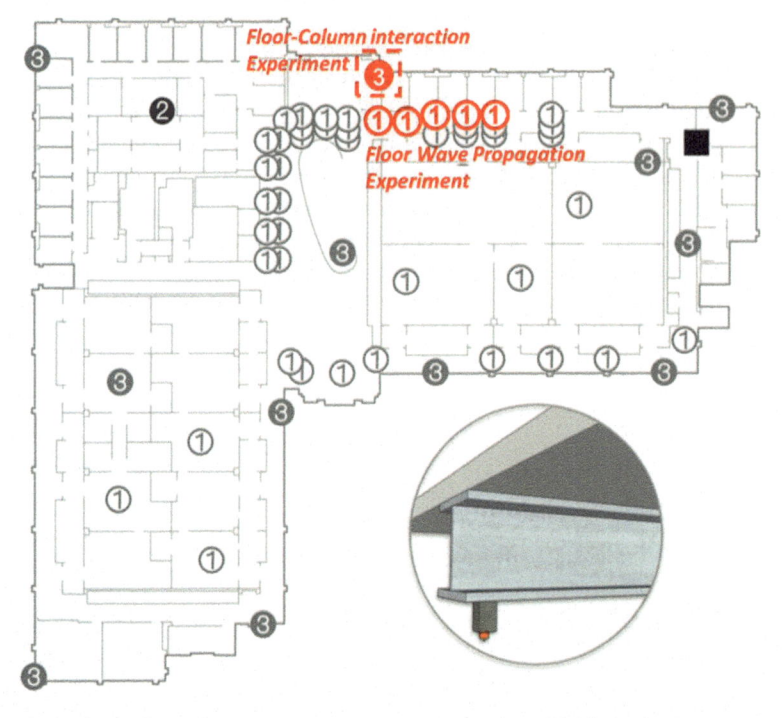

Fig. 33.1 Sensor deployment on the fourth floor. The sensors used in the current work are highlighted

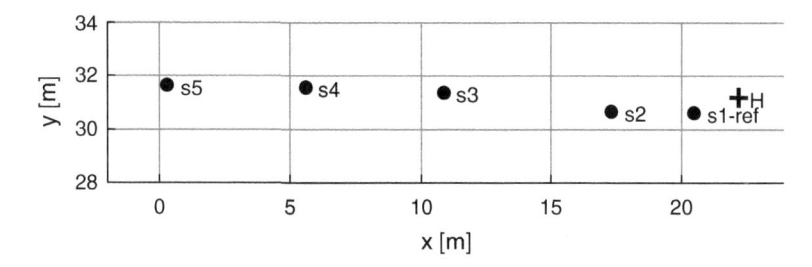

Fig. 33.2 Sensor layout used for the in-floor wave propagation experiment

Fig. 33.3 Sensor layout used for the floor-column interaction experiment

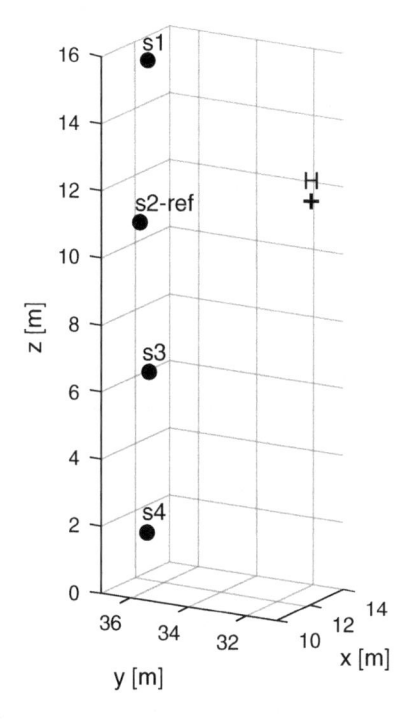

33.3 Results and Discussion

This section discusses the main results from the in-floor and the floor-column interaction experiments. For each experiment, the time trace of the measured signals along with their spectrogram are presented. The characteristics of the propagating waves are discussed qualitatively and wave speed estimates, based on the signal time-of-arrival (TOA), is calculated.

33.3.1 In-Floor Wave Propagation Experiment

The time history for the wave induced by the floor impact as measured by sensors S1–S5 are shown in Fig. 33.4a. The vertical red, dashed line indicates the time at which the signal is detected by sensor S1, which is nearest sensor to the event and thus is chosen to be the reference sensor in this experiment. The delay in the signal's TOA as a function of sensors' location can be clearly observed in the figure. Strong signal distortion along with reflections can be seen in the measured response. Figure 33.4b. shows the spectrograms of these signals. The frequency content of the excitation impact is found to be contained in the range of 0–4 kHz. The signal measured by the reference sensor, S1, shows no distinguishable variations in the TOA of the various frequencies constituting the signal. As the propagation distance increases, the high-frequency components of the propagating signal are found to arrive sooner than the low-frequency components. This is a clear sign of dispersion where wave speed varies as a function of frequency. The dispersive nature of the wave propagating through the concrete slabs of the floor indicates that the wave mode excited by the hammer impact is the first anti-symmetric (A_0) mode.

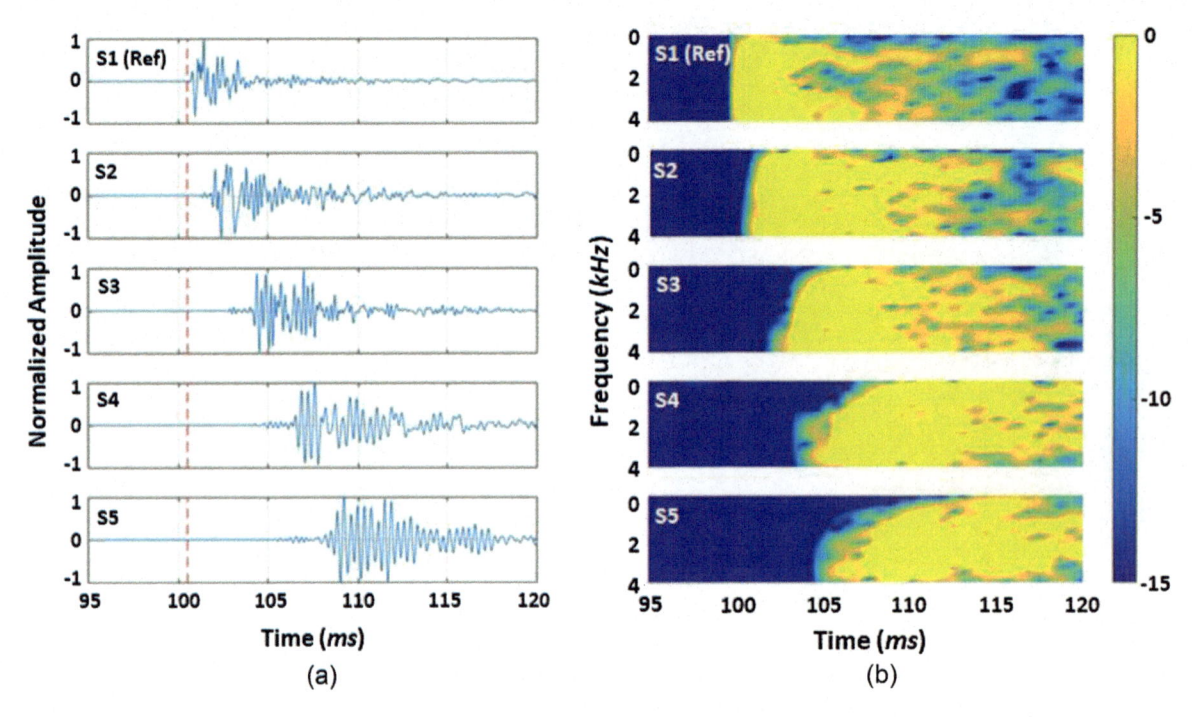

Fig. 33.4 (**a**) Wave forms and (**b**) spectrograms for the in-floor wave propagation experiment

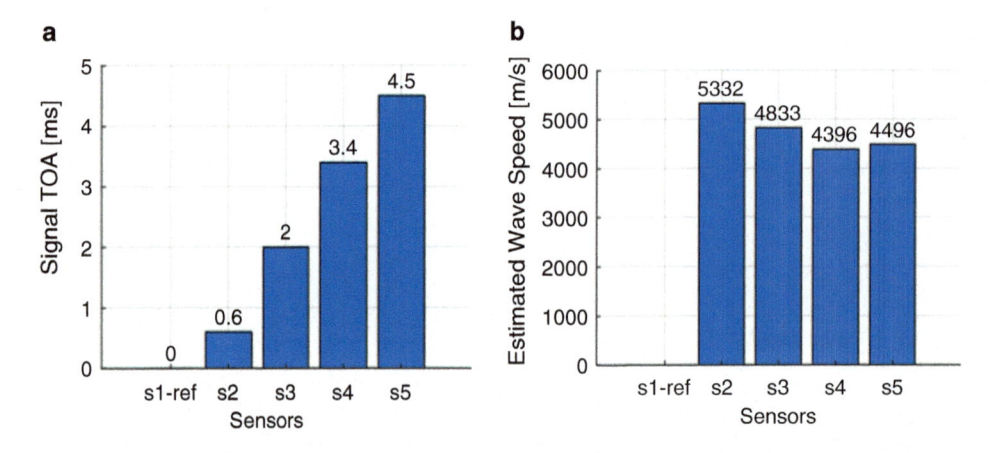

Fig. 33.5 (**a**) Signal time-of-arrival and (**b**) estimated wave speed, for the in-floor wave propagation experiment

While the dispersive nature of this wave mode renders traditional TOA-based calculations inaccurate, TOA has been used here to get rough estimates of the wave speed along the concrete floor. Figure 33.5a shows the TOA of the propagating wave, with sensor S1 being the reference. This TOA is defined as the time at which the measured response gets above the noise floor. With the known locations of these sensors, wave speed is calculated. The results are shown in Fig. 33.5b. Given the TOA definition adopted in this study, the calculated wave speed will reflect the speed of the highest-frequency component in the propagating wave as high-frequency components travel faster than the low-frequency ones, and thus, they are the first to arrive at the sensor location. Based on this, the wave speed estimates reported in Fig. 33.5b represent the speed value of the A_0 mode at 4 kHz. In the figure, It is clear that nearest sensor to the reference S1 provides the least accurate estimate. This is due to the limited time resolution of the measured signal, 0.03 ms in this experiment, which has a more profound impact on shorter time periods. As the sensor location gets further away from the reference, the speed estimate converges to about 4.5 km/s.

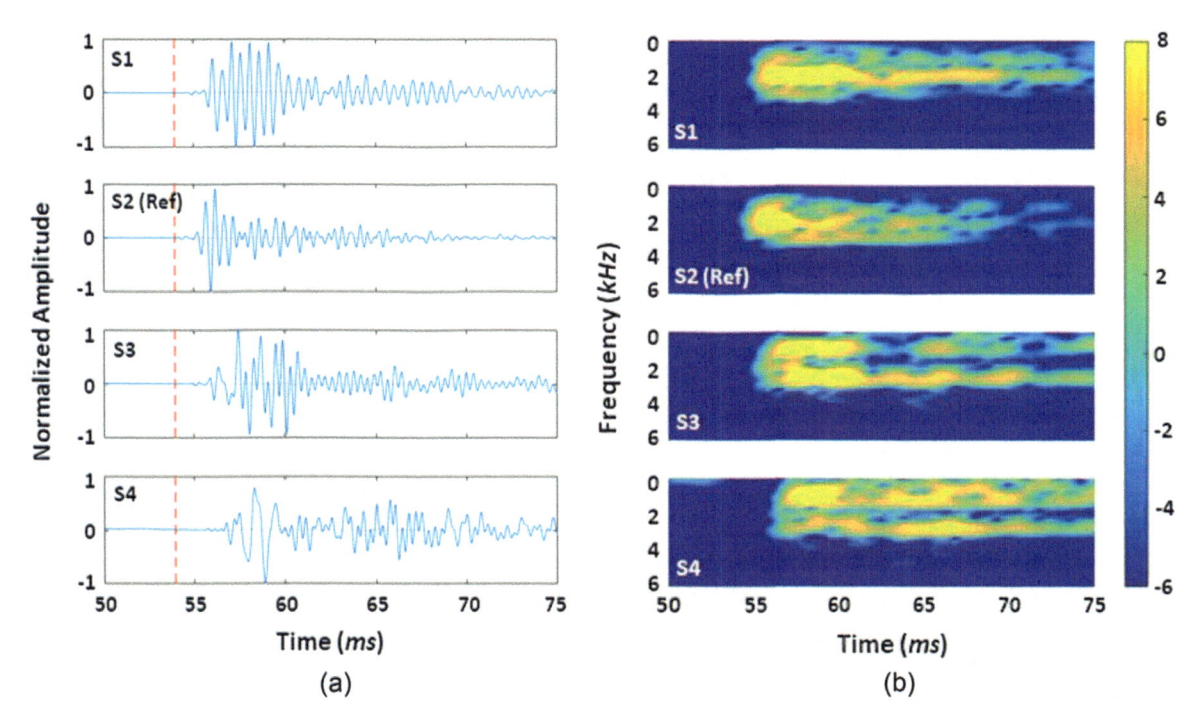

Fig. 33.6 (**a**) Wave forms and (**b**) spectrograms for the floor-column interaction experiment

33.3.2 Floor-Column Interaction Experiment

A similar analysis to the one presented in the previous subsection is repeated here for the floor-column interaction experiment. Figure 33.6a depicts the time history for the wave induced by the floor impact as measured by column sensors S1–S4. The floor impact induces an A_0 wave that propagate along the floor and this wave mode is dispersive in nature, as discussed earlier. The incident wave then interacts with the column and induces the wave forms shown in the figure. Sensor S2 is located on the fourth floor, where the impact excitation took place, and thus it is selected to be the reference for this experiment. Strong presence of reflections is noticed, especially in the signal measured by sensor S1, which is located near the building's roof. The spectrograms of these signals are shown in Fig. 33.6b. The frequency content of these signals is found to be contained in the range of 0–4 kHz, which agrees with what has been observed earlier. This is intuitive since floor and column waves are induced by similar events, i.e. floor impact. The signals measured by all sensors, S1–S4, show small variations in the TOA as a function of frequency. This indicates that wave speed is independent of frequency, a characteristic of the first symmetric (S_0) wave mode. Thus, wave mode conversion took place at the floor-column interface where the floor A_0 wave is converted to a column S_0 wave.

TOA-based calculations are carried out to get an estimate of wave speed, with sensor S2 being the reference. The results are summarized in Fig. 33.7. Sensor S1 is giving an erroneous estimate, which can be partly ascribed to the limited time resolution of the measurements. Sensors S2 and S3 signals yield a more accurate estimate of the wave speed at about 5.6 km/s. The estimated wave speed for the S_0 mode exceeds that obtained for the A_0 mode, which agrees with the fundamental characteristics of these wave modes. The accuracy of the reported values, however, is in question due to the limited sampling rate and the inherent limitations of the TOA technique. More accurate approaches for calculating wave speed of symmetric and anti-symmetric modes will be the focus of future studies.

33.4 Conclusions

In this study, the problem of wave propagation in a smart building, Virginia Tech's Goodwin Hall, is investigated. Elastic waves are induced in the building's floor using impact excitation and their propagation is monitored through a set of accelerometers mounted on the building's floors and columns. Two experiments have been presented and discussed in this study addressing in-floor wave propagation and floor-column interaction.

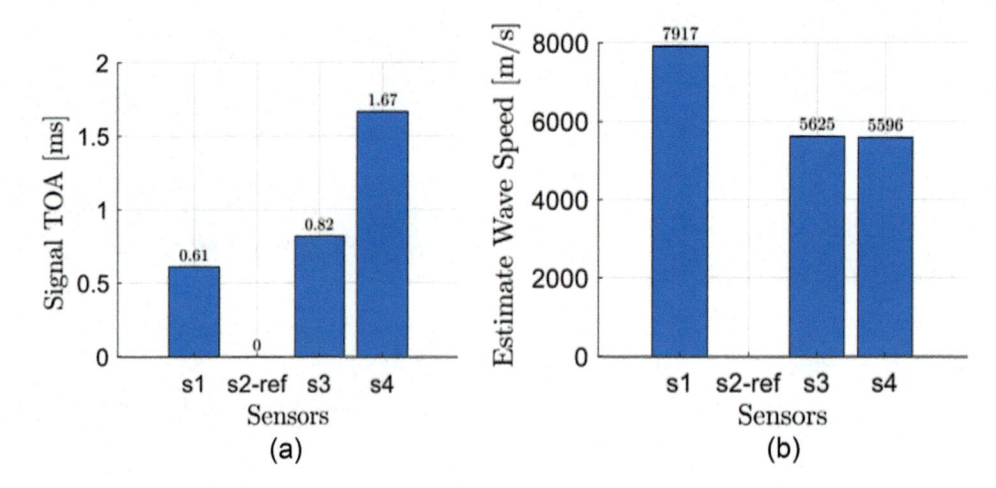

Fig. 33.7 (**a**) Signal time-of-arrival and (**b**) estimated wave speed for the floor-column interaction experiment

Experimental results revealed a strong dispersive nature of the waves propagating in the building's floor. This suggests that the waves induced in the floor due to impact excitation are of the A_0 mode. As floor waves interact with vertical columns, wave conversion takes place, and S_0 waves are found to propagate along the column. The nondispersive nature of the S_0 wave mode was clearly observed in the measured responses. Wave speed estimates based on TOA is also presented. While the correct trends are observed, when comparing the speeds of A_0 and S_0 waves, the limited sampling rate and the inherent limitations of the TOA technique adversely affect the accuracy of the reported values.

Future research will address these sources of inaccuracies, where dispersion curves for the concrete floor and columns will be constructed. The wave conversion phenomenon will also be addressed, where a series of floor-column/column-floor interactions will be studied to better understand how an event happening on a given floor is seen on other floors of the building.

Acknowledgement The authors wish to acknowledge the support as well as the collaborative efforts provided by our sponsors, VTI Instruments, PCB Piezotronics, Inc.; Dytran Instruments, Inc.; and Oregano Systems. The authors are particularly appreciative for the support provided by the College of Engineering at Virginia Tech through Dean Richard Benson and Associate Dean Ed Nelson as well as VT Capital Project Manager, Todd Shelton, and VT University Building Official, William Hinson. The authors would also like to acknowledge Gilbane, Inc. and in particular, David Childress and Eric Hotek. Dr. Tarazaga would also like to acknowledge the support provided by the John R. Jones III Faculty Fellowship. The work was conducted under the patronage of the Virginia Tech Smart Infrastructure Laboratory and its members.

References

1. Poston, J.D., Buehrer, R.M., Woolard, A.G., Tarazaga, P.A.: Indoor positioning from vibration localization in smart buildings. In: Position, Location and Navigation Symposium (PLANS), 2016 IEEE/ION, IEEE (2016)
2. Poston, J.D., Schloemann, J., Buehrer, R.M., Malladi, V.S., Woolard, A.G., Tarazaga, P.A.: Towards indoor localization of pedestrians via smart building vibration sensing. In: Localization and GNSS (ICL-GNSS), 2015 International Conference, IEEE (2015)
3. Alajouni, S., Albakri, M.I., Tarazaga, P.A.: Impact localization in dispersive waveguides based on energy attenuation of waves with the traveled distance. J. Mech. Syst. Signal Process. **105**, 361–376 (2018)
4. Todorovska, M., Ivanović, S., Trifunac, M.J.S.D., Engineering, E.: Wave propagation in a seven-story reinforced concrete building: I. Theoretical models. Soil Dyn. Earthq. Eng. **21**(3), 211–223 (2001)
5. Graff, K.F.: Wave Motion in Elastic Solids. Courier Corporation, North Chelmsford (1975)
6. Doyle, J.F.: Wave Propagation in Structures: Spectral Analysis Using Fast Discrete Fourier Transforms, 2nd edn. Springer, New York (1997)
7. Park, J.-H., Kim, Y.-H.: Impact source localization on an elastic plate in a noisy environment. J. Meas. Sci. Technol. **17**(10), 2757 (2006)

Chapter 34
Parameter Study of Statistics of Modal Parameter Estimates Using Automated Operational Modal Analysis

Silas S. Christensen and Anders Brandt

Abstract For any modal parameter estimation (MPE) method, there are a few control inputs that can have an impact on the modal parameter estimates. These control inputs are typically involving parameters like the maximum model order and how many time values or frequency lines that should be included in the MPE. In this paper, a comprehensive study on the influence of these parameters is conducted using the multi-reference Ibrahim Time Domain algorithm (similar to the cov-SSI method). Data from a laboratory Plexiglas plate are investigated, and an automated Operational Modal Analysis (OMA) algorithm is used to systematically select physical poles. The effect of each of the various control parameters are discussed in the paper.

Keywords Structural health monitoring (SHM) · Operational modal analysis (OMA) · Automated OMA (AOMA) · Modal parameter estimation (MPE) · Damping

34.1 Introduction

Conventional maintenance of civil-engineering structures relates to on-site inspections. For older structures or in case of recent failures that have been corrected, inspections are done more frequently. In most cases these structures are in good condition, hence the inspection is unavailing. Depending on accessibility, as well as the extent of the inspection, maintenance can be expensive. Furthermore, some faults are not detected through conventional maintenance due to human error. A novel approach to structural maintenance is Structural Health Monitoring (SHM), which refers to a system that allows for a systematic characterization of the health of a civil-engineering structure. The system is based on sensor technology, while the health of the structure relates to how much of its lifetime has already been used, and therefore how much of it is left. It is common to monitor modal parameters, as changes in these correspond to changes to the structure, i.e. boundary conditions, re-distributions of loads etc.

Operational Modal Analysis (OMA), is a popular concept that allows to determine modal parameters for structures in operation. It involves measuring the ambient response of structures, i.e. the forces that cause the response are unknown. The counterpart to OMA is Experimental Modal Analysis (EMA), in which the input of the system is measured. EMA is popular to use on smaller structures, i.e. objects that can fit into a laboratory, such that it is possible to isolate the structure from external forces, thereby controlling the input to the system by means of known excitation. For larger structures, i.e. most civil-engineering structures, the presence of external forces is inevitable, hence OMA is employed. A thorough comparison of EMA and OMA for measurements on a Plexiglass plate show that they perform equally, see [12].

It is crucial that the methods for estimating the modal parameters are consistent. There are many methods for Modal Parameter Estimation (MPE), and they can generally be divided into two subclasses, time domain and frequency domain methods, see [2]. Whether one method is better than the other is inconclusive, studies suggest that they perform equally well, see [11] and [16]. Transforming data from one domain to another and back again should not remove or add information. However, if power spectral densities are used, a bias error is introduced, due to the truncation of the measurement in the frequency domain, see [5]. In the time domain it is common to use free decay estimated as correlation functions or random decrement signatures. The former, correlation functions may be computed without bias [6].

In the following we will only consider time domain. Some of the most popular time domain methods are:

- the Ibramhim Time Domain (ITD) method, see [9] and Multireference ITD (MITD) method, see [8]

S. S. Christensen (✉) · A. Brandt
University of Southern Denmark, Odense, Denmark
e-mail: ssv@iti.sdu.dk

- the Polyreference Time Domain (PTD) method, see [19]
- the Eigensystem Realisation Algorithm (ERA) method, see [10]
- and Stochastic Subspace Identification (SSI) methods, see [18]

A numerical study [3] performed on a model of a 2DOF linear system with normal modes and proportional damping showed that the damping estimates found using ITD are more robust when it comes to closeness of modes and noise. This is in comparison to PTD and ERA. Another numerical study, see [14], that involved a lattice structure reported that ERA and SSI are performing equally well when it comes to determining modal parameters. The mentioned studies are of course limited to the structures that they were used on. Generally speaking, one MPE method may for one particular application work better than another. However, by varying control inputs for the desired MPE method, one may experience strong inconsistencies in the modal parameter estimates.

For any MPE method there are two control inputs that must be selected before modal parameters can be estimated. Those parameters are the model order and the amount of information that is fed into the MPE method. The first control input, the model order, denotes the number of equations to be solved, referring to the number of modes to be found in the data sets. If the model order is too low, crucial results could be omitted, while on the contrary, for a high model order, the output data may include computational estimates. The Akaike Information Criterion [1] or the Singular Value Criterion [4] are two examples of model order selection criteria. These procedures were established within the control engineering field, in which under- or over-fitting can yield wrong results. Within OMA, it is more common to use a high model order and then present the data using a stabilization diagram (consistency diagram), a plot showing the pole locations for different model orders as a function of frequency. For increasing model order, a new set of modes are found, and they tend to stabilize on the frequency axis suggesting the presence of a physical mode. For any given model order, there is no guarantee that all physical modes are found, but by using a sufficiently high model order, (yielding many poles), there is strong evidence whether a mode is physical or not.

The second control input, the amount of information used, is related to the number of frequency lines (when using transfer function) or time values (when using impulse responses or correlation functions) that should be used when estimating modal parameters. Since we are limiting the discussion here to time domain methods, the following will only address impulse responses and correlation functions, and therefore we may vary the number of time values or time lag values. The normalized random error of correlation functions increase with increasing time lags. Therefore, there is an optimum number of time lags to be used for any data set. Furthermore, the optimum number of time lags may depend on the mode, for example, higher modes die out faster, and thus require fewer lags. The number of lags that should be used is, at the moment, not well understood. As reported by [15], the influence of time blocks (related to time lags), is not discussed for SSI, but is known to have a great influence on the quality of the stabilization diagram.

A final control parameter, which is not strictly necessary for the system to be solvable, is related to the starting lag value. For correlation functions, it is recommended to remove the first few lags for the auto-correlation functions, for two reasons: first, that they are not pure free decays, and second that the auto-correlation function estimates are contaminated by sensor noise, see [13].

To summarize, the purpose of this paper is to investigate how the model order, the number of time lags and the starting time lag value affects the modal parameter estimates.

34.2 Theory

As per the introduction, a few of the most popular time domain MPE methods were listed, but for simplicity and due to the fact that these methods are very similar, only the MITD method is considered in the following. Starting off with Eq. (34.1), we have that the impulse response matrix for a system can be decomposed into a mode shape matrix $[\Psi]$, a diagonal pole matrix $\lceil e^{s_r t} \rfloor$, and a modal participation matrix $[L]$. The poles $\lceil s_r \rfloor$ are related to the frequency and damping of the system. The size of these matrices depend on the model order N, the number of references N_L and responses N_S.

$$[h(t)]_{N_L \times N_S} = [\Psi]_{N_L \times 2N} \lceil e^{s_r t} \rfloor_{2N \times 2N} [L]^T_{2N \times N_S} \qquad (34.1)$$

By repeating Eq. (34.1) a number of times, $\Delta t, 2\Delta t, \ldots, m$ and $\Delta t, 2\Delta t, \ldots, n$, the block Hankel matrix can be constructed, see Eq. (34.2). The block Hankel matrix has the size mN_L by nN_S, and in OMA, the most common is to either use $m = n$, or to make the block Hankel matrix approximately square.

$$[H_{mn}(t)] = \begin{bmatrix} [h(t)] & [h(t+\Delta t)] & \cdots & [h(t+(n-1)\Delta t)] \\ [h(t+\Delta t)] & [h(t+2\Delta t)] & \cdots & [h(t+n\Delta t)] \\ \vdots & \vdots & \ddots & \vdots \\ [h(t+(m-1)\Delta t)] & [h(t+m\Delta t)] & \cdots & [h(t+(m+n-2)\Delta t)] \end{bmatrix} \quad (34.2)$$

Equation (34.2), can be rewritten into the following

$$[H_{mn}(t)]_{mN_L \times nN_S} = [\widetilde{\Psi}]_{mN_L \times 2N} \lceil e^{s_r t} \rfloor_{2N \times 2N} [\widetilde{L}]^T_{2N \times nN_S} \quad (34.3)$$

In which $[\widetilde{\Psi}]$ and $[\widetilde{L}]$ are given by Eqs. (34.4) and (34.5)

$$[\widetilde{\Psi}] = \begin{bmatrix} [\Psi] \\ [\Psi]\lceil e^{s_r \Delta t} \rfloor \\ \vdots \\ [\Psi]\lceil e^{s_r(n-1)\Delta t} \rfloor \end{bmatrix} \quad (34.4)$$

$$[\widetilde{L}]^T = \begin{bmatrix} [L]^T & [L]^T\lceil e^{s_r \Delta t} \rfloor & \cdots & [L]^T\lceil e^{s_r(n-1)\Delta t} \rfloor \end{bmatrix} \quad (34.5)$$

By introducing the superscript $^+$ which denotes the Moore–Penrose inverse (or pseudo-inverse), Eq. (34.3) can be expressed as follows

$$[\widetilde{\Psi}]^+_{2N \times mN_L} [H_{mn}(t)]_{mN_L \times nN_S} = \lceil e^{s_r t} \rfloor_{2N \times 2N} [\widetilde{L}]^T_{2N \times nN_S} \quad (34.6)$$

There is no way to solve Eq. (34.6) from only knowing the impulse response function matrices. However, by formulating Eq. (34.2) for $t + \Delta t$ and rewriting it, one arrives at

$$[H_{mn}(t+\Delta t)]_{mN_L \times nN_S} = [\widetilde{\Psi}]_{mN_L \times 2N} \lceil e^{s_r \Delta t} \rfloor_{2N \times 2N} \lceil e^{s_r t} \rfloor_{2N \times 2N} [\widetilde{L}]^T_{2N \times nN_S} \quad (34.7)$$

By combining Eqs. (34.6) and (34.7), a system matrix $[A] = [\widetilde{\Psi}] \lceil e^{s_r \Delta t} \rfloor [\widetilde{\Psi}]^+$ can be computed from knowing $[H_{mn}(t)]$ and $[H_{mn}(t+\Delta t)]$, which may be seen in Eq. (34.8)

$$[A] = [H_{mn}(t+\Delta t)] [H_{mn}(t)]^+ \quad (34.8)$$

The system matrix A is an eigenvalue problem from which poles and mode shapes can be extracted. Next we employ the minimal realization approach with the intend to compress and suppress noise [10]. This uses a singular value decomposition

$$[H_{mn}(t)]_{mN_L \times nN_S} = [U]_{mN_L \times mN_L} \lceil S \rfloor_{mN_L \times nN_S} [V]^T_{nN_S \times nN_S} \quad (34.9)$$

The first $2N$ columns of the left singular matrix $[U]$, which we denote by $[U']_{mN_L \times mN_L}$, can then be used to compress the block Hankel matrix as follows

$$[H'_{mn}(t)]_{2N \times nN_S} = [U']^T_{2N \times mN_L} [H_{mn}(t)]_{mN_L \times nN_S} \quad (34.10)$$

Subsequent steps are to substitute the block Hankel matrix $[H_{mn}(t)]$ with the compressed block Hankel matrix $[H'_{mn}(t)]$ in Eq. (34.6) as well as doing the same steps for the time shifted block Hankel matrix $[H_{mn}(t+\Delta t)]$. After some algebraic manipulation, one arrives with two equations for the reduced system matrix $[A']$

$$[A']_{2N \times 2N} = [H'_{mn}(t+\Delta t)]_{2N \times nN_S} [H'_{mn}(t)]^T_{nN_S \times 2N} \left([H'_{mn}(t)]_{2N \times nN_S} [H'_{mn}(t)]^T_{nN_S \times 2N} \right)^{-1} \quad (34.11)$$

$$[A']_{2N \times 2N} = [H'_{mn}(t + \Delta t)]_{2N \times nN_S} [H'_{mn}(t + \Delta t)]^T_{nN_S \times 2N} \left([H'_{mn}(t)]_{2N \times nN_S} [H'_{mn}(t + \Delta t)]^T_{nN_S \times 2N} \right)^{-1} \quad (34.12)$$

Equations (34.11) and (34.12) refer to a low and high order coefficient normalization, see [2]. The size of the compressed system matrix, and therefore the number of mode shapes and poles, is only dependent on the model order. However the number of time lags used will have an impact on the block Hankel matrices that form the reduced system matrices.

In OMA applications we do not use impulse response functions, but (typically) correlation functions, due to the fact that correlation functions have the same characteristics of free decays as impulse responses [6].

34.3 Methodology

The experimental setup of the Plexiglas plate, on which measurements were taken, are shown in Fig. 34.1. The OMA test was done using 35 uni-axial accelerometers that were all mounted in a double symmetric pattern and in the same direction. The Plexiglas plate was suspended vertically in two springs, and excitation was done by scratching a pencil on the plate. 300 s responses were used, which corresponds to approximately 42500 periods of the lowest natural frequency. For more details on the data, the reader is referred to [12].

The correlation functions were produced using an unbiased estimator and a block-size of 512 samples. The interval, at which parameters were studied, is outlined in Table 34.1. A model order of 60 was used. By using a lower model order, fewer modal parameter estimates were output. Upon comparison of results originating from the same model order (but not having the same maximum model order), it was found that the modal parameter estimates were similar.

A total of 1225 correlation functions were computed. It is apparent that a total of 400 analyses were conducted at 2 different lag starting values and by varying the number of lags used in the correlation functions from 40 to 440 in intervals of 2.

As mentioned earlier it is common to use stabilization diagrams to select physical and well represented poles. Normally one pole for each mode shape is chosen, but in order to perform statistics, more information would be needed, hence all physical poles for each mode shapes must be added to the output data set. To overcome this time-consuming task, an

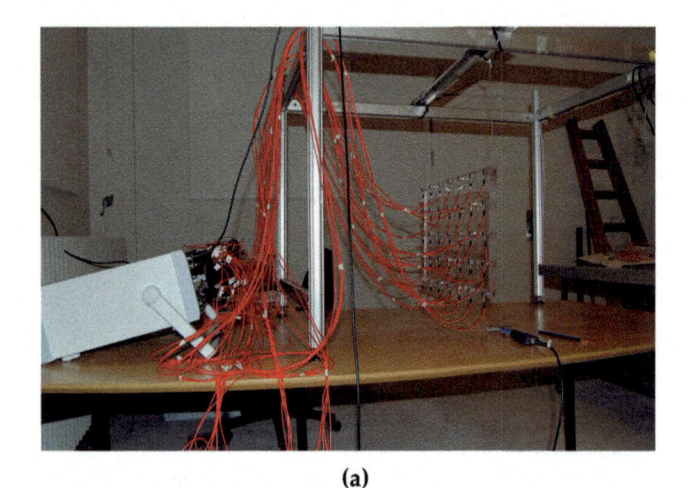

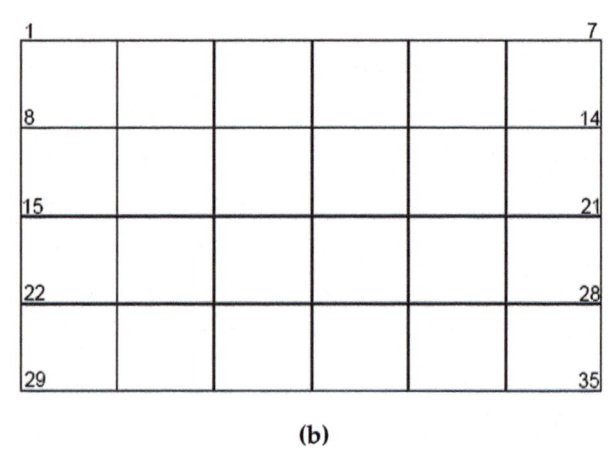

Fig. 34.1 Experimental setup and measurement grid of the Plexiglas plate. Figures are taken from [12]. (**a**) Experimental setup, showing suspension (free-free), as well as the 35 accelerometers. The pencil used for excitation is shown in the lower right corner. (**b**) Measurement grid of the Plexiglas plate, having 35 measurement points, each with a separation of 78 × 85 mm, and a full dimension of 533 × 321 × 20 mm (length × width × depth)

Table 34.1 Overview of the setup of control inputs in the analysis

Parameter	Description	First value	Last value	Interval
NStart	First value used in CF	5	35	30
NLines	No. of lags used in CF	40	440	2

algorithm that automatically detects physical poles have been applied in the analysis. This algorithm is based on employing probability analysis on the stabilization diagram, following a decision rule, that utilizes the Modal Assurance Criteria (MAC) to ensure consistency. The algorithm have shown to be successful in identifying well represented modes despite them being closely spaced or being weakly excited. For further reading, see [7].

34.4 Results

Since the number of lags used in the correlation function are of such importance in this paper, we will start by showing a selection of correlation functions computed from the measurements taken on the Plexiglas plate. In Fig. 34.2, all 35 auto-correlation functions are shown. Since the correlation functions include information about all modes and that the noise tail is unique for all correlation functions, it can be difficult, based on engineering intuition, to define an appropriate number of lags to be used in the further analysis. It can be seen that some auto-correlation functions decay rather fast, while others decay slowly. Most of the information in the correlation functions are retained in the first 300 time lag values with the exception of several auto-correlation functions that still carry some information past 500 time lag values.

For white noise excitation, the auto-correlation at time lag $\tau = 0$, correspond to the variance of the signal plus any extraneous noise in the signal. This value is significant in comparison to all other values of the correlation function. As mentioned in the Introduction and shown in [13], extraneous noise also contaminates some subsequent lags following $\tau = 0$. From Fig. 34.2 it is evident that the auto-correlation functions at $\tau = 0$ stand out. Since the excitation force is not strictly white, the first 5 lag values are removed.

In the following modal parameters for the first 6 mode shapes of the Plexiglas plate are reported. These modes are identified as:

- Mode 1—1st order bending around weak axis
- Mode 2—1st order torsion
- Mode 3—2nd order torsion
- Mode 4—2nd order bending around weak axis

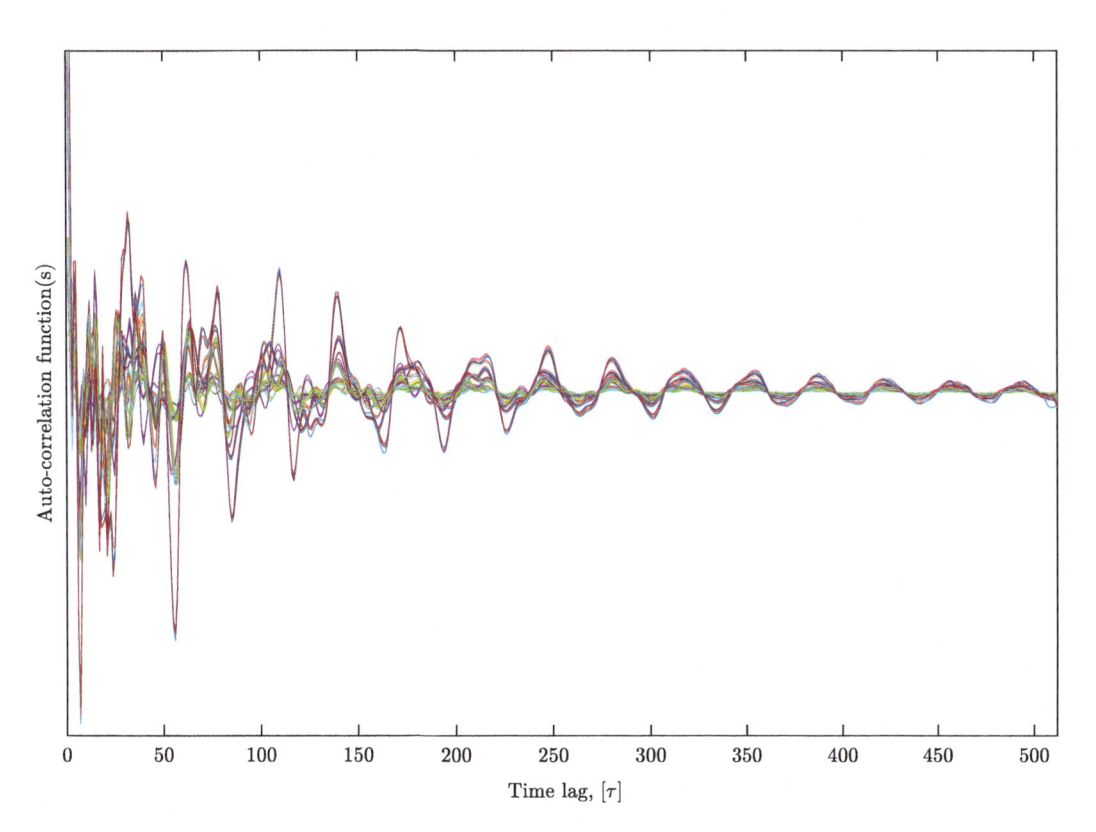

Fig. 34.2 Auto-correlation functions from measurements taken on the Plexiglass plate

- Mode 5—1st order bending around strong axis
- Mode 6—2nd order bending around strong axis

From the 6 mode shapes reported, there are 3 first order modes and 3 s order modes. Only damping ratio estimates are presented. The frequency estimates are less susceptible to changes in the control input, and therefore of less interest. Figure 34.3 shows the estimated mean damping ratio as a function of time lags used in the correlation function for all 6 mode shapes, i.e. sub-figures a to f. For each mode, the damping ratio estimates are subdivided into 3 classes, as seen below:

- Damping ratio estimates for $0 < \text{MOrder} \leq 20$
- Damping ratio estimates for $20 < \text{MOrder} \leq 40$
- Damping ratio estimates for $40 < \text{MOrder} \leq 60$

Furthermore, the damping ratio estimates are partitioned into 4 intervals based on the time lag value they belong to. The time lag value intervals are as follows:

- Damping ratio estimates for $40 < \text{NLines} \leq 140$
- Damping ratio estimates for $140 < \text{NLines} \leq 240$
- Damping ratio estimates for $240 < \text{NLines} \leq 340$
- Damping ratio estimates for $340 < \text{NLines} \leq 440$

In Table 34.2 some statistics for these 12 groups of damping ratio estimates can be found. The number of estimates, their mean value and normalized error are shown. Important to note is that the mean value is the damping ratio given in %, while the normalized error is defined as one standard deviation of the estimates divided by 100.

For mode 1, it is evident that the damping ratio estimates for the 3 model order intervals are different when using less than 150 time lag values. Above 150 time lag values there are significantly less estimates at all model order intervals, particularly at MOrder > 20. An explanation for this lies in the automated OMA algorithm used. As mentioned the algorithm reports well represented modes that have similar MAC characteristics. By manually inspecting the stabilization diagrams for time lag values 160 to 180 in intervals of 2, it was found that the stabilization diagram at first glance looked fine. However, strong variation in the MAC values were the cause that mode 1 was not well represented in this range of time lag values. Generally speaking for mode 1, the most damping ratio estimates are obtained at $0 < \text{MOrder} \leq 20$ while the lowest normalized error is found for the estimates between $240 < \text{NLines} \leq 340$ at 0.77% based on 1132 estimates. It can also be seen that the normalized error of the damping ratio estimates are significantly higher for estimates found for MOrder > 20.

For mode 2, a similar picture as for mode 1 is seen. A small bias is seen on the mean value across all model orders for increasing number of time lag values. For estimates between $340 < \text{NLines} \leq 440$ a normalized error of 1.26% is found based on 1253 estimates. Several normalized errors are below 1%, but with less than 205 estimates.

For mode 3, there seem to be variations in the mean damping estimates for the 3 intervals of model order. For $0 < \text{MOrder} \leq$ it appears that the estimates are steady. Particularly for time lag values below 175 and above 320 the estimates seem to be reasonable stable. The mean value increases from 2.67% to 2.74% depending on whether the first 100 of the last 100 time lag values are used, while the normalized error varies from 0.99% to 1.74% across the 4 time lag value intervals used. The most estimates that also have a low normalized error is either the first 100 time lag value or the last 100, respectively having 1037 and 1253 estimates.

For mode 4, the variation in the mean value for different model orders at increasing number of time lags is less profound in comparison to the previous mode. Especially for $0 < \text{MOrder} \leq 20$, the mean value is stable and only has a small inclination for the first 50 time lag values. The averaged mean value for the last 100 time lags is slightly lower than for the first 100 time lag values. A complete opposite observation as for the previous mode. The lowest normalized error is found for $0 < \text{MOrder} \leq 20$ between $140 < \text{NLines} \leq 240$ at 1.28% based on 1084 estimates.

For mode 5, the mean value for the damping estimates are the most stable than for all previous modes, but some variation is present at time lag values above 200. Furthermore, the estimates for model orders below 20 are best, i.e. the mean value is stable and the normalized error is low. The number of estimates for the higher model orders are concentrated at the time lag values below 150. The lowest normalized error is found for $0 < \text{MOrder} \leq 20$ between $240 < \text{NLines} \leq 340$ at 0.57% based on 1009 estimates.

For mode 6, the same observation for the first 150 time lag values as for mode 5 is made. However, above 150 time lag values, the mean damping ratio estimates at $20 < \text{MOrder} \leq 40$ attain value above that of estimates at $0 < \text{MOrder} \leq 20$, while estimates for $40 < \text{MOrder} \leq 60$ does the complete opposite. After around 300 time lag values this is reversed and more profound. However, for $0 < \text{MOrder} \leq 20$ the mean value damping estimates are stable for all time lag values used. The lowest normalized error is found for $0 < \text{MOrder} \leq 20$ between $140 < \text{NLines} \leq 240$ at 0.39% based on 1279 estimates.

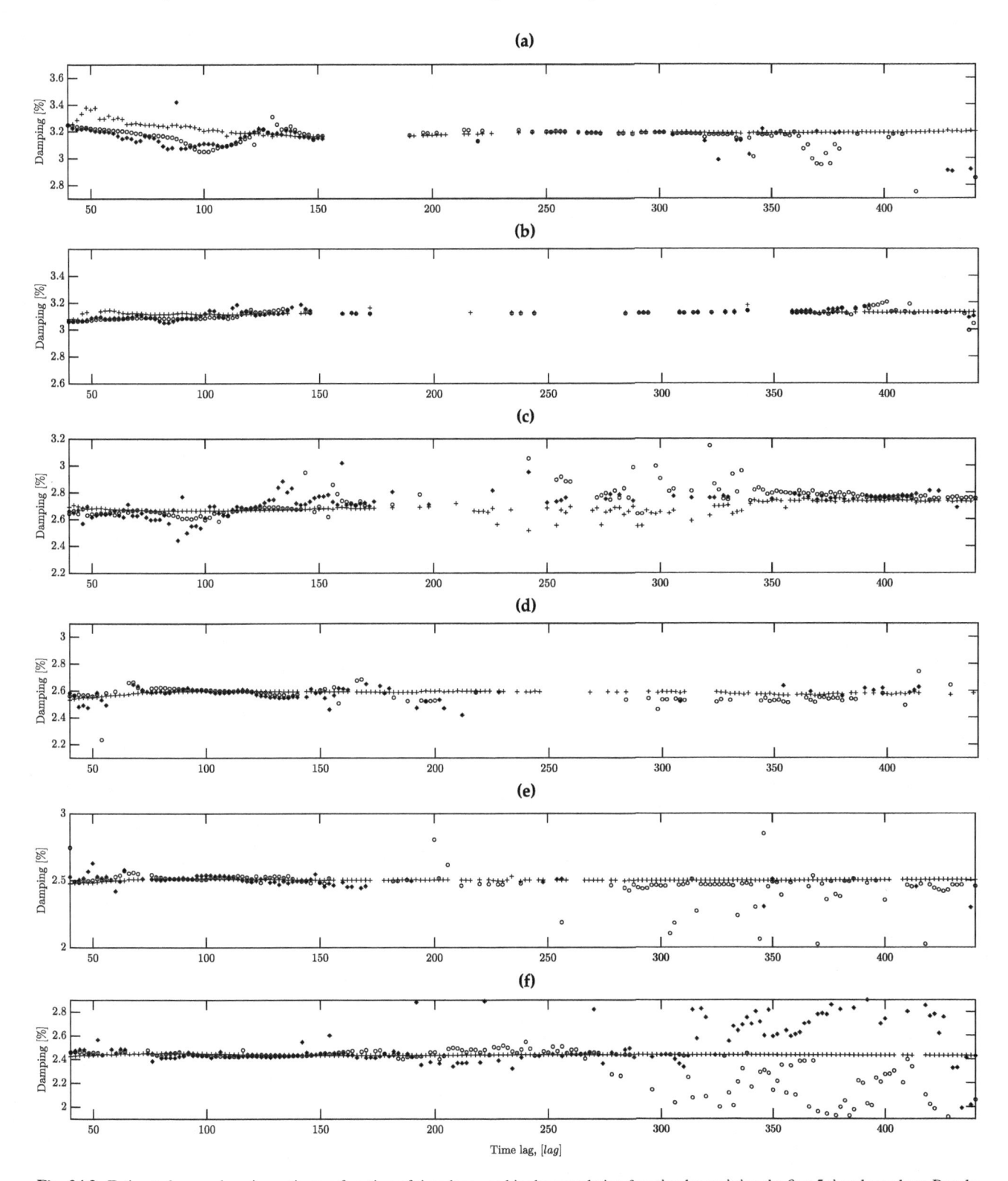

Fig. 34.3 Estimated mean damping ratio as a function of time lags used in the correlation function by omitting the first 5 time lag values. Results for the first 6 modes of the Plexiglas plate are shown. Damping ratio estimates for $0 < \text{MOrder} \leq 20$ are represented using a black +. Damping ratio estimates for $20 < \text{MOrder} \leq 40$ are represented using a black ○. Damping ratio estimates for $40 < \text{MOrder} \leq 60$ are represented using a black ∗. (**a**) Mode 1—1st order bending around weak axis. (**b**) Mode 2—1st order torsion. (**c**) Mode 3—2nd order torsion. (**d**) Mode 4—2nd order bending around weak axis. (**e**) Mode 5—1st order bending around strong axis. (**f**) Mode 6—2nd order bending around strong axis

Table 34.2 Statistics on damping ratio estimates based on model order intervals and time lag value intervals, see Fig. 34.3 for details

		Mode 1			Mode 2			Mode 3			Mode 4			Mode 5			Mode 6		
		No. of estimates	Mean value	Normalized error	No. of estimates	Mean value	Normalized error	No. of estimates	Mean value	Normalized error	No. of estimates	Mean value	Normalized error	No. of estimates	Mean value	Normalized error	No. of estimates	Mean value	Normalized error
$0 < MOrder \leq 20$	$40 < Nlines \leq 140$	1106	3.23	7.19	1037	3.12	1.70	1634	2.67	1.51	1314	2.59	1.98	1325	2.51	1.18	1011	2.44	0.93
	$140 < Nlines \leq 240$	526	3.17	1.34	205	3.12	0.99	681	2.67	6.61	1084	2.59	1.28	1025	2.50	2.13	1279	2.43	0.39
	$240 < Nlines \leq 340$	1132	3.19	0.77	426	3.13	1.74	861	2.64	15.06	530	2.59	1.81	1009	2.50	0.57	1389	2.44	0.48
	$340 < Nlines \leq 440$	1673	3.19	2.43	1253	3.13	1.26	1012	2.74	4.30	891	2.57	2.80	1303	2.50	0.55	1434	2.43	1.07
$20 < MOrder \leq 40$	$40 < Nlines \leq 140$	1269	3.17	6.50	1243	3.09	20.03	908	2.64	9.44	681	2.60	7.11	663	2.52	18.23	709	2.44	5.64
	$140 < Nlines \leq 240$	48	3.17	2.96	44	3.12	0.78	55	2.92	154.22	23	2.57	6.14	28	2.53	17.52	213	2.54	55.84
	$240 < Nlines \leq 340$	336	3.18	2.30	132	3.12	1.18	85	2.84	21.22	33	2.48	22.64	92	2.42	47.23	479	2.22	78.85
	$340 < Nlines \leq 440$	121	3.00	37.37	125	3.13	3.32	640	2.78	4.23	58	2.51	46.44	88	2.42	29.92	1660	2.07	79.67
$40 < MOrder \leq 60$	$40 < Nlines \leq 140$	1667	3.16	33.78	1596	3.10	4.07	1463	2.66	26.10	1123	2.58	3.86	1119	2.52	7.92	1185	2.43	5.36
	$140 < Nlines \leq 240$	105	3.15	0.82	111	3.13	2.24	150	2.75	28.38	77	2.56	8.39	88	2.54	26.17	387	2.56	77.66
	$240 < Nlines \leq 340$	26	3.10	15.09	16	3.13	0.79	27	2.95	67.51	1	2.52	0.00	3	2.50	1.33	132	2.73	127.73
	$340 < Nlines \leq 440$	16	3.03	15.30	40	3.14	2.74	126	2.97	105.39	10	2.60	2.56	6	2.20	54.54	189	2.77	60.86

Figure 34.4 shows the estimated mean damping ratio as a function of time lags used in the correlation function for all 6 mode shapes, i.e. sub-figures a to f. The first 35 time lag values are omitted. For each mode shapes, all damping ratio estimates are subdivided into three classes, as seen below. Table 34.3 presents the averaged damping ratio estimates and normalized errors at different model orders, but also in different intervals of time lag values.

For mode 1; 1st order bending around weak axis, see (a), there are significantly more estimates at all model order intervals. Particularly at time lag values between 200 and 350 the mean damping ratio estimates are in agreement across all 3 model order intervals. From Table 34.3 at $240 < \text{Nlines} \leq 340$ it is evident that the averaged mean values for the 3 model order intervals are very similar and have low normalized error. Obviously from Fig. 34.4 the mean value is lower for the two highest order intervals. The lowest normalized error is found for $0 < \text{MOrder} \leq 20$ between $240 < \text{NLines} \leq 340$ at 1.18% based on 1674 estimates.

For mode 2; 1st order torsion, see (b), there are also significantly more estimates. Small discrepancies in mean value damping estimates across all time lag values are observed. The most stable results are found at $0 < \text{MOrder} \leq 20$ using an intermediate number of time lag values. The lowest normalized error is found for $0 < \text{MOrder} \leq 20$ between $240 < \text{NLines} \leq 340$ at 0.30% based on 1644 estimates.

For mode 3; 2nd order torsion, see (c), there are inconsistencies in the mean damping ratio estimates at time lag values below 280. Despite $0 < \text{MOrder} \leq 20$ yields the best results, the normalized error for the estimates found at the first 100 time lag values are rather high. The lowest normalized error is found for $0 < \text{MOrder} \leq 20$ between $340 < \text{NLines} \leq 440$ at 1.64% based on 1159 estimates. The mean value on the other hand is higher in comparison to the first 100 time lag values. This was also observed in Fig. 34.3c.

For mode 4; 2nd order bending around weak axis, see (d), the estimates at the 2 higher model order intervals seem to generally underestimate the mean damping ratio estimate. Also, very few estimates are available for time lag values above 350. The mean damping ratio estimates for $0 < \text{MOrder} \leq 20$ are most stable, but show some variation. The lowest normalized error is found for $0 < \text{MOrder} \leq 20$ between $40 < \text{NLines} \leq 140$ at 4.19% based on 1366 estimates.

For mode 5; 1st order bending around strong axis, see (e), there are some variation in the mean damping ratio estimates between 100 and 200 time lag values. Particularly for the two higher model order intervals. Otherwise the 3 model order intervals are in some agreement. The lowest normalized error is found for $0 < \text{MOrder} \leq 20$ between $340 < \text{NLines} \leq 440$ at 5.83% based on 1097 estimates.

For mode 6; 2nd order bending around strong axis, see (f), there are rather large dependencies in the estimates for the 2 higher model order intervals. However, the estimates for $0 < \text{MOrder} \leq 20$ yield good results and appear to be stable for all time lag values. The lowest normalized error is found for $0 < \text{MOrder} \leq 20$ between $240 < \text{NLines} \leq 340$ at 1.99% based on 1517 estimates.

34.5 Discussion

Based on the results in the previous section there is no doubt that the damping ratio estimates are very good and that the data used are of high quality. Achieving a mean damping ratio using more than 1000 estimates that have a normalized error below 1% is far beyond expected. However, there are some element of interest that should be addressed. The estimates for mode 1 seen in Fig. 34.3a and in Table 34.2 shows a peculiar behavior for the first 150 time lag values. The 3 intervals of model order yielded different trends, and the mean damping ratio estimates does not follow a leveled straight line, which is the general case for all the other modes. This was almost gone when using a time lag starting value of 35 instead of 5. Also when more information of the correlation functions were included, this effect vanished. The cause for this must lie with the 30 time lag values that are omitted in the second data processing, since when using a time lag starting value of 35 this effect is almost gone.

Another important observation is that for mode 3 a higher mean damping ratio estimate were reported when using more time lag values. This is explained by accounting for the noise tail, a bias error is added, which is present in the estimates, see [17]. Notice that the bias is positive, yielding damping estimates that are higher than supposed to. However, only mode 3 is affected by this, which could indicate that by using the noise tail, it is not granted that all modes get a positive bias.

Furthermore, the auto-correlation function as seen in Fig. 34.2 appear to be heavily dominated by many modes for the first 100 time lag values, while only a few mode shapes remain present in the last 100 time lag values. For the modal parameter estimation method used, this is not strictly the case. By including more time lag values, the estimates are for most modes unchanged, hence some information about all modes were still there.

The perhaps most important finding is that in order to obtain the best damping ratio estimates for the Plexiglass plate, a single time lag value should not be used. Since some modes may be well represented over the course of all time lag values,

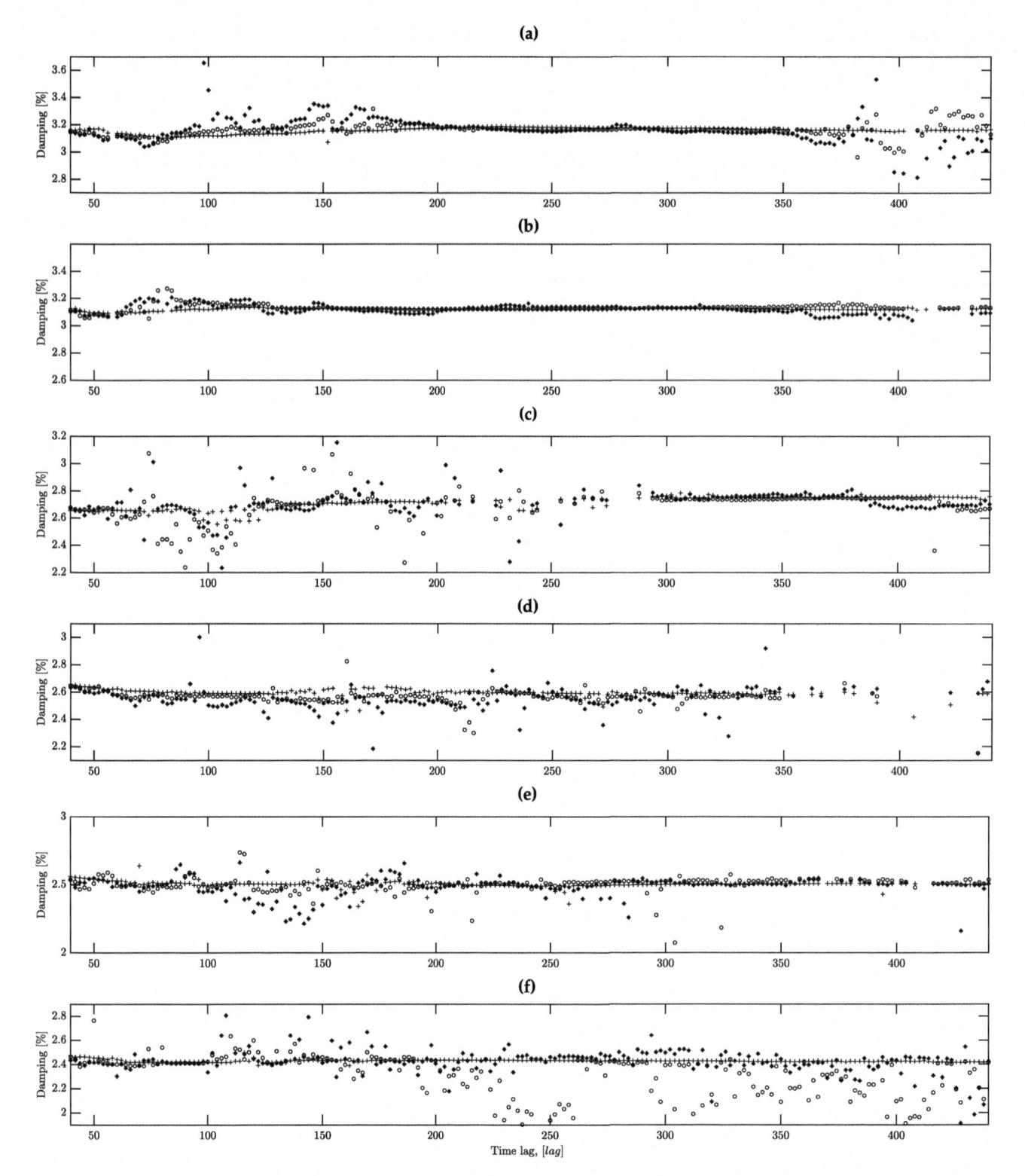

Fig. 34.4 Estimated mean damping ratio as a function of time lags used in the correlation function by omitting the first 35 time lag values. Results for the first 6 modes of the Plexiglas plate are shown. Damping ratio estimates for $0 < \text{MOrder} \leq 20$ are represented using a black +. Damping ratio estimates for $20 < \text{MOrder} \leq 40$ are represented using a black ∘. Damping ratio estimates for $40 < \text{MOrder} \leq 60$ are represented using a black ∗. (**a**) Mode 1—1st order bending around weak axis. (**b**) Mode 2—1st order torsion. (**c**) Mode 3—2nd order torsion. (**d**) Mode 4—2nd order bending around weak axis. (**e**) Mode 5—1st order bending around strong axis. (**f**) Mode 6—2nd order bending around strong axis

Table 34.3 Statistics on damping ratio estimates based on model order intervals and time lag value intervals, see Fig. 34.4 for details

		Mode 1			Mode 2			Mode 3			Mode 4			Mode 5			Mode 6		
		No. of estimates	Mean value	Normalized error	No. of estimates	Mean value	Normalized error	No. of estimates	Mean value	Normalized error	No. of estimates	Mean value	Normalized error	No. of estimates	Mean value	Normalized error	No. of estimates	Mean value	Normalized error
0 < MOrder ≤ 20	40 < Nlines ≤ 140	1472	3.13	2.92	1391	3.12	1.65	1433	2.65	16.12	1366	2.60	4.19	1317	2.52	7.54	1205	2.43	7.78
	140 < Nlines ≤ 240	1673	3.17	6.97	1621	3.12	0.30	838	2.71	4.01	1311	2.60	14.04	1331	2.49	16.51	1483	2.43	3.04
	240 < Nlines ≤ 340	1674	3.18	1.18	1644	3.12	0.30	654	2.75	5.49	1290	2.59	7.09	1458	2.49	8.33	1517	2.43	1.99
	340 < Nlines ≤ 440	1481	3.16	1.59	1417	3.12	1.15	1159	2.76	1.64	358	2.54	30.09	1097	2.50	5.83	1434	2.42	3.12
20 < MOrder ≤ 40	40 < Nlines ≤ 140	824	3.14	3.86	683	3.15	32.32	812	2.59	34.44	699	2.58	4.90	769	2.51	18.10	1199	2.45	30.12
	140 < Nlines ≤ 240	1113	3.19	8.91	1212	3.12	0.74	316	2.71	56.97	534	2.56	25.23	611	2.48	16.62	1385	2.30	52.86
	240 < Nlines ≤ 340	1070	3.16	2.12	1103	3.13	0.79	638	2.74	12.93	438	2.56	6.69	552	2.51	11.42	1144	2.19	63.49
	340 < Nlines ≤ 440	514	3.15	17.14	477	3.14	7.10	963	2.73	7.63	32	2.46	38.34	137	2.52	9.38	1538	2.20	53.76
40 < MOrder ≤ 60	40 < Nlines ≤ 140	781	3.19	51.85	588	3.12	4.84	875	2.66	39.62	866	2.57	14.99	852	2.49	19.35	1163	2.45	36.47
	140 < Nlines ≤ 240	1081	3.21	6.52	1070	3.12	2.74	366	2.70	36.39	572	2.52	21.01	594	2.50	16.23	1275	2.43	38.52
	240 < Nlines ≤ 340	1150	3.16	2.54	1139	3.13	1.66	284	2.74	10.68	391	2.56	16.22	624	2.48	18.26	862	2.46	18.82
	340 < Nlines ≤ 440	501	3.07	29.24	513	3.09	3.73	292	2.72	4.64	39	2.60	32.18	119	2.49	21.58	1072	2.34	45.33

some modes are not well represented at other time lag values. However, for model orders, the best estimates was by a large margin found in the low model order regime, $0 < \text{MOrder} \leq 20$. It is common to over-fit models to obtain more poles, but in this study it is shown that the estimates originating from the higher orders, are generally less credible.

34.6 Conclusion

The MITD method and an automated OMA algorithm have consistently extracted damping ratio estimates from measurement responses of a Plexiglass plate. Damping ratio estimates were generally best when using a low model order. Furthermore, one time lag value should not be used to obtain the optimal damping ratio estimates for all modes. The time lag starting value can have an effect on the estimates, especially when using few time lag values.

Acknowledgements The work presented is supported by the INTERREG 5A Germany-Denmark program, with funding from the European Fund for Regional Development.

References

1. Akaike, H.: A new look at the statistical model identification. IEEE Trans. Autom. Control **19**(6), 716–723 (1974)
2. Allemang, R.J., Brown, D.L.: A unified matrix polynomial approach to modal identification. J. Sound Vib. **221**(3), 301–322 (1998)
3. Bajric, A., Brincker, R., Georgakis, C.T.: Evaluation of damping using time domain OMA techniques. In: Proceedings of 2014 SEM Fall Conference and International Symposium on Intensive Loading and Its Effects. Society for Experimental Mechanics, Inc., Bethel Island (2014)
4. Bauer, D.: Order estimation for subspace methods. Automatica **37**(10), 1561–1573 (2001)
5. Brandt, A.: Noise and Vibration Analysis – Signal Analysis and Experimental Procedures. Wiley, Chichester (2011)
6. Brincker, R., Ventura, C.: Introduction to Operational Modal Analysis. Wiley, Hoboken 2015.
7. Christensen, S.S., Brandt, A.: Automatic Operational Modal Analysis Using Statistical Modelling of Pole Locations. ISMA43 (2018)
8. Fukuzono, K.: Investigation of Multiple-Reference Ibrahim Time Domain Modal Parameter Estimation Technique. Master's thesis, University of Cincinnati, Cincinnati (1986)
9. Ibrahim, S.R., Mikulcik, E.C.: A method for the direct identification of vibration parameters from the free response. Shock Vibration Bull. **47**, 183–198 (1977)
10. Juang, J., Richard, P.S.: An eigensystem realization algorithm for modal parameter identification and model reduction. J. Guid. Control Dynam. **8**(5), 620–627 (1984)
11. Ljung, L., Glover, K.: Frequency domain versus time domain methods in system identification. Automatica **17**(1), 71–86 (1981)
12. Orlowitz, E., Brandt, A.: Comparison of experimental and operational modal analysis on a laboratory test plate. Measurement **102**, 121–130 (2017)
13. Orlowitz, E., Brandt, A.: Influence of noise in correlation function estimates for operational modal analysis. In: Proceedings of 37th International Modal Analysis Conference (IMAC). Springer, Cham (2019)
14. Peeters, B., Roeck, G.D.: Stochastic system identification for operational modal analysis: a review. J. Dyn. Syst. Meas. Control. **123**(4), 659–667 (2001)
15. Rainieri, C., Fabbrocino, G.: Influence of model order and number of block rows on accuracy and precision of modal parameter estimates in stochastic subspace identification. Int. J. Lifecycle Performance Engineering **1**(4), 317–334 (2014)
16. Schoukens, J., Pintelon, R., Rolain, Y.: Time domain identification, frequency domain identification. Equivalencies! Differences? In: Proceedings of the 2004 American Control Conference. IEEE, Boston (2004)
17. Tarpo, M., Olsen, P., Amador, S., Juul, M., Brincker, R.: On minimizing the influence of the noise tail of correlation functions in operational modal analysis. Protein Eng. **199**, 1038–1043 (2017)
18. Van Overschee, P., De Moor, B.: Subspace Identification for Linear Systems–Theory–Implementation–Applications. Springer, Berlin (1996)
19. Vold, H., Kundrat, J., Rocklin, G.T., Russell, R.: A multi-input modal estimation algorithm for mini-computers. SAE Int. Congress and Exposition **91**(1), 815–821 (1982)

Chapter 35
Dynamic Bridge Foundation Identification

Nathan Davis and Masoud Sanayei

Abstract The reuse of existing bridge foundations during superstructure replacement is increasing in popularity due to the potential for time and cost savings. In many cases, an existing foundation will need to be reanalyzed for new loading or design codes for it to be suitable for another 75-year design life. While these foundations have a demonstrated history of performance, there is often limited data available from their original construction, and important details like depth or type may not be available. There is interest in the bridge engineering community for solutions to determining the details of unknown foundations, and to verify the performance of existing foundations. A method is presented that uses dynamic strain and acceleration measurements taken from substructure elements during operational loading to characterize the behavior of the underlying foundations. These measurements are used to assemble frequency response functions (FRFs) that describe the load-response behavior of the tested foundation element as a function of frequency. This method only requires that a limited number of measurements be performed on exposed portion of the bridge substructure (piers, columns piles, etc.), and can be performed during daily operational conditions. The direct result of this analysis is an updated boundary condition that be used during modeling of the superstructure in place of the fixed conditions typically used. Furthermore, this method can be used to identify foundation depth and type, verify design parameters, and update models of pile and drilled shaft behavior. A case study is presented where dynamic measurements were taken from an in-service drilled shaft foundation during operational loading.

Keywords Bridge · Dynamic · Foundation · Strains/accelerations · Frequency response function (FRF)

The Powder Mill Bridge is a three-span continuous steel girder bridge with two interior pier bents that each consist of three reinforced concrete pier columns. The columns are 920 mm wide and are each supported by a drilled shaft 1.07 m in diameter. The drilled shafts are adjacent to a dam-controlled river and the top 11.5 m of soil consist of a medium to stiff overconsolidated clay underlain by bedrock. The bridge experiences moderately frequent truck traffic, in part due to a nearby landfill. The PMB was chosen as a test case due to its accessibility, the long pier columns that allow for easy instrumentation, and the commonality of the bridge type. Strain transducers were connected to a dynamic strain gauge signal amplifier and conditioner and mounted at four points on the pier column. Four accelerometers, two on either side (one vertically and one horizontally oriented) were also mounted to the column, as shown in Fig. 35.1.

Readings were obtained from the sensors during approximately 8 h of operational traffic usage of the bridge. From these recordings, strain and acceleration time histories were retained when data exceeded a threshold level of excitation. The spectral density of these recordings were determined from the data to identify the frequencies most excited during the recorded events. The frequency response function (FRF) of the foundation was estimated at those frequencies, and averages from all the events were determined, as shown in Fig. 35.2.

The observed FRFs were compared to analytical FRFs for pile foundations calculated using equations originally derived by Novak [1]. Parameters were varied until the difference between the analytical FRF and the calculated FRF was minimized. Three parameters that were identified to most heavily affect the foundation stiffness and be differentiable from the other were identified: the pile length, the pile stiffness, and the shear modulus of the soil. These parameters were identified to be relatively close or in the range of their predicted values, as shown in Table 35.1.

N. Davis (✉) · M. Sanayei
Department of Civil and Environmental Engineering, Tufts University, Medford, MA, USA
e-mail: nathan.davis@tufts.edu; masoud.sanayei@tufts.edu

Fig. 35.1 Instrumentation setup at the PMB for two sides of the same column (**a, b**)

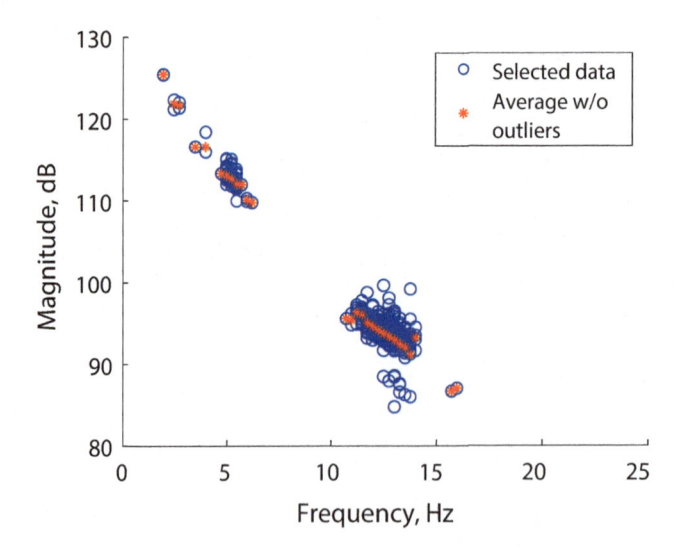

Fig. 35.2 Foundation FRFs measured at the PMB

The case study presented in this paper was intended to identify the ability of the proposed method to identify foundation parameters. The identified parameters appear to be reasonable but were not confirmed with measurements taken from the bridge. Further research is expected to identify the sensitivity of the proposed method to bridge hazards such as scour or structural deterioration.

Table 35.1 Converged parameters for PMB foundation

Parameter	Expected value	Converged value
Pile length (Lp)	11.5 m	12.0 m
Pile stiffness (Ep)	25.7 GPa	20.6 MPa
Soil shear modulus (G)	8–20 MPa	11.1 MPa

Reference

1. Novak, M.: Dynamic stiffness and damping of piles. Can Geotech. J. **11**(4), 574–598 (1974)

Chapter 36
Damping Ratios of Reinforced Concrete Structures Under Actual Ground Motion Excitations

Dan Lu, Jiayao Meng, Songhan Zhang, Yuanfeng Shi, Kaoshan Dai, and Zhenhua Huang

Abstract Structural damping ratio which quantifies the energy dissipation of civil structures under external excitations plays a critical role in the seismic design and assessment of civil structures. In existing building design provisions and guidelines, however, the structural damping ratio is only suggested either as a single fixed value or as an optional value for the general structure type adopted. For example, damping ratio 5% is commonly recommended for all reinforced concrete (RC) structures in practical seismic design, which may not be sufficient to represent the realistic damping features of different RC structures under ground motions with different amplitudes. This research explored deeper understandings on the structural damping features of different RC structures under actual ground motion excitations. A series of seismic response records of RC structures were collected from the "Center for Engineering Strong Motion Data" (CESMD) database. These records were then categorized into three typical lateral resisting systems: moment-resisting frame systems, shear wall systems, and moment-resisting frame plus shear wall systems. The equivalent structural damping ratios for different systems of RC structures were then estimated based on the categorized response records with different amplitudes. Finally, an empirical statistical relationship was established, offering a refined basis for civil engineers to reasonably choose the equivalent damping ratios during the design and post-earthquake assessment of the RC structures.

Keywords Damping ratio · Modal parameter identification · Nonlinear · Strong motion observation · Empirical formulas

36.1 Introduction

In recent years, the number of aged buildings suffering from natural disasters has shown an increasing trend. This increasing trend has stimulated a rapid development of the state-of-the-art devices for vibration control. The application of such devices requires the reliable information of structural damping ratio, which is a commonly used physical parameter to characterize the energy dissipation of structural dynamic response. However, the actual structural damping ratio for existing and new-designed buildings subject to earthquake and wind loading is difficult to be precisely quantified because of its complicated mechanism, which has become a troublesome issue in optimizing structural designs [1].

Unlike characterizing the structural stiffness and mass properties, which can be intuitively evaluated from the material and geometric features, quantifying the structural damping ratio is much more complicated. Currently, the structural damping ratio of civil structures is usually obtained by experimental/field tests, which measure the responses of structures to different excitations [2], such as shake tables, wind loadings, environment loadings, or strong earthquakes [3]. Among these experiments, the shake table test uses excessively idealized models, therefore it cannot exactly represent real structures.

D. Lu · S. Zhang · Y. Shi
Department of Civil Engineering, Institute for Disaster Management and Reconstruction, Sichuan University, Chengdu, China

J. Meng
State Key Laboratory of Disaster Reduction in Civil Engineering, Tongji University, Shanghai, China

K. Dai (✉)
Department of Civil Engineering, Institute for Disaster Management and Reconstruction, Sichuan University, Chengdu, China

State Key Laboratory of Disaster Reduction in Civil Engineering, Tongji University, Shanghai, China
e-mail: kdai@scu.edu.cn

Z. Huang
Department of Engineering Technology, University of North Texas, Denton, TX, USA
e-mail: zhenhua.huang@unt.edu

Since the 1970s, a large number of dynamic response records of real civil structures have been collected for the analysis of structural damping ratios. The influences of construction materials, structure dimensions, foundation types, vibration orientations, vibration amplitudes, excitation types, non-structural components, etc. on the structural damping ratio have been studied. These studies resulted in the currently recommended design values of structural damping ratios in building design provisions of different countries. For example, in the United States, the structural damping ratio of steel and RC (reinforced concrete) structures can be calculated as an inverse proportion to the structural height [4]. In China, the structural damping ratio is given as a few fixed numbers for several cases, such as 0.05 for normal case, 0.04 for steel structures less than 50 m, 0.03 and 0.02 for structures less and higher than 200 m, respectively, 0.05 for plastic analysis of earthquake [5], 0.04 for concrete high-rise buildings under multiple earthquakes, 0.02 for wind vibration of high-rise buildings, and 0.01–0.02 for mixed structures [6].

Even though the above studies and design provisions have involved many influence factors for structural damping ratios. There are still many concerns such as: (1) the mechanism of damping energy dissipation in real structure remains unclear. Limited research has been conducted in this field. Wyatt [7] and Davenport [8] proposed Stiction and Stick-Slip theories, respectively, to explain the damping energy dissipation mechanism, which depicted a close relationship between the deformation of structural members and the damping energy dissipation. Based on these theories, Spence and Kareem [9] proposed a joint-probability amplitude-dependent damping model, which considered the viscous damping in structural materials and the Coulomb friction damping between components for the low-amplitude motion. (2) The tested structural damping ratios showed high dispersion. The relative errors for the estimated damping ratios are more than 50 times higher than that for the estimated frequencies [10]. (3) The structural damping ratio is very sensitive to identification methods. Previous studies have proved that the reliability of the half-power bandwidth method, a widely used method in early studies, is low. Therefore, the modal analysis and system identification method have been widely adopted in recent years.

Apparently, the study of structural damping ratio is still far from comprehensive and satisfactory. This research explores deep understandings on the structural damping features of different RC structures under actual ground motion excitations. The field response data of several RC structures collected by the CESMD are analyzed. An empirical formula for the equivalent structural damping ratio of RC structures is proposed statistically for practical use, which including the three common types of RC structures (the RC frame structures, the RC shear wall structures, and the RC frame-shear wall structures).

36.2 Parameter Identification

In this section, the theoretical analysis procedure of this study is introduced. Figure 36.1 showed the analytical model of a multi-story building under ground motion $v_b(t)$.

This model made the following assumptions:

1. The number of modes contributing to responses is no more than the number of sensors.
2. No external loading is applied to the building in addition to the ground motion.
3. The damping forces are linear combinations of the absolute velocity of each floor and the relative velocity between adjacent floors.

The equilibrium of each floor (Fig. 36.1b–d) yields equation (36.1a, 36.1b, 36.1c: from top to bottom):

$$\begin{cases} -m_1\ddot{v}_1^{(m)} - c_1^{(m)}\dot{v}_1^{(m)} + k_1\left[v^{(b)} - v_1^{(m)}\right] + c_1^{(s)}\left[\dot{v}^{(b)} - \dot{v}_1^{(m)}\right] + k_2\left[v_2^{(m)} - v_1^{(m)}\right] + c_2^{(s)}\left[\dot{v}_2^{(m)} - \dot{v}_1^{(m)}\right] = 0 \\ -m_j\ddot{v}_j^{(m)} - c_j^{(m)}\dot{v}_j^{(m)} + k_j\left[v_{j-1}^{(m)} - v_j^{(m)}\right] + c_j^{(s)}\left[\dot{v}_{j-1}^{(m)} - \dot{v}_j^{(m)}\right] + k_{j+1}\left[v_{j+1}^{(m)} - v_j^{(m)}\right] + c_{j+1}^{(s)}\left[\dot{v}_{j+1}^{(m)} - \dot{v}_j^{(m)}\right] = 0 \\ -m_n\ddot{v}_n^{(m)} - c_n^{(m)}\dot{v}_n^{(m)} + k_n\left[v_{n-1}^{(m)} - v_n^{(m)}\right] + c_n^{(s)}\left[\dot{v}_{n-1}^{(m)} - \dot{v}_n^{(m)}\right] = 0 \end{cases} \quad (36.1)$$

Physical interpretations of each notation in Eq. (36.1) are described in Fig. 36.1b–d. Equation (36.1) represents the motion of the building floors. The unique solution of Eq. (36.1) implies that the equation has an inhomogeneous form. Dividing Eq. (36.1a) by m_1 and moving $\ddot{v}_1^{(m)}$ to the right-hand side of the equation yields:

$$-\frac{c_1^{(m)}}{m_1}\dot{v}_1^{(m)} + \frac{k_1}{m_1}\left[v^{(b)} - v_1^{(m)}\right] + \frac{c_1^{(s)}}{m_1}\left[\dot{v}^{(b)} - \dot{v}_1^{(m)}\right] + \frac{k_2}{m_1}\left[v_2^{(m)} - v_1^{(m)}\right] + \frac{c_2^{(s)}}{m_1}\left[\dot{v}_2^{(m)} - \dot{v}_1^{(m)}\right] = \ddot{v}_1^{(m)} \quad (36.2)$$

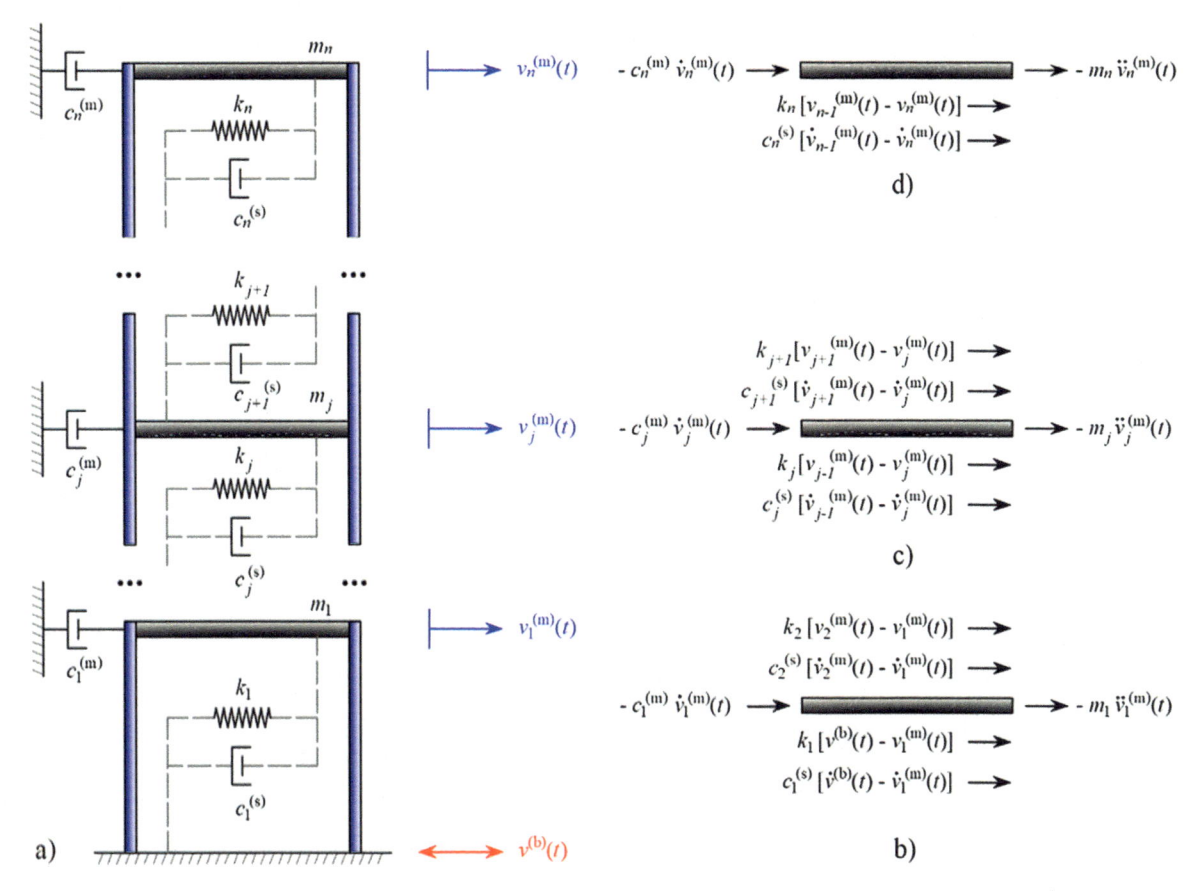

Fig. 36.1 A shear-type building under ground motion: (**a**) model of the building; equilibriums of (**b**) the first floor, (**c**) the ith floor and (**d**) the top floor

The unknown coefficients in Eq. (36.2) can be calculated by the Least Square Method as:

$$\overline{\mathbf{C}}^{(1)} = \left\{ \overline{c}_1^{(m)}/m_1 \ \overline{k}_1/m_1 \ \overline{c}_1^{(s)}/m_1 \ \overline{k}_2/m_1 \ \overline{c}_2^{(s)}/m_1 \right\}^{\mathrm{T}} \tag{36.3}$$

Dividing Eq. (36.1b) by m_1 on both sides yields its inhomogeneous form, Eq. (36.4), for $j = 2, 3, \cdots, n-1$:

$$-\frac{m_j}{m_1}\ddot{v}_j^{(m)} - \frac{v_j^{(m)}}{m_1}\dot{v}_j^{(m)} + \frac{k_{j+1}}{m_1}\left[v_{j+1}^{(m)} - v_j^{(m)}\right] + \frac{c_{j+1}^{(s)}}{m_1}\left[\dot{v}_{j+1}^{(m)} - \dot{v}_j^{(m)}\right] = -\frac{k_j}{m_1}\left[v_{j-1}^{(m)} - v_j^{(m)}\right] - \frac{c_j^{(s)}}{m_1}\left[\dot{v}_{j-1}^{(m)} - \dot{v}_j^{(m)}\right] \tag{36.4}$$

By substituting the calculated values of parameters $\overline{k}_2/m_1$ and $\overline{c}_2^{(s)}/m_1$ from Eq. (36.2) into the right-hand side of Eq. (36.4) for $j = 2$, the remaining parameters in Eq. (36.4) can be calculated iteratively until $j = n - 1$. For the jth floor, the unknown coefficients, by the Least Square Method, are:

$$\overline{\mathbf{C}}^{(j)} = \left\{ -\overline{m}_j/m_1 \ -\overline{c}_j^{(m)}/m_1 \ \overline{k}_{j+1}/m_1 \ \overline{c}_{j+1}^{(s)}/m_1 \right\}^{\mathrm{T}} \tag{36.5}$$

For the top floor ($j = n$), dividing both sides of Eq. (36.1c) by m_1 yields:

$$-\frac{m_n}{m_1}\ddot{v}_n^{(m)} - \frac{v_n^{(m)}}{m_1}\dot{v}_n^{(m)} = = -\frac{k_n}{m_1}\left[v_{j-1}^{(m)} - v_j^{(m)}\right] - \frac{c_n^{(s)}}{m_1}\left[\dot{v}_{n-1}^{(m)} - \dot{v}_n^{(m)}\right] \tag{36.6}$$

The parameters $\bar{k}_n/m_1$ and $\bar{c}_n^{(s)}/m_1$ can be calculated by substituting Eq. (36.4) for $j = n - 1$ into the right-hand side of Eq. (36.6). The unknown coefficients in Eq. (36.6), by the Least Square Method, are:

$$\overline{\mathbf{C}}^{(n)} = \left\{ -\overline{m}_n/m_1 \quad -\overline{c}_n^{(m)}/m_1 \right\}^{\mathrm{T}} \tag{36.7}$$

Using Eq. (36.3), Eq. (36.5) and Eq. (36.7), all the unknown physical parameters of the building model in Fig. 36.1 can be obtained. Then, the normalized mass matrix $\mathbf{M}$, damping matrix $\mathbf{C}$, and stiffness matrix $\mathbf{K}$ of the model can be reconstructed as:

$$\overline{\mathbf{M}} = m_1 \mathbf{M} = m_1 \begin{bmatrix} \overline{m}_1 & & & \\ & \overline{m}_2 & & \\ & & \ddots & \\ & & & \overline{m}_n \end{bmatrix} \tag{36.8}$$

$$\overline{\mathbf{C}} = m_1 \mathbf{C} = m_1 \begin{bmatrix} \bar{c}_1^{(s)} + \bar{c}_2^{(s)} & -\bar{c}_2^{(s)} & 0 & \cdots & 0 \\ -\bar{c}_2^{(s)} & \bar{c}_2^{(s)} + \bar{c}_3^{(s)} & -\bar{c}_3^{(s)} & & \vdots \\ 0 & \ddots & \ddots & \ddots & 0 \\ \vdots & & -\bar{c}_{n-1}^{(s)} & \bar{c}_{n-1}^{(s)} + \bar{c}_n^{(s)} & -\bar{c}_n^{(s)} \\ 0 & \cdots & 0 & -\bar{c}_n^{(s)} & \bar{c}_n^{(s)} \end{bmatrix} + m_1 \begin{bmatrix} \bar{c}_1^{(m)} & & & \\ & \bar{c}_2^{(m)} & & \\ & & \ddots & \\ & & & \bar{c}_n^{(m)} \end{bmatrix} \tag{36.9}$$

$$\overline{\mathbf{K}} = m_1 \mathbf{K} = m_1 \begin{bmatrix} k_1 + k_2 & -k_2 & 0 & \cdots & 0 \\ -k_2 & k_2 + k_3 & -k_3 & & \vdots \\ 0 & \ddots & \ddots & \ddots & 0 \\ \vdots & & -k_{n-1} & k_{n-1} + k_n & -k_{n-1} \\ 0 & \cdots & 0 & -k_{n-1} & k_n \end{bmatrix} \tag{36.10}$$

The free vibration of the building model can then be expressed using these reconstructed matrices as Eq. (36.11):

$$\mathbf{M}\ddot{\mathbf{X}} + \mathbf{C}\dot{\mathbf{X}} + \mathbf{K}\mathbf{X} = 0 \tag{36.11}$$

Usually, the reconstructed damping matrix is not proportional, i.e. $\mathbf{C} \neq \alpha \mathbf{M} + \beta \mathbf{K}$. In order to solve for the modal parameters, a state-space form of Eq. (36.11) is constructed as:

$$\mathbf{P}\dot{\mathbf{U}} + \mathbf{Q}\mathbf{U} = 0 \tag{36.12}$$

where $\mathbf{P} = \begin{bmatrix} \mathbf{C} & \mathbf{M} \\ \mathbf{M} & 0 \end{bmatrix} \in \mathbb{R}^{2n \times 2n}$, $\mathbf{Q} = \begin{bmatrix} \mathbf{K} & 0 \\ 0 & -\mathbf{M} \end{bmatrix} \in \mathbb{R}^{2n \times 2n}$, $\mathbf{U} = \begin{bmatrix} \mathbf{X} \\ \dot{\mathbf{X}} \end{bmatrix} \in \mathbb{R}^{2n \times 1}$

Assuming that the solution of Eq. (36.12) is in the form of $\mathbf{U} = \lambda \mathbf{\Phi}$, then Eq. (36.11) is written as:

$$\mathbf{P}\mathbf{\Phi} = -\lambda \mathbf{Q}\mathbf{\Phi} \tag{36.13}$$

The eigenvalue of Eq. (36.13) can be solved as:

$$\lambda_j = -\xi_j \omega_j \pm i\omega_j \sqrt{1 - \xi_j^2}, \quad (j = 1, 2, \cdots, n)$$

where ω_j is the undamped circular frequency of the jth mode, and the structural damping ratio can be calculated as:

$$\bar{\xi}_j = \frac{|\mathrm{Re}(\lambda_j)|}{|\lambda_j|} \tag{36.14}$$

The eigenvector of Eq. (36.13) is:

$$\Phi_j = \left\{ \lambda_j \varphi_j \ \varphi_j \right\}^T \in \mathbb{C}^{2n \times 1} \tag{36.15}$$

where $\varphi_j \in \mathbb{C}^{n \times 1}$ represents the jth complex mode.

36.3 Case Example

In this section, a typical case example is illustrated for the modal analysis method introduced in Sect. 36.2. This case example analyzed a 10-story shear wall structure (CSMIP station No. 24385) as shown in Fig. 36.2. Precast concrete shear walls are the main transverse resistance system of the structure. The dynamic responses of the building to earthquakes are measured by accelerometers. The tag numbers and locations of these accelerometers are shown in Fig. 36.2. Earthquake records for 4.4 Mw Encino Earthquake, on 17th March 2014 measured by accelerometers No.10, No.11, No.12, and No.16 were selected for this study. No.16 accelerometer measured the excitation signal (Fig. 36.3a) and the other three measured the responses of the building (Fig. 36.3b–d).

By integrating the measured acceleration responses in a frequency domain, velocity and displacement responses of the building were obtained as shown in Figs. 36.4 and 36.5, respectively. During this integrating process, the components of the acceleration responses below 0.16 Hz frequency, which is much lower than the natural frequency of the building, were filtered out in order to eliminate trend effects.

By applying these acceleration, velocity, displacement responses to the analysis procedure described in Sect. 36.2, the modal parameters of the building (frequency, modes, damping) were identified. While analyzing, each time-domain signal was divided into several sub-signals based on different time windows to study the time-varying effect on the damping

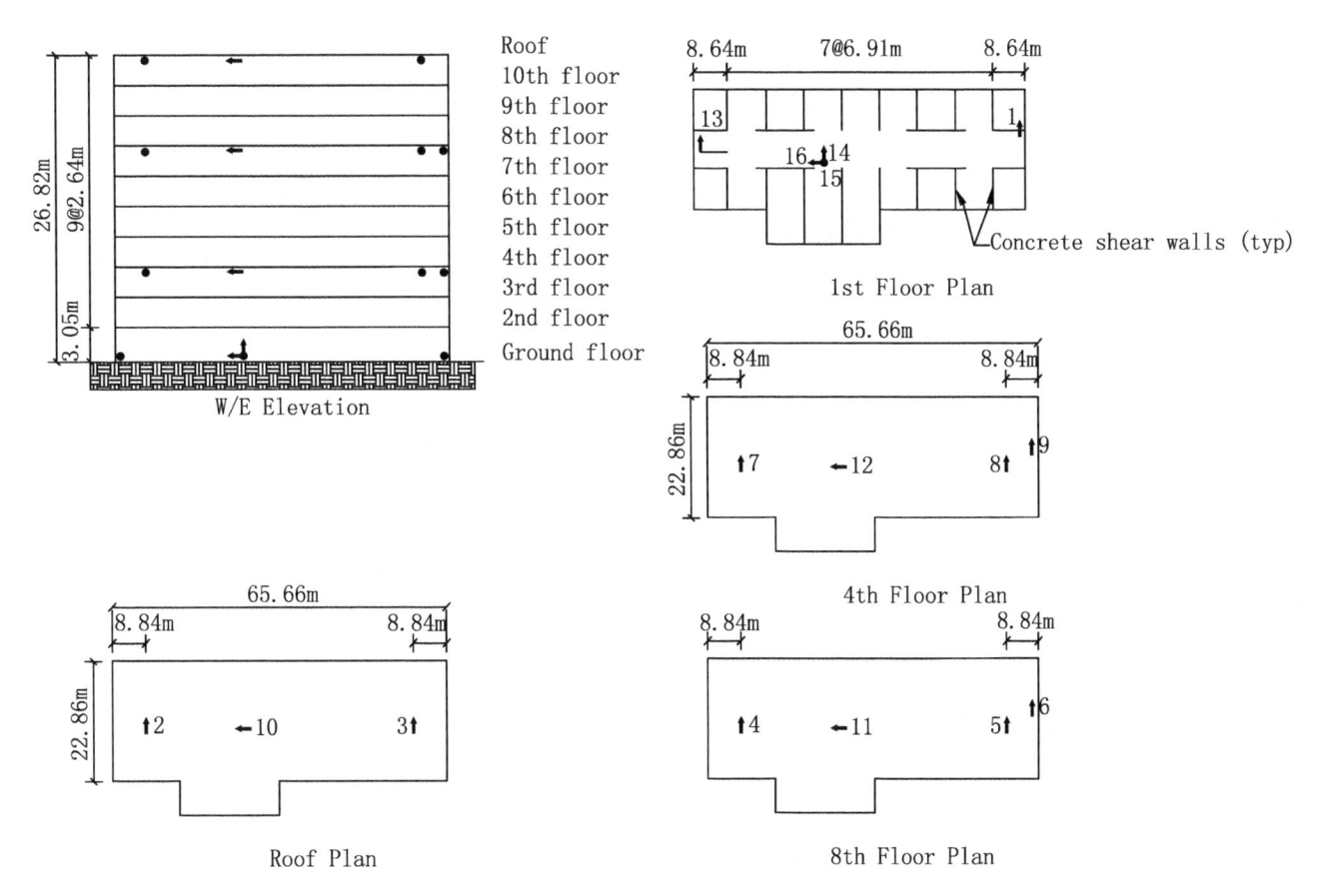

Fig. 36.2 Location of the accelerometers [11]

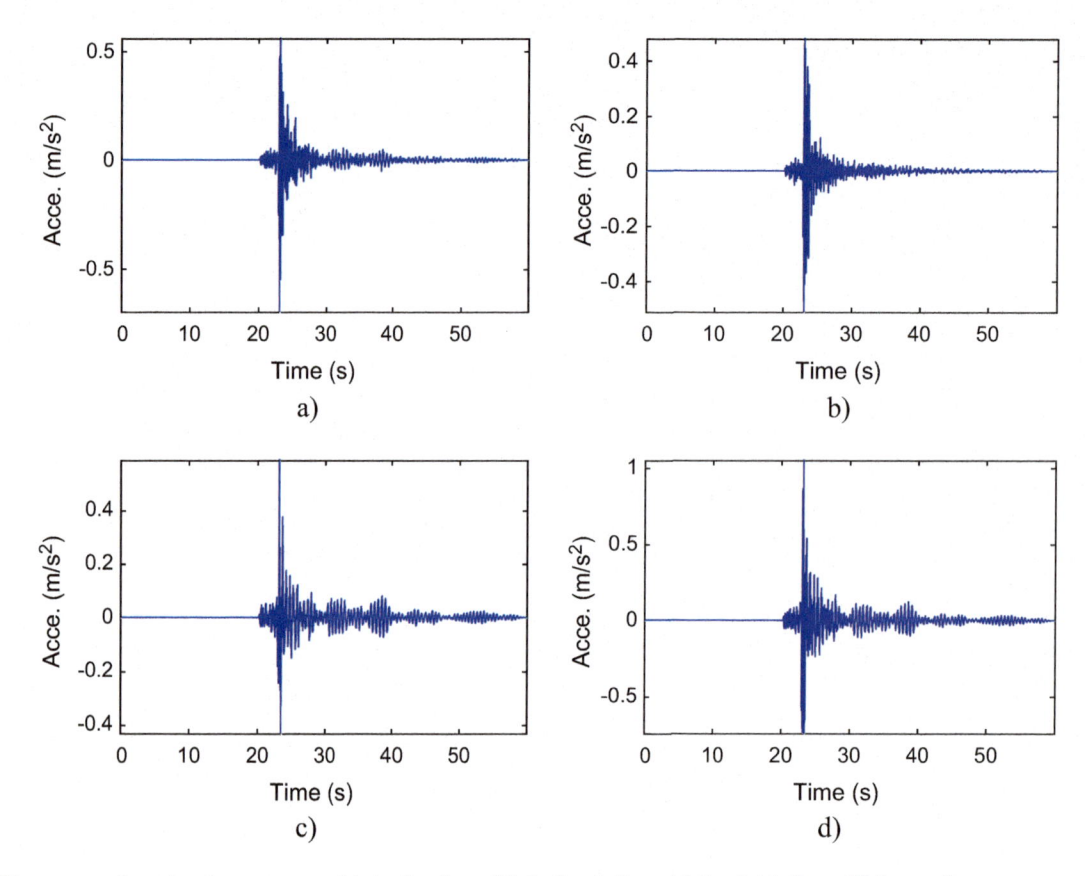

Fig. 36.3 The measured acceleration responses: (**a**) the first floor; (**b**) the fourth floor; (**c**) the eighth floor; (**d**) the top floor

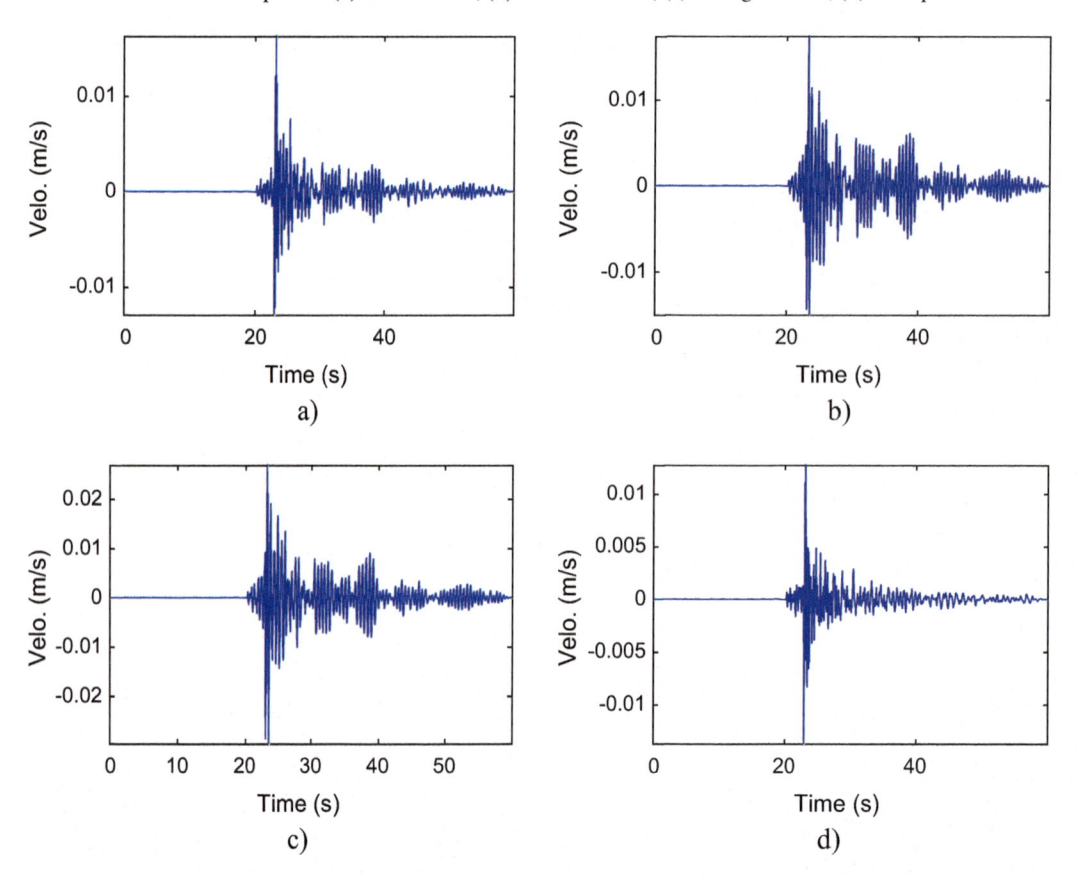

Fig. 36.4 The velocity responses obtained by the frequency-domain integral: (**a**) the first floor; (**b**) the fourth floor; (**c**) the eighth floor; (**d**) the top floor

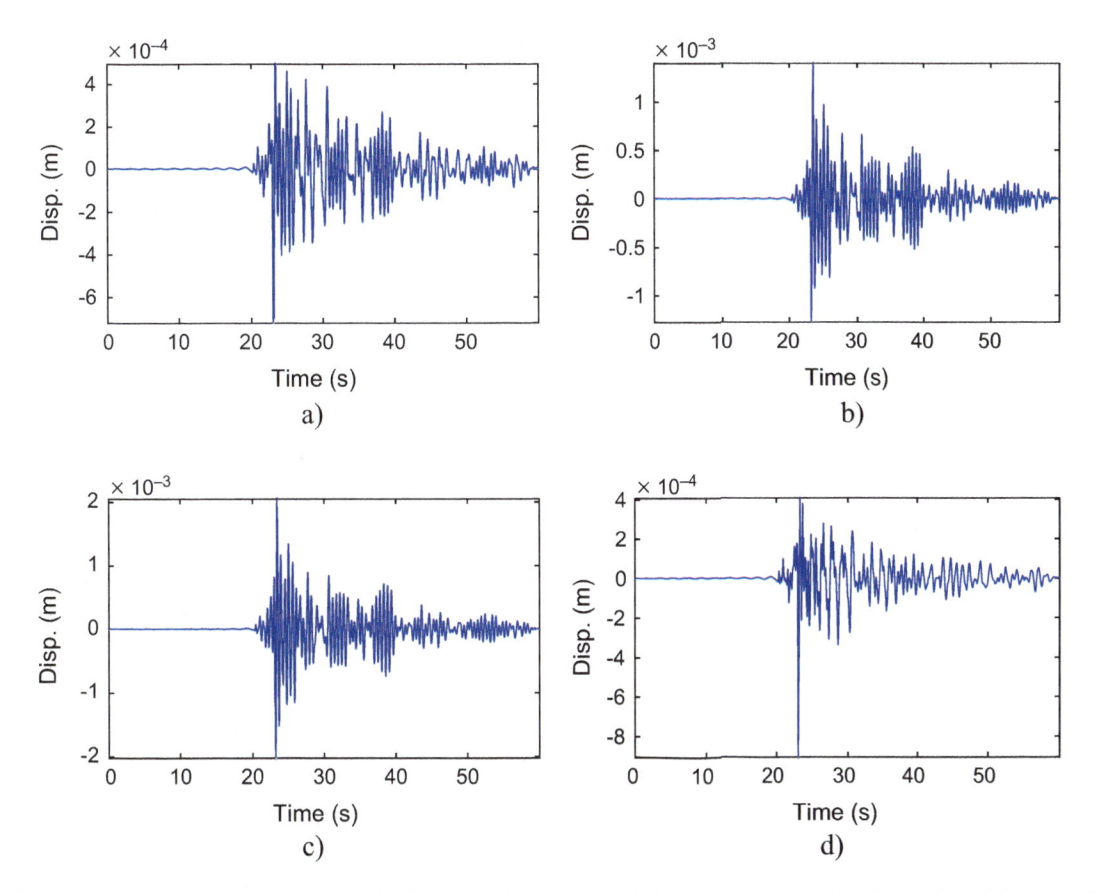

Fig. 36.5 The displacement responses obtained by the frequency-domain integral: (**a**) the first floor; (**b**) the fourth floor; (**c**) the eighth floor; (**d**) the top floor

Table 36.1 The mode frequencies and damping ratios

Mode	f (Hz)	ξ (%)
1	2.486	4.307
2	6.909	2.508
3	9.484	2.432

behavior with different PGAs (Peak Ground Accelerations). Special attention needs to be paid on two issues when defining the length of the time window function: (1) the length of the time window function should be as short as possible in order to reveal the time-varying effect with a high resolution, (2) the length of the time window should yield $t_d \geq 20/f_1$ (where f_1 is the fundamental frequency of the structure) in order to ensure that the characteristics of the fundamental mode can be captured. Therefore, the length of the time window as $t_d = 10$ s was used, which includes approximately 2000 sampling points. Table 36.1 and Fig. 36.6 showed the analysis results of natural frequencies, damping ratios, and mode shapes of the building for one example sub-signal (20 30 s) with PGA of 0.512 m/s^2.

Each sub-signal can result in a pair of damping ratio and PGA data. With a large group of response records on a variety of buildings, the relationship between the structural damping ratio and PGA can be revealed statistically. The procedure is introduced as follows.

36.4 Statistical Investigation of Damping Ratio

Based on analyses of a variety of buildings under various earthquakes, an empirical relationship between the first mode structural damping ratio and the PGA for RC structures was revealed statistically. In order to obtain a reasonable statistic relationship, a large database of dynamic response records on different buildings under different earthquakes is required. Therefore, this study selected 258 dynamic response records on 45 RC structures from the CESMD data center. Among the

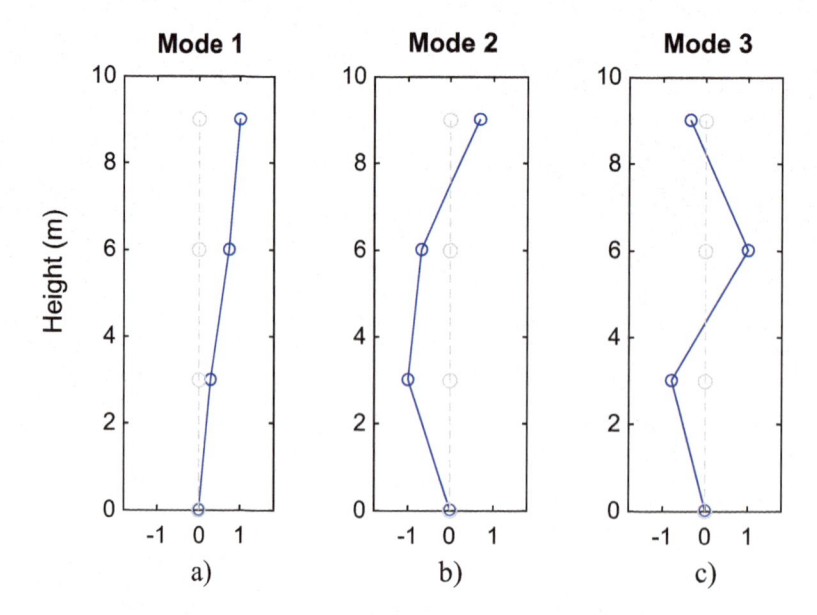

Fig. 36.6 The identified mode shapes: (**a**) the first mode; (**b**) the second mode; (**c**) the third mode

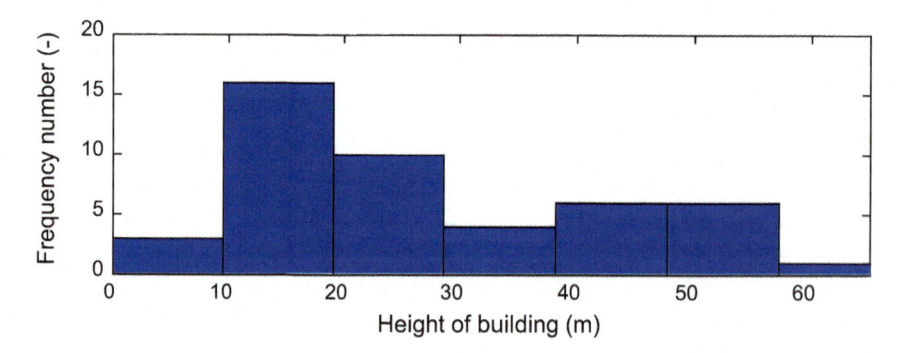

Fig. 36.7 Height distribution histogram for of RC buildings

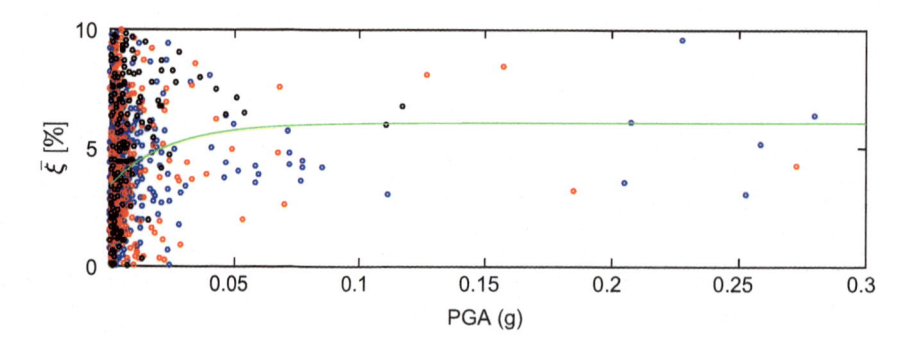

Fig. 36.8 The estimated damping ratios in terms of different PGA values

258 records, 162 are on shear-wall structures (25), 30 are on moment-frame structures (7), and 66 are on moment-frame plus shear-wall structures (14). Figure 36.7 showed the height histogram of the selected buildings. All these selected buildings belong to low to medium height buildings. Because of this, the variation trend of the structural damping ratio corresponding to different building heights cannot be obtained in this study. Therefore, this study focused on the relationship between the structural damping ratio and the PGA for RC buildings with relatively similar height.

Figure 36.8 showed the identified results of the structural damping ratio vs. PGA. The blue, red, and black dots in the figure represent the shear wall structures, the frame-shear wall structures, and the frame structures, respectively. As shown, the dots for PGA less than $0.01g$ are randomly distributed. The possible reasons are: (1) the building dynamic responses for these dots are weak and significantly polluted by noises before earthquake arrives, and (2) the accelerometers used in this study are designed for measuring high-amplitude responses which have wide measurement range but coarse resolution. These accelerometers are unable to capture weak dynamic responses precisely and result in unacceptable errors in the calculated damping ratios. Therefore, the data dots in Fig. 36.8 with PGA less than 0.0015 m/s^2 were filtered out in this study. Besides, because this paper is just a starting point for this damping ratio study, statistical analyses based on different structure types are not demonstrated here. Only the relationship between the structural damping ratio and PGA for generalized RC structures are reported in this paper.

To obtain a regression curve for the damping ratio vs. PGA relationship, this study selected the exponential function, based on existing literature and constant experiments, which can be written as:

$$\xi = c_1 - c_2 e^{\frac{-PGA}{c_3}} \tag{36.16}$$

The fitting error of the regression curve was defined as square root sum square of all errors. By minimizing the fitting error, the parameters c_1, c_2, c_3 in Eq. (36.16) were calculated as 6.05, 2.65, 0.023, respectively. The green curve in Fig. 36.8 represents the regression results. Based on Fig. 36.8, it can be concluded that the damping ratio of RC structures increases with respect to PGA increases when the PGA value is small, and it yields a constant value (6.05%) when the PGA value reaches approximately 0.1 g. This result is very different from the fixed values recommended in current RC building design provisions.

36.5 Conclusion

This research explored deeper understandings on the structural damping features of different RC structures under actual ground motion excitations. A time-domain Least Square Method was developed and applied for estimating the physical parameters of RC structures, which allows identifying the structure's modal parameters by reconstructing the normalized stiffness and mass matrices. The developed method requires few sampling points, which allows studying the time-vary effect of damping behavior by shortening the window function.

As a case study, this research analyzed an extensive large number of real building dynamic response records. A reliable statistical relationship between the equivalent damping ratio and the peak ground motions (PGA) was validated. An empirical statistic curve describing the structural damping ratio of RC buildings with respect to PGA was proposed based on nonlinear regression analyses.

The proposed empirical results can be applied not only in a post-earthquake assessment of building structures, but also in a dynamic analysis for structural seismic design. However, the study of structural damping behavior is far from comprehensive and satisfactory. The improvement of damping models and the refinement of statistical analyses will be the future focus.

Acknowledgements The authors would like to acknowledge the support from the International Collaboration Program of Science and Technology Commission of Ministry of Science and Technology, China (Grant No. 2016YFE0105600), the International Collaboration Program of Science and Technology Commission of Shanghai Municipality and Sichuan Province (Grant No. 16510711300, 18GJHZ0111), the 111 Project (Grant No. B18062), the National Natural Science Foundation of China (Grant No. U1710111, 51878426), and the Fundamental Research Funds for Central Universities of China.

References

1. Mao, W.: Research on structural damping in a shaking table test of a high-rise building. J Wuhan Univ Technol. **32**(6), 1671–2431 (2010). (in Chinese)
2. Cruz, C., Miranda, E.: Evaluation of damping ratios for the seismic analysis of tall buildings. J Struct Eng. **143**(1), 04016144 (2016)
3. Wang, Z.: Study and application on damping model of reinforced concrete frame-shear wall structures. PhD thesis, Hunan University (2016) (in Chinese)
4. Pacific Earthquake Engineering Research Center/Applied Technology Council: Interim Guidelines on Modeling and Acceptance Criteria for Seismic Design and Analysis of Tall Buildings: PEER/ATC 72-1[S]. PEER/ATC, Redwood City (2010)
5. China Academy of Building Research: Code for seismic design of buildings: GB50011-2010[S]. China Architecture & Building Press, Beijing (2010). (in Chinese)

6. Ministry of Housing and Urban-Rural Development of the People's Republic of China: Technical Specification for Concrete Structures of Tall Building: JGJ3-2010[S]. China Architecture & Building Press, Beijing (2010). (in Chinese)
7. Wyatt, T.A.: Mechanisms of damping. In: Symposium on Dynamic Behaviour of Bridges, vol. 05, pp. 10–21 (1997)
8. Davenport, A.G., Hill-Carroll, P.: Damping in Tall Buildings: Its Variability and Treatment in Design, pp. 42–57. Building Motion in Wind ASCE, New York (1986)
9. Spence, S.M.J., Kareem, A.: Tall buildings and damping: a concept-based data driven model. J Struct Eng. **140**(5), 155–164 (2014)
10. Bernal, D., Döhler, M., Kojidi, S.M., et al.: First mode damping ratios for buildings. Earthq. Spectra. **31**(1), 367–381 (2015)
11. https://strongmotioncenter.org/NCESMD/photos/CGS/bldlayouts/bld24385.pdf

Chapter 37
Launching Semi-automated Modal Identification of the Port Mann Bridge

A. Mendler, C. E. Ventura, L. Nandimandalam, and Y. Kaya

Abstract Southwest British Columbia (B.C.) is home to many of the largest bridges in North America, and the majority of its *lifeline* bridges are cable-stayed. It is also one of the most active seismic zones in Canada and the world. To help face the obvious challenges that this poses for structural engineers, the Ministry of Transportation embarked on a province-wide monitoring program in 2009, called the B.C. Smart Infrastructure Monitoring System (BCSIMS). It combines the vibration data from an urban strong motion network and a structural health monitoring network, and pursues two objectives: (a) to enable an efficient emergency response by implementing a post-earthquake damage assessment module, and (b) to install a long-term monitoring module for cost-effective operation over the entire lifespan of structures. This paper contributes to the post-earthquake damage assessment module and focuses on the modal identification of the Port Mann Bridge using the stochastic subspace identification (SSI) method. Moreover, a hierarchical clustering approach is used to automatically select stable modes of vibration. The envisaged damage assessment module is based on subspace-based methods, and the modal parameter estimation is merely a side product of the ongoing studies. Nonetheless, it is an integral part of every structural health monitoring network and allows first insights into the influence of environmental/operational variables on the dynamic behaviour of the bridge when, for example, combined with regression analysis. No bridge weigh-in-motion system is installed, but a strong correlation can be found between a traffic index estimated from toll cameras and the fluctuation of modal frequencies. This highlights the dominant effect of traffic loads over other environmental variables.

Keywords BCSIMS · Port Mann Bridge · Stochastic subspace identification · Environmental/operational variables · Regression

37.1 Introduction

For the Southwest of British Columbia (B.C.), the development of a reliable structural health monitoring (SHM) network for bridges is of utmost importance. Geologic evidence has shown that the tectonic fault lines about 150 km off the coast of B.C. generate earthquakes with Richter magnitudes up to 9.0 every 300–700 years, with the last one occurring in 1700 [1]. At the same time, 45% of B.C.'s population concentrates in Metro Vancouver, a sea port city that is cut off from the rest of Canada, with the only infrastructural links being the two bridges across the Pacific inlet to the North, and the five bridges across the Fraser River delta to the South. In the event of an earthquake, the success and safety of emergency response and evacuation services depend on the structural health state of each of these bridges, and the ability to assess them quickly and reliably. This way, bridge closures can be minimized, and time-consuming inspections can be prioritized based on their performance.

Vibration-based structural health monitoring is a promising technology which helps to realize the vision of a real-time monitoring system. However, numerous vibration sensors on each bridge are required to be able to record and interpret the vibrational behavior of bridges. In 2009, the Ministry of Transportation & Infrastructure embarked on a monitoring program called the B.C. Smart Infrastructure Monitoring System (BC-SIMS) [2]. The resulting online platform[1] combines two networks. On one hand, it makes available the data of a strong motion network including 162 earthquake stations,

---
[1]BCSIMS homepage: www.bcsims.ca.

A. Mendler (✉) · C. E. Ventura · L. Nandimandalam
Department of Civil Engineering, University of British Columbia, Vancouver, BC, Canada
e-mail: alexander.mendler@ubc.ca

Y. Kaya
B.C. Ministry of Transportation and Infrastructure, Coquitlam, BC, Canada

and offers several tools to track historical earthquakes, simulate future events, or issue seismological reports right after an earthquake has occurred. On the other hand, it incorporates a structural health monitoring network connecting 16 bridges to a central server.

An efficient modal identification algorithm is an integral part of every structural health monitoring network. It evaluates the characteristic vibration behavior of bridges in regular intervals and stores all modal parameters, *i.e.* natural frequencies, mode shapes, damping ratios, and (relative) modal participation factors etc. Changes in the modal parameters can be correlated with the presence of damage [3]. However, research has shown that natural frequencies can be very insensitive to minor damage scenarios [4], and the fluctuation of environmental and operational variables (EOVs), such as temperature or traffic fluctuations, can impact the dynamic parameters to an equal extent [5]. This paper aims to do three things: (a) to validate the developed modal identification algorithm, (b) to define the baseline modal parameters of the Port Mann Bridge, and (c) to indicate which environmental/operational factors influence the vibrational response most significantly.

37.2 The Port Mann Bridge

37.2.1 Importance Categorization

The Port Mann Bridge is an extraordinary bridge from both a political and engineering perspective. As per the Canadian Highway Bridge Design Code, it is classified as a *lifeline* bridge, meaning it is deemed vital to the integrity of the regional transportation network and stands out due to its importance with respect to social, economic and security aspects. Moreover, this means that its seismic design is extremely robust. In the event of a major earthquake (with a return period of 2475 years), any post-earthquake damage has to be repairable and the serviceability may be limited but must still be guaranteed [6].

37.2.2 Structural System

The structural system is complex and appears heavy. With a width of ten traffic lanes (about 65 m) and a total length of 2020 m, the Port Mann Bridge was the widest bridge worldwide at the time of its inauguration in 2012. The structure is divided into the North approach, the 850 m-long cable-stayed bridge from (Fig. 37.1) with a main stay of 470 m, and the South approach. The composite deck follows a hierarchical design (with longitudinal girders, transverse floor beams and a concrete slab), and is split up into the West deck and the East deck, which are partially connected through median struts. Two cable planes in fan-arrangement suspend each deck from the two 158 m-tall towers that are located between the decks; in total, 288 cables are installed. The supports consist of tie down piers at the end of each backstay, as well as wind shoes. Numerous foundation piles with diameters up to 1.8 m support the towers and piers.

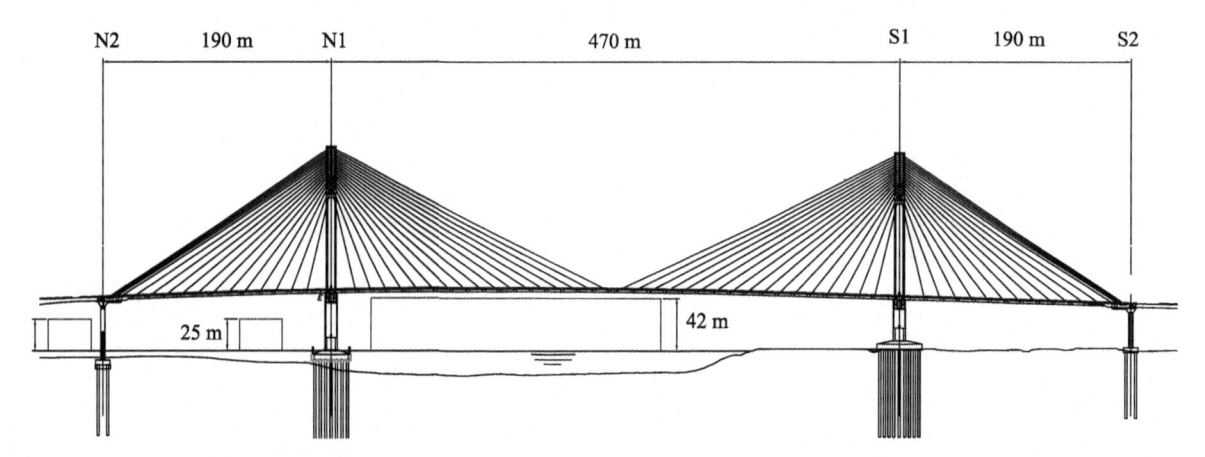

Fig. 37.1 Side elevation of the Port Mann Bridge

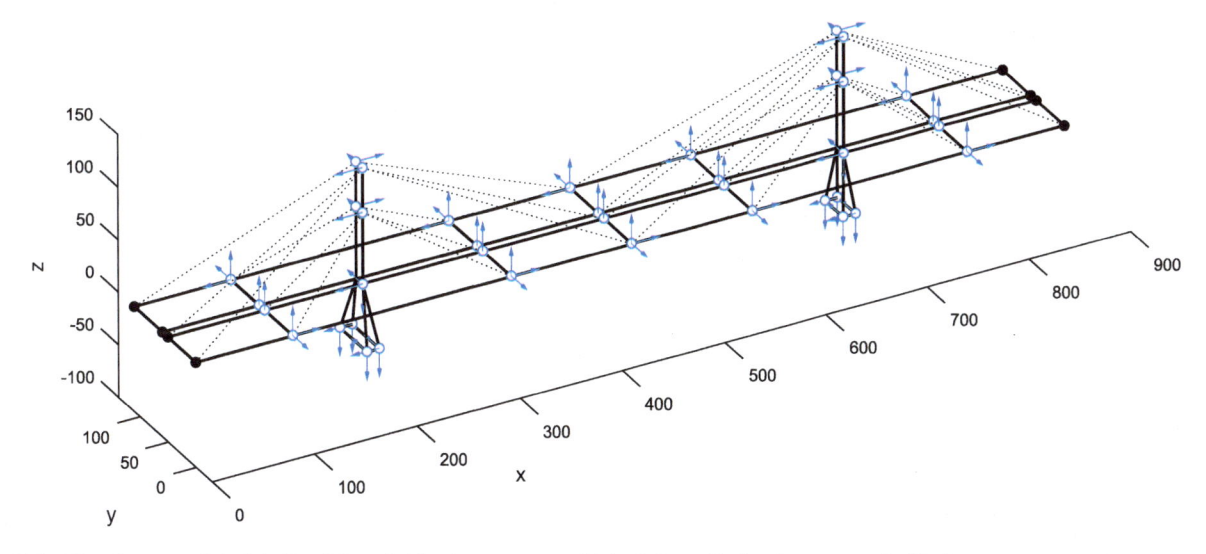

Fig. 37.2 Wire-frame model of the Port Mann Bridge (not true to scale) including all vibration sensors (in blue)

37.2.3 Instrumentation

The instrumentation is elaborate, and a total of 340 measurement channels are connected to the online server of the BCSIMS. The cable-supported part is equipped with 34 vibration sensors (70 channels) and 12 displacement transducers at the expansion joints (Fig. 37.2). Moreover, two weather stations are mounted on top of the towers to continuously measure the ambient temperature, relative humidity, wind direction and speed. Ultimately, a toll station is located at the South approach and records both the number and length of crossing vehicles in both directions.

37.3 Operational Modal Analysis

The implemented algorithm is the data-driven stochastic subspace identification principal component (SSI-PC) [5, 6] combined with an automated clustering approach for mode selection [7].

37.3.1 Dynamic System Model

Stochastic state space models are used to approximate the ambient vibrations. They are based on control theory and are split up into two separate equations. The first line, the state vector $\dot{x}_{k+1}$, describes the system change over time, which depends on the mass, damping and stiffness matrices M, C, K, all of which are included in the discrete state transition matrix $A \in \mathbb{R}^{n \times n}$, with n being the model order. Moreover, a noise term w_k is present, which includes the ambient excitations. The second line, the observation vector y_k, describes the system feedback, and provides information on the current vibration state stored in the output matrix C. It contains all observable state matrices, *i.e.* the output matrices for displacement, velocity or acceleration. Again, a noise vector v_k is present, which models the noise components that are, for example, introduced through electrical measurement equipment. Both noise vectors are unknown but modelled as zero-mean white noise [8].

$$\dot{x}_{k+1} = Ax_k + w_k \tag{37.1}$$

$$y_k = Cx_k + v_k.$$

The matrices A and C are sufficient to fully describes the vibration of structures, and the remaining question is how they can be approximated from ambient vibrations, and how to choose an appropriate model order n.

37.3.2 Block Hankel Matrix

The covariance function[2] is a helpful means to describe the similarity of wave pattern between multiple sensors. It is the first engineering unit that can be associated with structural information. However, in the data-driven version, the direct calculation of covariance functions is circumvented and geometrical projections are used instead [8]. To do so, the discrete measurements $Y_i \in \mathbb{R}^r$ at time the lag $i = 1, \ldots, N$ measured at all sensor locations r are arranged in two matrices also called the "the past" $\mathcal{H}_{(-)}$ and the "the future" $\mathcal{H}_{(+)}$.

$$\mathcal{H}_{(-)} = \frac{1}{\sqrt{N}} \begin{bmatrix} Y_1 & Y_2 & \ldots & Y_{N-2i+1} \\ Y_2 & Y_3 & \ldots & Y_{N-2i+2} \\ \vdots & \vdots & & \vdots \\ Y_i & Y_{i+1} & \cdots & Y_{N-1} \end{bmatrix} \quad \mathcal{H}_{(+)} = \frac{1}{\sqrt{N}} \begin{bmatrix} Y_{i+1} & Y_{i+2} & \ldots & Y_{N-i+1} \\ Y_{i+2} & Y_{i+3} & \ldots & Y_{N-i+2} \\ \vdots & \vdots & & \vdots \\ Y_{2i} & Y_{2i+1} & \cdots & Y_N \end{bmatrix} \quad (37.2)$$

The projection matrix P can be calculated by projecting the row space of future measurements onto the row space of past measurement. The information in the resulting matrix is closely related to covariance functions, although the direct calculation has been avoided [9].

$$\mathcal{H}_i = W_1 P_i W_2; \quad P_i = \mathcal{H}_{(+)} \mathcal{H}_{(-)}^T \bullet \left(\mathcal{H}_{(-)} \mathcal{H}_{(-)}^T \right)^* \bullet \mathcal{H}_{(-)} \quad (37.3)$$

The matrices W_1, W_2 are weighting matrices. Depending on the weighting matrices, the algorithm is subdivided into the unweighted principal component (UPC), the principal component (PC), or the canonical variate algorithm (CVA). In this paper, the PC version is chosen. All matrices in Eqs. (37.2) and (37.3) are block Hankel matrices $\mathcal{H}$, i.e. matrices with identical blocks on the main diagonal from the bottom left to the top right.

37.3.3 Subspace Identification

The next task is to link the estimated block Hankel matrix $\mathcal{H}$ to the state space model, or more precisely to the system and output matrices A, C. Again, the block Hankel matrix can be associated with auto-covariance functions which, according to control theory, factorizes into the output matrix, the system matrix, and the cross-covariance between the state and the measurement vectors $R_{yy,i} = (CA^{i-1})R_{xy,i}$, provided the measurements are long enough [8]. In other words, the auto-covariance functions can be interpreted as some kind of matrix unit response to white noise excitations [10]. The system specific (bracket) term CA^{i-1} is called the observability matrix and the other term is load dependent and called the controllability. When applied to the block Hankel matrix in Eq. (37.3), the equation unfolds as follows

$$\mathcal{H} = \mathcal{O}_i Z_q; \quad \mathcal{O} = \left(C^T \ (CA)^T \ (CA^2)^T \ \cdots \ (CA^p)^T \right)^T. \quad (37.4)$$

This decomposition is done using thin singular value decomposition $\mathcal{H} = USV^T$. Since singular values and vector are sorted in descending order with respect to their significance, and the modes of vibration are assumed to dominate the information stored in these matrices, the matrices can be truncated at the desired model order. Making use of Eq. (37.4), the matrix of observability yields $\mathcal{O}_i = U_1(S_1)^{1/2}$. When looking at the different components in $\mathcal{O}_i$, in Eq. (37.4), it also becomes clear how the dynamic system matrices can be derived from the subspace matrices: the observation matrix C is taken from the first block-row, and the state transition matrix A can be obtained using the following shift invariance property $\mathcal{O}_i^\uparrow A = \mathcal{O}_{i\downarrow}$ [8]. This equation can be solved by least-square regression using the pseudo inverse.

$$A = \left(\mathcal{O}_i^\uparrow \right)^* \mathcal{O}_{i\downarrow} \quad (37.5)$$

---

[2] The relation between the covariance and correlation functions writes $cor(y, y) = Cov(y, y)/(\sigma_y \sigma_y)$.

The equations only hold true for an infinitely long set of data ($N \to \infty$), which is why the system matrices estimated from measurement records with N samples are merely approximations.

37.3.4 Modal Parameter Estimation

Once the dynamic system model is approximated, see Eq. (37.1), it is straightforward to solve the well-known eigenvalue problem, and derive poles as well as the corresponding mode shapes.

$$\det(A - \lambda I) = 0 \quad A\phi_\lambda = \lambda \phi_\lambda \quad \phi = C\phi_\lambda \tag{37.6}$$

The eigenvalues can now be transformed from modal and discrete coordinates into continuous physical coordinates indicated by a subscript "c". The natural frequencies and damping ratios yield

$$f = \frac{|\lambda_c|}{2\pi}; \quad \zeta = -\frac{\mathrm{Re}(\lambda_c)}{|\lambda_c|}; \quad \lambda_c = \frac{\ln(\lambda)}{\Delta t}. \tag{37.7}$$

37.3.5 Stabilizing Diagram

More challenging than solving the problem is to stabilize the obtained solution, that is, to find accurate estimates of the physical modes of vibration, and to separate them from spurious noise modes. The presence of noise modes is inevitable. They are created by noise components in the signal, and the reason why they are identified is that the model order of the underlying state space model generally has to be chosen much higher than the actual number of modes of vibration. However, in many cases, they show dynamic properties that are very different from physical modes, which is why they can be filtered out. If the identifying characteristics enjoy universal validity, they are called "hard" noise mode rejection criteria. For example, structural modes of vibration always occur in complex conjugate pairs. If the criteria employ user-defined threshold values, which might vary from structure to structure, they are called "soft" mode rejection [11]. For brevity, all employed criteria are summarized in Table 37.1. For bridge structures, the mode shape complexity appears to be one of the most powerful criteria. It verifies whether all locations vibrate exactly in-phase or out-of-phase, which is the case for physical mode shape vectors.

The accuracy of the modal parameters depends on two user-defined parameters. One parameter is the number of time lags i used to estimate the block Hankel matrix. It can be associated with the maximum wave length that can be evaluated. That is why it could be set based on the first natural frequency and the sampling frequency $i \geq f_s/(2f_{n,1})$ [12]. A large value increases the computational effort, but guarantees that low frequency modes can be captured. The number of time lags also limits the maximum model order n of the underlying state space model, which is why some authors prefer another criterion that includes n as well as the number of sensors r, so $n \leq ri$ [10]. Some authors started to evaluate i for each mode individually, to minimize the noise contamination and maximize the accuracy of modal parameters [13]. The second user-defined parameter is the model order n. The value is generally unknown and even for repeated ambient vibration tests on the same structure,

Table 37.1 Automated SSI-PC

Hard criteria		Soft criteria	
Complex conjugate pairs	Discard modes that do not occur in complex conjugate pairs	Mode shape complexity	Set an upper bound for the modal phase collinearity $MPC < thres$ [14]
Stability	Require a positive damping ratio $\zeta > 0$ [11]	Modal contribution	Require a minimal relative modal participation factor $\Pi > thres$
Realistic damping ratio	Set upper bound $\zeta < 0.15$	Statistical accuracy	Discard modal frequencies that have very wide confidence intervals, as noise modes show to have greater uncertainties [15]

Noise mode rejection criteria

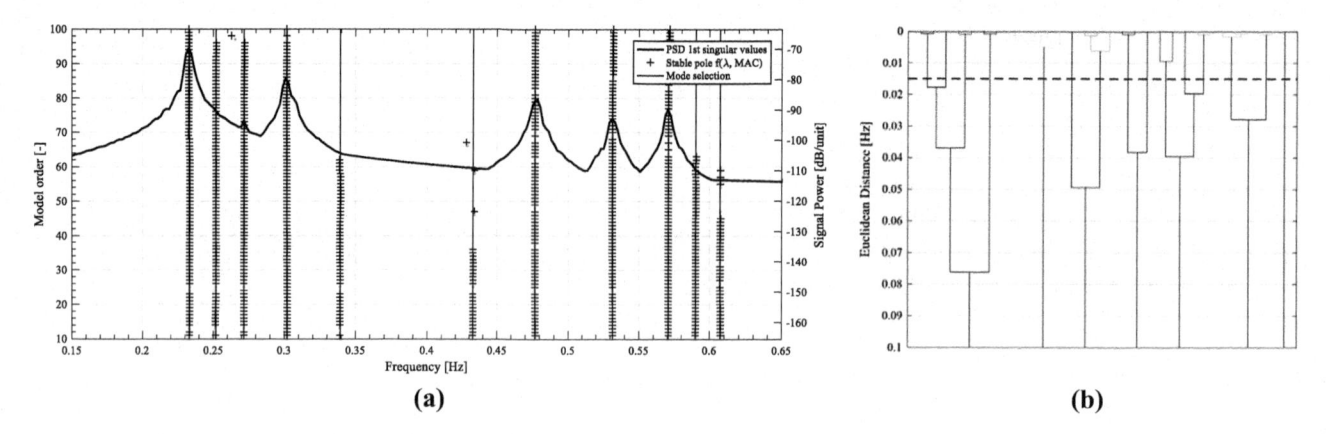

Fig. 37.3 Automated SSI-PC: (**a**) automated stabilizing diagram, and (**b**) hierarchical cluster tree

the optimal n can vary, because the excitation and noise characteristics can vary and not all modes of vibration are always sufficiently excited.

A stabilizing diagram is a good way to discriminate physical modes from noise modes, and to decrease the uncertainties of the estimates. Hereby, the modal parameters are estimated for a whole range of model orders, while keeping the number of block rows sufficiently large. Subsequently, the estimates of two successive calculations can be compared, and 'stable' modes of vibration can be identified. A mode of vibration is understood to be stable if the deviation of its modal parameters is within certain threshold limits dP_{lim}. To include modal frequencies, mode shapes and damping ratios, the following stabilizing criterion was implemented [7].

$$dP = d\lambda + (1 - dMAC) \leq dP_{lim} \tag{37.8}$$

Figure 37.3 shows a typical stabilizing diagram, where stable modes are indicated by a "+" symbol, and form vertical lines. For visualization purposes, the first singular value of the power spectral density is also plotted, although it does not affect the analysis.

37.3.6 Automated Mode Selection

The last step of the automated mode selection is a procedure that groups all stable modes, and evaluates the mean modal parameters over all model orders, as well as the standard deviations. To do that, a hierarchical clustering approach is applied, which divides the groups based on the horizontal distance between the data points in the stabilizing diagram from Fig. 37.3 [7]. First, the data points in the stabilizing diagram above are re-organized with respect to their horizontal distance measured in [Hz], and the result is the hierarchical tree which is plotted in Fig. 37.3. Secondly, the number of clusters is determined by trimming the tree at a user-defined distance value, indicated by a dashed horizontal line. To avoid misclassification, all modes within each cluster are eliminated that show inconsistent damping ratios or mode shapes. Moreover, the 99.7% confidence intervals of each cluster frequency are evaluated ($\mu_{f,\,clust} \pm 3\sigma_{f,\,clust}$), and cluster with overlapping confidence bounds are combined. This way, modes that are highly dispersed over the model order are not falsely interpreted as two or more independent modes [7]. Ultimately, the number of clusters is reduced to the most dominant ones, meaning the clusters with the greatest number of modes are selected. This step is based on the assumption that the more "stable" a pole is, the more likely it is a physical one.

37.3.7 Continuous Mode Tracking

Since some modes of vibration are not always sufficiently excited, they cannot always be identified, and it is necessary to assign the estimated modal parameters to the user-defined reference modes of vibration. In this paper, the modal parameters are allocated to pre-defined modes if the natural frequency falls within a certain frequency bin, and for verification the mode shapes are compared by means of the modal assurance criterion. The limitations of such a non-adjusting, or "static" mode tracking procedure are apparent, as a continuous off-set from the reference value will not be tracked. Such an off-set could be caused by altering material stiffness over the lifespan of the bridge, or accumulating damage scenarios, *e.g.* corrosion or loss in pre-tension. Having said that, this study sets out to define the healthy state of the Port Mann Bridge observed over 6 months and no large deviations are expected.

37.3.8 Summary

First, the similarities of wave signals measured at different measurement locations are evaluated through geometrical projections and stored in the block Hankel matrix. Subsequently the matrix is decomposed into its subspace using SVD, and state space matrices are approximated. By solving the eigenvalue problem of the dynamic system model, natural frequencies, mode shapes and damping ratios can be retrieved. The model order n of the optimal model is unknown a-priori, but affects the quality of the fit. Therefore, the calculation is repeated for a range of model orders, and stable solutions are identified by means of a hierarchical clustering approach.

37.4 Regression Models

37.4.1 Polynomial Model

Multiple-linear regression analysis is a common way to describe the relationship between output and input variables. Essentially, a system response variable u is reproduced through a superposition of input terms p_i that are multiplied by a constant coefficient β_i. Input terms are also referred to as *predictors*, and they can be quadratic or cubic, or a product of multiple input variables, and do not have to be linear. Moreover, a constant term β_0 is used, which in combination with all linear terms yields the system response. It is called the *intercept* and should not be mistaken for a mean value, as it carries no physical meaning. Ultimately, an error term ε has to be introduced to describe all effects that cannot be explained through predictors. The predictors are given, and the coefficients β are typically derived so the sum of squares of the error term is minimized.

$$u = \beta_0 + \beta_1 p_1 + \beta_2 p_2 + \cdots + \beta_j p_n + \varepsilon \tag{37.9}$$

37.4.2 Goodness of Fit

To quantify the goodness of fit, several statistical criteria are available. In this paper, the adjusted coefficient of determination (R^2-value) is used. It describes the proportion of the variance in the response variable that is predictable, *i.e.* the ratio of the variance explained by the model to the total variance. More precisely, the adjusted R^2 value is used, which normalizes the sum of squares of the regression model SS_{reg} not only by the number of samples N (this term would be proportional to the variance) but also considering the number of predictor variables k in the model. Another way to assess the goodness of fit is to observe the residual ε and to make sure that it does not follow the same trends as the original response variable.

$$R^2_{adj} = 1 - \left(\frac{N-1}{N-k}\right) \cdot \frac{SS_{reg}}{SS_{tot}} \tag{37.10}$$

37.4.3 Challenges

The quality of the modelled system response depends on the proper choice of predictor variables p_i, and appropriate polynomials. If an essential input variable is missing, the algorithm will not be able to reproduce the system response

Table 37.2 Estimation of a traffic index by means of mass multipliers

Bin	Length [m]	Vehicle class	Vehicle description	GVW (t)	Mass multiplier
1	<6.0	Class 1–3	Motorcycles, cars, light single unit trucks	2.35	1.000
2	6.0–12.5	Class 4–7	Buses, single unit trucks (2–4 axles)	6.67	2.840
3	12.5–22.5	Class 8–10	Single trailer trucks	18.77	8.317
4	22.5–35.0	Class 11–13	Multi trailer trucks and B-trains (5–7 axles)	21.55	
5	>35.0	Class 13	Multi-trailer trucks and B-trains (>7 axles)	28.35	

adequately. Another source of error is co-linearity in the input variables; It occurs if two input variables exhibit a linear relationship to each other, and consequently, it cannot be distinguished which one actually influences the system response. A typical example would be temperature and humidity, which often follow opposite trends, so only one of the two variables should be used for regression analysis. Ultimately, overfitting should be avoided. It occurs if noise processes are modelled and, typically, the chance of overfitting increases as the complexity of the model increases. In other words, it is desirable to keep the predictor terms as simple as possible and avoid higher order terms.

37.5 Data Processing

37.5.1 Traffic Load Estimation

Traffic load is an operational variable. No bridge weight-in-motion system is installed, so no information on the velocity of vehicles is available. However, the records from toll cameras can be used to estimate traffic loads. The toll station is located at the end of the South approach and measures both the number of passing vehicles as well as their length, and calculates the tolls based on three length bins. The length bins are related to the vehicle classes of the Federal Highway Administration (FHWA), and are characterized by an average gross vehicle weight (GVW) rating. Since several vehicle classes fall into one bin, their gross weight was averaged, taking into account the annual number of vehicles on highways for each class, see Table 37.2.

37.5.2 Normalization

Environmental variables are recorded at the weather stations and directly used for regression analysis. Having said that, all input and response variables are normalized, *i.e.* the mean values are subtracted divided by their standard deviation.

37.6 Application

37.6.1 Modal Parameter Accuracy

To validate the modal analysis results of the Port Mann Bridge, they are juxtaposed with documented modal parameters from the literature [16], where both field tests and former SHM data analyses were collected.[3] The measurement duration is set according to the rule of thumb $T = 1000/f_{0,\,min} \leq 90$ min [10], so the damping ratio estimates converge. The first nine modes of vibration could be estimated accurately. The fundamental frequency of the bridge is 0.233 Hz (natural period 4.29 s), and the mean deviation of all frequency estimates from the target values is 0.13%. For damping ratios, on the other hand, the mean deviation amounts up to 24.7%, see Table 37.3. Mode 5 and 6 deserve special attention as they are not always sufficiently excited.

---

[3] Target values have been created in ARTeMIS [17].

Table 37.3 Natural frequencies, damping ratios and mode shape descriptions of the first nine modes of vibration

No.	Comment	Natural frequency [Hz]			Damping ratio [% crit.]		
		Target	Result	Error	Target	Result	Error
1.	Deck 1st bending	0.233	**0.233**	0.04%	0.68	**0.54**	20.6%
2.	Tower 1st transverse	0.251	**0.251**	0.24%	0.84	**0.53**	36.9%
3.	Deck 1st torsional	0.272	**0.272**	0.15%	0.64	**0.51**	20.3%
4.	Deck 2nd bending	0.302	**0.302**	0.10%	0.66	**0.49**	25.8%
5.	Deck 1st transverse	N/A	**0.339**	N/A	N/A	**1.07**	N/A
6.	Mainstay local torsional	0.432	**0.432**	0.07%	1.14	**1.82**	37.4%
7.	Deck 3rd bending	0.478	**0.477**	0.15%	0.62	**0.45**	27.4%
8.	Backstay local bending	0.532	**0.532**	0.06%	0.68	**0.52**	23.5%
9	Deck 4th bending	0.572	**0.571**	0.23%	0.53	**0.37**	30.2%

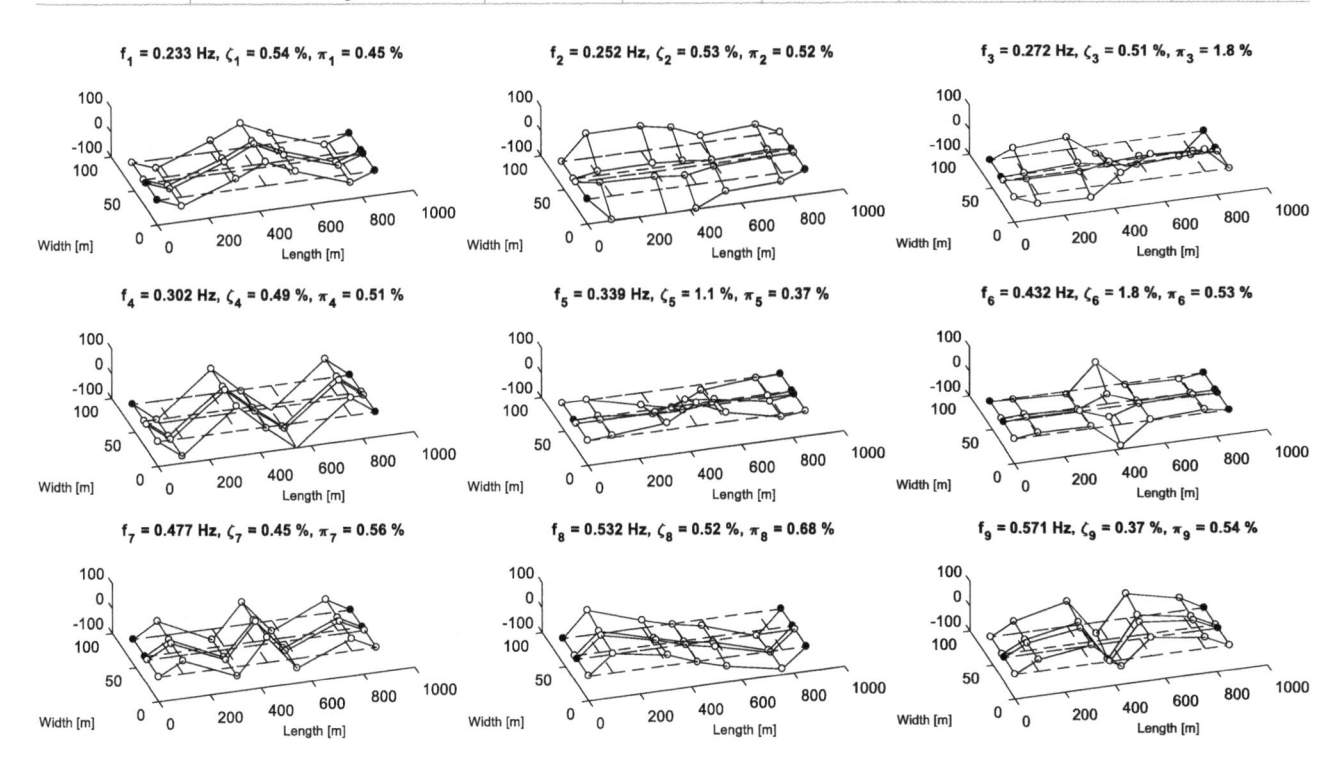

Fig. 37.4 Modes shapes of the Port Mann Bridge deck

37.6.2 Mode Shapes

For plotting reasons, the mode shapes of the first nine modes of vibration are plotted in Fig. 37.4 with focus on the bridge deck. It can be seen that Mode 5, 6 and 8 are local modes of individual spans. Each mode shape is further commented on in Table 37.3.

37.6.3 Influence of Environmental Variables on Natural Frequencies

The first nine global and local modes of vibration exhibit large vibration amplitudes of the bridge deck. That means the vibration behaviour may as well be approximated through a theoretical formula that only considers the deck's vibration. For simplification, the stay cables can be replaced by an elastic bedding with spring stiffness k, and the entire deck can be described through a simply supported beam of length l with distributed mass μ and a cross-sectional bending stiffness EI. For simplification, the constant factor that can be used to describe the effect of pretension forces or the elastic bedding is not

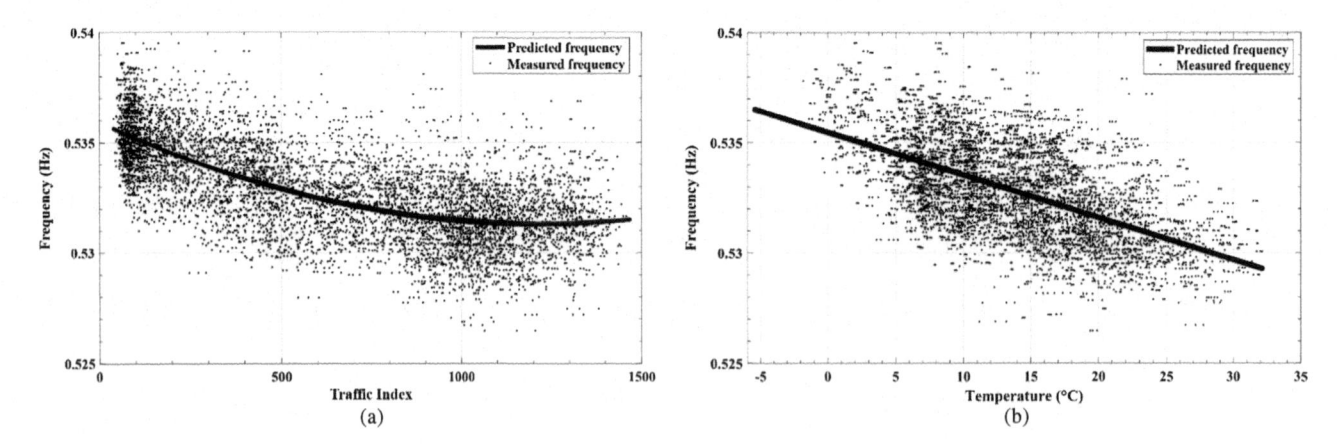

Fig. 37.5 Relation between the natural frequency of Mode 8 and (**a**) the estimated traffic index, and (**b**) ambient temperatures

described here. The parameter n describes the mode number, or the number of sinus half-waves, and ζ is the modal damping ratio in percent of the critical damping ratio.

$$\omega_d = \omega_e\sqrt{1-\zeta^2}; \quad \omega_e = \frac{(n\pi)^2}{l^2}\sqrt{\frac{EI}{\mu}} \cdot const \quad (37.11)$$

Traffic Fluctuations

The traffic index evaluated based on toll cameras appears to have a very strong correlation with the natural frequencies. Preliminary studies showed a weak correlation between the number of vehicles and modal frequencies [18], however, the new traffic index presented in this paper seems to be a very meaningful parameter to explain the fluctuation of modal frequencies. The regression model with the best overall fit includes a linear and a quadratic term of the traffic index, see Fig. 37.5. In fact, a very similar fit can be achieved by including the traffic index as a inverse square root term. This model seems to comply with the theoretical relation between the natural frequency and a uniformly distributed load μ from Eq. (37.11). For Mode 8, for example, traffic fluctuation accounts for up to 47.8% of the total variance. This makes the traffic index the most dominant environmental/operational variable.

Temperature Fluctuations

Fluctuations in ambient temperature also show a clear correlation with modal frequencies. With reference to Eq. (37.11), temperature fluctuations have an effect on the dimensions of the bridge components and the material stiffness, so l and EI. The physical relations are much more complex, as the decisive factor is not the ambient temperature but the material temperature, and temperature gradient could have a more significant effect than a uniform change in temperature [5]. However, shifting the temperature records to account for a warming-up phase did not increase the accuracy of the regression model, and due to the location of temperature sensors (on top of the towers) and the limited number of sensors, the temperature gradient cannot be considered. To conclude, a quadratic regression term does not significantly improve the approximation, and thus, the temperature fluctuations are only considered through a linear term. In combination with the regression terms for traffic loads, up to 69.0% of the variance in modal frequency could be explained, see Fig. 37.6.

Other Variables

Other environmental variables, such as the wind speed and direction, or statistical values such as the mean vibration level were taken into account. In addition, multiple artificial variables were created that consider the wind loads in specific wind directions, for example, lateral wind loads on the bridge deck and towers [19]. However, the increase in accuracy of the regression model was insignificant (<3.0% of the total variance).

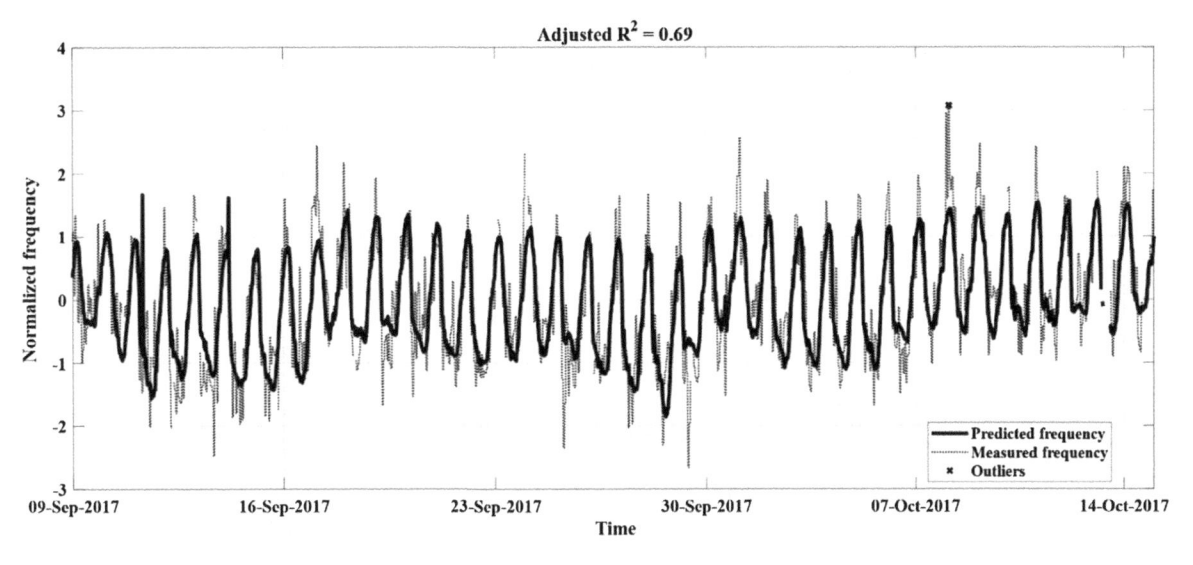

Fig. 37.6 Optimal regression model explaining up to 69% of the total variance of the natural frequency of Mode 8

37.6.4 Influence of Environmental Variables on Damping Ratios

Damping ratio estimates are subject to high uncertainties in operational modal analysis, but their fluctuation seems to follow a distinct daily and weekly pattern very much like the natural frequencies. Having said that, the optimal regression model was able to explain about 20% of the variance in the fluctuation, even when combined with principal component analysis. Nonetheless, the influence of the traffic volume on the damping ratio estimates appears to be significant. On average, the magnitude of the damping ratio estimates under high traffic load is twice as large.

37.7 Discussion

A dynamical bridge profile including the first nine modes of vibration could be established based on structural health monitoring data obtained through the B.C. Smart Infrastructure Monitoring System (BC-SIMS). The obtained modal parameters can serve as reference values and a baseline for all future studies on the Port Mann Bridge.

The accuracy of the modal frequencies and damping ratios has been evaluated in comparison to an extensive study that took into account the results of field measurements and online tests [16], and maximum deviations of 0.24% and 37.4% have been found. Possible reasons for the deviation are differences in the signal pre-processing, as well as the employed sub-versions of the SSI algorithm and environmental conditions. To verify the accuracy, another study on a laboratory test specimen, the Yellow Frame, was conducted, and the maximum deviation could be reduced to 0.07% for natural frequencies and 22.22% for damping ratios [20].These results were deemed satisfying, as the average deviations were merely 0.03% and 13.5%, respectively, and it is well-documented that damping estimations from ambient vibrations are subject to high uncertainties [10].

Despite the comparatively simple mode selection procedure, the implemented modal identification algorithm showed to be suitable for automation and long-term monitoring within the BC-SIMS. Over a period of 6 months, the first nine modes of vibration could be tracked reliably, although a very simple mode tracking procedure was implemented. In this paper, the horizontal distance in the stabilizing diagram is used to form a metric for mode clustering, instead of using both natural frequency and mode shape information [7]. The results are satisfactory. However, the algorithm is not fully automated, because two parameters have to be known a priori, including the minimal distance of adjacent modal frequencies and the number of most dominant modes. This is why it is more precise to speak of "semi-automated" modal identification. Having said that, the bridges within a monitoring system are well-studied, and the required information is available.

Correlation analysis revealed initial insight into the influence of environmental and operational variables on the modal frequencies of the bridge. In summary, traffic loads appear to have the most dominant effect and can be used to explain large parts of the total variance in the modal parameters, such as the natural frequencies of Mode 8. This result is consistent with other studies in the literature that were conducted on a suspension bridge [19]. Temperature fluctuations are the second most influential variable, and all other variables appear to have a minor effect. The estimated traffic index should be treated carefully, however, as no information on the velocity of passing vehicles is available from toll cameras. In case of a traffic jam, for example, the traffic index becomes meaningless, as it leads to a very low traffic index when the real load is at a maximum. Having said that, traffic jams on the Port Mann Bridge are rare, and thus, can be handled as outliers in a statistical sense.

37.8 Conclusion

This paper documents the implementation of an automated modal identification algorithm (data-driven SSI-PC) in the online server of the British Columbia Smart Monitoring Network (BCSIMS). The results showcase the accuracy and suitability of the implemented algorithm, and summarize modal parameters of the healthy structure, which can be used as a baseline for future damage detection and localization studies on the Port Mann Bridge. Ultimately, and due to the dominant effect of traffic fluctuations on the "normal" vibration behaviour, it may be beneficial to implement a bridge weight-in-motion system that takes into consideration the velocity of the passing vehicles.

Acknowledgements The financial support from the German Academic Exchange Service (DAAD) and the Canadian Natural Sciences Engineering Research Council (NSERC) is gratefully acknowledged.

References

1. Goldfinger, C., Nelson, C.H., Morey, A.E., et al.: Turbidite event history—methods and implications for Holocene Paleoseismicity of the Cascadia Subduction Zone. US Geological Survey, U.S. Department of the Interior, Paper 1661–F (2012)
2. Kaya, Y., Ventura, C., Huffman, S., et al.: British Columbia smart infrastructure monitoring system. Can. J. Civil Eng. **44**(8), 579–588 (2017)
3. Farrar, C.R., Worden, K.: Structural Health Monitoring: A Machine Learning Perspective. Wiley, Oxford (2012)
4. Doebling, S.W., Farrar, C.R., Prime, M.B.: A summary review of vibration-based damage identification methods. Shock. Vib. Digest. **30**(2), 91–105 (1998)
5. Farrar, C.R., Doebling, S.W., Straser, E.G., et al.: Variability of modal parameters measured on the Alamosa Canyon Bridge. In: Proceedings of the IMAC—15th International Modal Analysis Conference (1997)
6. Canadian Highway Bridge Design Code (S6-14) (2014)
7. de Almeida Cardoso, R., Cury, A., Barbosa, F.: A clustering-based strategy for automated structural modal identification. Struct. Health Monitor. **17**(2), 201–217 (2017)
8. van Overschee, P., de Moor, B.: Subspace Identification for Linear Systems: Theory-Implementation-Application. Springer, Berlin (2012)
9. Peeters, B.: System identification and damage detection in civil engineering. Doctoral dissertation, Katholieke Universiteit Leuven (2000)
10. Brincker, R., Ventura, C.E. (eds.): Introduction to Operational Modal Analysis. Wiley, New York (2015)
11. Reynders, E., Houbrechts, J., de Roeck, G.: Fully automated (operational) modal analysis. Mech. Syst. Signal Process. **29**, 228–250 (2012)
12. Rainieri, C., Fabbrocino, G.: Operational Modal Analysis of Civil Engineering Structures. Springer, New York (2014)
13. Tarpø, M., Olsen, P., Amador, S., et al.: On minimizing the influence of the noise tail of correlation functions in operational modal analysis. Proc. Eng. **199**, 1038–1043 (2017)
14. Richard, S.P., Elliott, K.B., Axel, S.: A consistent-mode indicator for the eigensystem realization algorithm. J. Guid. Contr. Dynam. **16**(5), 852–858 (1993)
15. Verboven, P., Parloo, P., Guillaume, P., et al.: Autonomous structural health monitoring-part I: modal parameter estimation and tracking. Mech. Syst. Signal Process. **16**(4), 637–657 (2002)
16. McDonald, S.: Operational modal analysis, model updating, and seismic analysis of a cable-stayed bridge. Master thesis, University of British Columbia (2016)
17. Structural Vibration Solutions A/S, ARTeMIS Modal, Denmark (2015)
18. Kaya, Y., Mendler, A., Ventura, C.E.: Structural health monitoring network in British Columbia, Canada. In: Proceedings of the EWSWH—9th European Workshop on Structural Health Monitoring (2018)
19. Cross, E.J.: On SHM in changing environmental and operational conditions. Doctoral dissertation, University of Sheffield (2012)
20. Mendler, A., Ventura, C.E., Allahdadian, S.: The Yellow frame: experimental studies and remote monitoring of the structural health monitoring benchmark structure. In: Proceedings of the IMAC—36th International Modal Analysis Conference, vol. 2, pp. 233–244 (2018)